Fundamental Theories of Physics

Volume 191

The international monograph series “Fundamental Theories of Physics” aims to stretch the boundaries of mainstream physics by clarifying and developing the theoretical and conceptual framework of physics and by applying it to a wide range of interdisciplinary scientific fields. Original contributions in well-established fields such as Quantum Physics, Relativity Theory, Cosmology, Quantum Field Theory, Statistical Mechanics and Nonlinear Dynamics are welcome. The series also provides a forum for non-conventional approaches to these fields. Publications should present new and promising ideas, with prospects for their further development, and carefully show how they connect to conventional views of the topic. Although the aim of this series is to go beyond established mainstream physics, a high profile and open-minded Editorial Board will evaluate all contributions carefully to ensure a high scientific standard.

More information about this series at http://www.springer.com/series/6001

Dieter Schuch

Quantum Theory from a Nonlinear Perspective

Riccati Equations in Fundamental Physics

Springer

Dieter Schuch
Institut für Theoretische Physik
Goethe-University Frankfurt am Main
Frankfurt am Main
Germany

ISSN 0168-1222 ISSN 2365-6425 (electronic)
Fundamental Theories of Physics
ISBN 978-3-319-65592-5 ISBN 978-3-319-65594-9 (eBook)
https://doi.org/10.1007/978-3-319-65594-9

Library of Congress Control Number: 2017955648

Printed on acid-free paper

This Springer imprint is published by Springer Nature
The registered company is Springer International Publishing AG
The registered company address is: Gewerbestrasse 11, 6330 Cham, Switzerland

Dedicated to the memory of my parents
Anneliese and Franz
and my grandmother Katharina.

Preface

This book is based on courses on “Riccati, Ermakov and the Quantum-Classical Connection”, “Nonlinearities and Dissipation in Classical and Quantum Physics” and “Is Quantum Theory Intrinsically Nonlinear?” that I taught at Goethe University in Frankfurt am Main. Contact with Springer publishing house was made at the opening reception of the Conference DICE 2012 in Castiglioncello, Italy where I met Dr. Angela Lahee, Senior Editor Physics at Springer. After talking with her about Plato, Pythagoras and their relation to an unconventional view on quantum theory, the idea was born to write a monograph on this topic and publish it in the Springer Series “Fundamental Theories of Physics”.

The contract was signed in July 2013 and the original plan was to deliver the manuscript within 2 years. However, the field was increasingly expanding since then and I obtained my own new results that I definitely wanted to include in the book. So, various parts of Chaps. 2, 3, 5 and 7, as well as Appendix D, represent results that were found and published only after the contract was signed. Still, it is impossible to cover all aspects of the field, particularly the effective description of dissipative systems and work on the Ermakov invariant and related equations of motion. Therefore, the references presented in this book represent a concise selection of all the material that is published in this field. Further references can be found in the papers cited but a complete bibliography on this subject is beyond the scope of this book. My apologies to all authors who contributed significant work in this field and are not cited. Please be reassured that this is without contempt. It is likely that some papers escaped my knowledge, while others known to me were omitted as I had to select a reasonable number on a subjective basis.

Many of the new results mentioned above have been obtained in collaboration with colleagues (some of whom have become very dear friends), particularly in Mexico, Spain and Italy.

The joint efforts with Marcos Moshinsky at Instituto de Física at UNAM in Mexico City made a significant impact on my work. It started in the latter years of the last century and developed into a deep friendship that lasted until his death in 2009. Some of the papers that I consider my best or most influential ones originated from this collaboration with Marcos or were inspired by him. Fortunately, the

Mexican connection was not severed after he had passed away but continued with one of his colleagues, Octavio Castaños, at Instituto de Ciencias Nucleares at UNAM who then made contact with Oscar Rosas-Ortiz at CINVESTAV in Mexico City. The projects undertaken with my Mexican colleagues were always generously supported by their respective Institutes and in part by CONACyT for which I am immensely grateful.

Work on the Bateman model and its connections with my earlier research stemmed from "Damping in Granada", a conference organized by Victor Aladaya. This expanded into my Spanish collaboration to also include Julio Guerrero and Francisco López-Ruiz. The connection with 't Hooft's idea of be-ables was then established while visiting with Massimo Blasone of the University of Salerno in Italy.

Links to my chemistry roots still exist via Robert Berger of Marburg University in Germany and his former student Joonsuk Huh, now a Professor at Sungkyunkwan University in South Korea.

Certainly, my teachers deserve sincere thanks and, notably, my supervisor Hermann Hartmann had lasting influence on my attitude towards research, teaching and academia. Not only was he (officially) a Professor of Physical and Theoretical Chemistry, he was also a theoretical physicist in disguise who published his first paper with Arnold Sommerfeld in Munich. His broad knowledge in numerous academic fields was also quite impressive and, even in his latter years, he was still very open-minded to unconventional approaches of youngsters like me and supported my heretic ideas of nonlinear modifications of the Schrödinger equation in my Ph.D. thesis; at that time, not something that could be taken for granted.

Actually, I got infected with the idea of nonlinear Schrödinger equations by his collaborator and my dear Korean colleague K.-M. Chung who introduced me to a Korean paper on the subject. Despite countless personal misfortunes, he followed my progress in this field until he died in 2005.

Being essentially an orphan in the Theoretical Chemistry Department after my supervisor's death in 1984, another collaborator of Prof. Hartmann, Prof. Karl Hensen, supported me in my metamorphosis from a theoretical chemist to a theoretical physicist. To this day, he never doubted that I would succeed and encouraged me in many a dark period, something very important to me. From the very beginning of my studies, just for fun, I attended lectures on Theoretical Physics given by Prof. Rainer Jelitto. This later proved advantageous in my move from chemistry to physics. Then, it happened that said Prof. Jelitto took over the role of a kind of godfather when I switched to physics and followed my work with interest until he died in 2011.

I wish to thank my Ph.D. student Hans Cruz Prado for his assistance in the final stage of the manuscript; for integrating the figures and tables and putting together all the parts into one opus.

Thanks also to Angela Lahee of Springer Publishers for her patience and trust that this project would come to a successful end.

However, all of this would not have been possible without the support of my family. My parents and my grandmother stimulated my interest in the beauty, aesthetics and elegance of nature, numbers and their relations. They also supplied the mental, moral and material support that allowed me to pursue my ideas.

Special thanks go to my wife, Yvette. She is a constant source of encouragement for my work, was (and still is) my most meticulous critic (not only on matters linguistic), though also my biggest fan. Her proficient advice for this book was not always made use of as I insisted that the international language of scientists is bad English (to quote Bogdan Mielnik) and it is not my intention to win the Pulitzer Prize. Without Yvette's relentless support and technical know-how, this book would not have been written.

Frankfurt am Main
Mai 2017

Dieter Schuch

Contents

Chapter 1
Introduction

An attempt to look at quantum theory from a different perspective also leads to the question of where the idea of "quantization", i.e., dividing our material world into "smallest building blocks", actually originates. In the western hemisphere this takes one back to ancient Greece and its philosophers. In this case, one might first think of Democritus (ca. 460–ca. 371 B.C.) who coint the term "atomos" for the smallest building blocks of matter which cannot be divided any further. The word "atom" remains today. However, the atom turned out to be composed of even smaller constituents and, at present we are at the level of quarks and gluons. Is this the end, or will it be possible to detect yet smaller components with higher energized accelerators? Maybe we are even searching for the wrong answer and should not be looking for the smallest material particle but rather for some general elementary structure(s) that are ubiquitous in the universe and not depending on the size!

Another Greek philisopher even more famous than Democritus was probably thinking more along this line. In his book "Timaeus", Plato (428/29–348/47 B.C.) gives his view of how the world is made up in terms of right-angled triangels. Werner Heisenberg, who was equally fascinated with and puzzeled by this text summarizes this idea in his book "Der Teil und das Ganze" [1]. The general idea is that matter is made up of right-angled triangles which, after being paired to form isosceles triangles or squares, are simply joined together to construct the regular bodies of stereometry: cube, tetrahedron, octahedron and icosahedron. These four solids then represent the basic units of the four elements: earth, fire, air and water. Plato makes no statement about the size of the triangles, only about their form and resulting properties.

Some 2000 years later, the idea of Platonic solids (polyhedra) fascinated Johannes Kepler (1571–1630) so much that, in his quest for harmony in nature, he tried to explain the orbits of the planets in our solar system by fitting one polyhedron onto another so that the radii of spheres enveloping these polyhedra would correspond to the mean distances of the planets from the sun.

An aspect of "quantization" was brought into this picture by Titius von Wittenberg (1729–1796) and Johann Elert Bode (1747–1826) who proposed a series of numbers

D. Schuch, *Quantum Theory from a Nonlinear Perspective*,
Fundamental Theories of Physics 191, https://doi.org/10.1007/978-3-319-65594-9_1

(integers!) that describes the (relative) distance of the planets from the sun (Titius–Bode law) similar to Bohr's model of the atom. (Remarkably, a new formulation of this law has been found by Reinisch ([2] and literature quoted therein) using a nonlinear (NL) formulation of a (formal) macroscopic Schrödinger equation (SE) to describe the solar system. This NL formalism is equivalent to the treatment of a NL complex Riccati equation as presented in Sect. 2.3 in the context of time-dependent (TD) quantum mechanics and a NL formulation of time-independent (TI) quantum mechanics [3] as discussed in Sect. 3.2.

Kepler tried to connect the geometry of the planetary orbits and movement of the planets with some kind of imaginary sounds – the music of the spheres. This takes us back to the ancient Greeks and the right-angled triangles.

The Greek philosopher Pythagoras lived around 570–500 B.C. Today, even if (almost) nothing survived from our early mathematics lessons, most people can recollect the theorem named after him and some might even be able to quote it as $a^2 + b^2 = c^2$ where a and b are the catheti and c the hypotenuse of a right-angled triangle. Pythagoras and his pupils were well-known for their dogma "everything is number"; number meaning *integer*. They applied it to develop a musical scale (see Kepler's music of the spheres) and also to the right-angled triangle. So, the Pythagorean triples are three *integers* denoting the length of the three sides of a right-angled triangle thus fulfilling Pythagoras' theorem. The most common example is $(3, 4, 5)$ with $9 + 16 = 25$. Asked for a few more examples of this kind, even individuals affiliated with mathematics have difficulty providing some (or even one) though infinitely-many triples exist! Moreover, there is even a rather simple rule to find these triples. This rule, or something similar, was probably already known in Mesopotamia around 2000 B.C. but certainly Diophantus of Alexandria (around the year 250) knew it.[1]

Why do I mention this here? And what does this have to do with the topics stated in the title of this book? In Chap. 2 it is shown that a complex nonlinear evolution equation, in particular a Riccati equation, can be obtained from the dynamics of Gaussian wave packet (WP) solutions of the TD Schrödinger equation (TDSE) that also provides the key to answering the above question of obtaining Pythagorean triples.

Returning to a more recent era of physics, around the beginning of the 20th century, physicists were puzzled by what is called wave-particle duality. For instance light, that (after Maxwell) was finally considered to be a continuous wave, behaved like discrete particles in certain experiments such as the photoelectric effect. Contrarily, electrons that (in the meantime) were assumed to be particles, displayed wave-like behaviour in some experiments and produced interference patterns. The dichotomy of light versus matter, or continuous versus discrete, was only resolved in the mid-

[1] Fermat claimed to have found an elegant proof showing that if a, b and c are integers, the relation $a^n + b^n = c^n$ cannot be fulfilled for n larger than 2; but the margin of Diophantus' book "Arithmetica" was too small to write it down. This so-called Fermat's conjecture was proven only recently by A.J. Wiles.

twenties of the last century by Schrödinger and Heisenberg (and finally Dirac) with the development of quantum theory [4–6].

Though physically equivalent, Schrödinger's wave mechanics turned out to be more successful and receptive to the physics community than Heisenberg's matrix mechanics that used a less familiar mathematical description than Schrödinger's partial differential equation. (The Schrödinger picture is also preferred in this book.) As both formulations are closely-related to classical Hamiltonian mechanics, they also have similar properties. In particular, there is no direction of time in the evolution of the system and energy is a constant of motion (at least in the cases that are usually discussed in textbooks and can be solved analytically in closed form).

However, as everyone can observe daily in the surrounding world, nature actually behaves quite differently. There is a direction of time in almost every evolutionary process (and we usually cannot reverse it directly even if we would sometimes like to do so). Also mechanical energy is not a conserved quantity but dissipated into heat by effects like friction. There are ways of explaining and including these phenomena into the theories mentioned earlier. However, for ordinary people, concepts like (Poincaré's) recurrence time that is longer than the age of the universe are not really convincing. Nevertheless, quantum theory (with all its technological developments) is undoubtedly the most successful theory so far, and not only in physics but also from an economic viewpoint.

The situation in physics took a different twist near the end of the 20th century with the development of *Nonlinear Dynamics*. This theory is able to describe evolutionary processes like population growth with limited resources or weather patterns and other such complex systems as they occur in real life. At the same time, it can also take into account phenomena like irreversibility of evolution and dissipation of energy.

So why not combine the two theories to get the best of both worlds? In order to answer this question one must specify what the essential elements of these theories are and which aspects can be abandoned in order to have, at least, a chance for unification.

Starting with quantum mechanics, the *quantization* of *action* introduced by Max Planck around the year 1900 in order to explain the black-body radiation is obviously the most fundamental concept of all quantum theory. Interestingly enough, action usually cannot be measured directly as it is the product of two physical quantities (like position times momentum or energy times time) that both cannot be measured exactly at the same time. (Something that Heisenberg later on proved to be impossible in principle.)

So it is actually not energy, represented by the Hamiltonian, that is the most fundamental quantity in this theory but action, an aspect that will become of interest when dynamical invariants are discussed in Chap. 2.

The second indispensible ingredient of quantum physics has been specified by C.N. Yang in his lecture on the occasion of Schrödinger's centennial celebration [7]. In his opinion, the major difference between classical and quantum physics is the occurrence of the imaginary unit $\mathrm{i} = \sqrt{-1}$ in quantum mechanics since it enters physics here in a fundamental way (not just as a tool for computational convenience) and "complex numbers became a conceptual element of the very foundations of physics". The very meaning of the fundamental equations of wave mechanics and matrix mechanics (the TDSE and the commutator relations) "would be totally destroyed if one tries to get rid of i by writing them in terms of real and imaginary parts". Also E.P. Wigner stresses the important role of complex numbers in quantum mechanics in his article "The Unreasonable Effectiveness of Mathematics in the Natural Sciences" [8]. This, however, should not be a problem in the unification of quantum theory and NL theories like Nonlinear Dynamics as, in the latter one, also complex numbers can play an eminent role (e.g., in the complex quadratic family leading to the Mandelbrot set).

The contrast between reversible time-evolution in quantum theory and irreversible evolution as possible in Nonlinear Dynamics appears more problematic. Even if one would restrict the unification only to systems with reversible evolution, the major problem would seem to be that quantum theory in its conventional form is a linear one, i.e., it is essentially based on linear differential equations. On the other hand, Nonlinear Dynamics, by definition, uses NL differential equations (or discretized versions of such). Why then should linearity be so important for quantum theory? As mentioned above, quantum theory can explain the wave properties of material systems such as diffraction patterns that can essentially be considered a superposition of different solutions of the same equation. However, this superposition principle is usually only attributed to linear differential equations!

Is there a way out of this dilemma?

In principle, Nonlinear Dynamics (or NL theories in general) cannot exactly be linearized otherwise it would lose not only its *linguistic* meaning, but also its *physical* properties. Therefore, the only solution seems to be a NL version of quantum theory that takes into account all of the conventional properties of this theory (including a kind of superposition principle) while displaying formal compatibility with NL theories like Nonlinear Dynamics. One might ask why we should give up the nice mathematical properties of a linear theory; what do we gain? We might gain additional information that cannot be obtained easily (or even be expected) from the linear form of quantum theory. The sensitivity of NL theories to initial conditions is an example that demonstrates an advantage of a NL formulation over the linear one. Also the NL treatment of complex quantities mixes real and imaginary parts, or phase and amplitude, of these quantities. As it turns out, this mixing is not arbitrary but related to some conservation laws that are not at all obvious in the linear version of quantum mechanics. Furthermore, a formal link between a NL version of quantum mechanics and other NL theories, e.g., soliton theory (as shown in Chap. 7), enables analogies to be drawn between them and knowledge gained in one field to be transferred and applied in another.

What then should this nonlinear version of quantum mechanics look like? There are several modifications adding NL terms to the Schrödinger equation (particularly in the dissipative case, examples are mentioned in Chap. 4, but also in a general context [9, 10]) and the resulting problematic aspects of some of these attempts are discussed in the literature ([11–14] and literature cited therein). In the approach presented in this work quantum theory is not considered incomplete in its present form

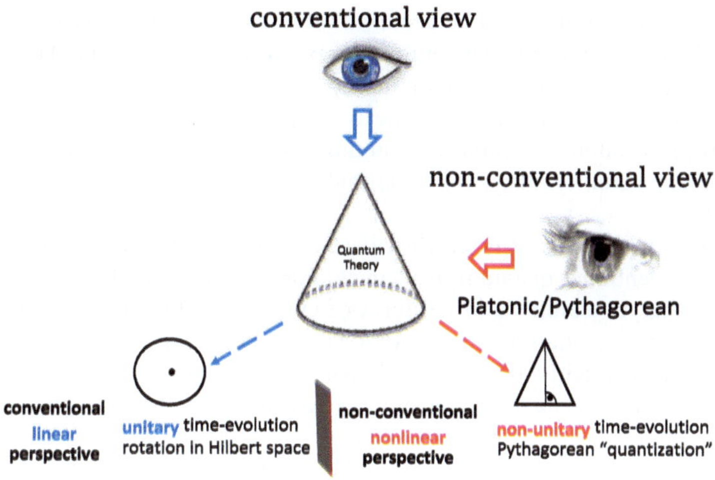

Fig. 1.1 Quantum theory from different perspectives

(at least for non-dissipative[2] systems); that means it does not require any (eventually NL) additions. The conventional view is regarded rather as a particular projection of quantum theory but this theory contains more information and further properties that are lost by looking at it from this perspective. If one would consider quantum theory as a three-dimensional object, like a cone (see Fig. 1.1), the conventional view could be compared to looking at it from the top, thus having the impression of viewing a circle. This corresponds to the conventional linear perspective with unitary time-evolution as a rotation in Hilbert space and other established properties.

The non-conventional view presented in the following could be compared to looking at the same cone from the side, giving the impression of viewing an isosceles triangle that can be divided into two right-angled triangles, as it were, a Platonic/Pythagorean viewpoint. This then leads to a non-conventional, NL perspective that can also include non-Hermitian Hamiltonians, non-unitary time-evolution (and, with a simple extension, dissipative open systems) and finally can be linked to the Pythagorean "quantization" in terms of the above-mentioned triples.

In particular, the search is for a NL reformulation of quantum mechanics that can be exactly linearized. It thus retains the property of a kind of superposition principle but still exhibits properties of NL systems like sensitivity to the choice of initial conditions, scale invariance (i.e., only relative changes like logarithmic derivatives

[2]Usually, "dissipative" is set against "conservative", the latter describing systems where the energy, and therefore also the Hamiltonian function, is a constant of motion. However, already the parametric oscillator with $\omega = \omega(t)$ leads to a time-dependent Hamiltonian $H(t)$, i.e., non-conservative but without dissipation due to friction forces. To distinguish this type of systems from dissipative ones the first type is referred to as "non-dissipative" systems.

matter) etc. How can this goal be achieved? It is known that the Riccati[3] equation provides all necessary properties for this purpose. It is a NL differential equation but can be linearized (involving a logarithmic derivative), therefore still preserving a kind of superposition principle. However, due to its (quadratic) nonlinearity, it is sensitive to the choice of initial conditions and, in the case of complex functions, mixes real and imaginary parts, or, phase and amplitude of these functions, respectively, in a unique way.

Therefore, in Chap. 2, it is shown where a (complex) Riccati equation already occurs in conventional quantum mechanics. For this purpose, the TDSE for cases with exact analytic solutions in the form of Gaussian WPs will be considered; in other words, for potentials at most quadratic (or bilinear) in position and momentum variables, here explicitly for the one-dimensional harmonic oscillator (HO) with $V = \frac{m}{2}\omega^2 x^2$ and constant frequency $\omega = \omega_0$, its generalization, the parametric oscillator with $\omega = \omega(t)$ and, in the limit $\omega \to 0$, for the free motion $V = 0$. It will be shown that the information about the dynamics of the systems can not only be obtainable from the TDSE but equally well (and even more) from a complex, quadratically-nonlinear, inhomogeneous Riccati equation.

The direct solution of this equation (by transformation to a homogeneous Bernoulli equation, once a particular solution of the Riccati equation is known) shows the sensitivity of the dynamics to the choice of the initial conditions. This characteristic feature of a NL differential equations is not at all obvious in the linearized form. In an alternative treatment of the complex Riccati equation, it is transformed into a real (but still NL) so-called Ermakov equation. This equation, together with the Newtonian equation describing the motion of the WP maximum, allows for the definition of a dynamical invariant that is still a constant of motion even if the Hamiltonian of the system no longer has this property (e.g., if $\omega = \omega(t)$). This invariant turns out to be important for the formulation of generalized creation and annihilation operators, corresponding coherent states and the Wigner function of the system and can also be derived by an algebraic method.

Finally, the complex Riccati equation can be linearized to a complex Newtonian equation. The relation between the amplitude of the complex variable fulfilling this equation and the WP width, as well as its relation to the phase of this variable, are explained. The latter relation, representing a conservative law, is due to the nonlinearity of the Riccati equation. Real and imaginary parts of the linearized complex variable are also sufficient for defining the time-dependent Green function (or Feynman kernel) and the representation of canonical transformations (in terms of the symplectic group) in quantum mechanics for the systems under consideration.

The variable fulfilling the Ermakov equation also allows for rewriting the quantum mechanical contribution of the mean value of the Hamiltonian calculated using the WP solution, i.e., the ground state energy, in a form that fits into the Lagrange/Hamilton formalism of classical mechanics but now for typical quantum mechanical properties like position and momentum uncertainties.

[3]Jacopo Riccati lived from 1676 to 1754. He wrote on philosophy and physics and is mainly known for the differential equation named after him.

In Chap. 3, similarities with supersymmetric (SUSY) quantum mechanics in the TI case are shown. Furthermore, a NL reformulation of TI quantum mechanics in the form of a real NL Ermakov equation or a complex NL Riccati equation for arbitrary potentials are given.

In Chap. 4 different approaches for an effective description of open systems with dissipative friction forces and irreversible dynamics are considered. In particular, models are discussed in detail where the environmental degrees of freedom do not appear explicitly or can be eliminated by certain constraints. Those models that provide a correct description for the classical aspect, as represented by the motion of the WP maximum, and the quantum aspect, as represented by the dynamics of the WP width, can be interrelated classically by a combination of canonical and non-canonical transformations, quantum mechanically by a corresponding combination of unitary and non-unitary transformations. The crucial point is the non-canonical or non-unitary transformation between a formal canonical and the physical level of description.

Choosing a NL modification of the TDSE with complex logarithmic nonlinearity for the description of irreversibility and dissipation (that can be related to other established methods in the above-mentioned way), the WP dynamics dealt with in Chap. 2 is further examined in Chap. 5 including a dissipative, linear velocity-dependent friction force.

Chapter 6 explains how the results of Chap. 3 for the TISE can be extended to include the afore-mentioned friction effect. The results turn out to be consistent with the TD treatment.

In Chap. 7 examples are given where NL Riccati equations also occur in other fields of physics thus allowing results obtained in the quantum mechanical context to be transferred, where possible, to these systems. Starting from examples with real Riccati equations (as they occur in statistical thermodynamics, systems with Nonlinear Dynamics and solitons), the discussion is extended to complex Riccati equations (as they occur in classical optics, Bose–Einstein condensates and cosmological models) and finally to the more abstract problem of determining the Pythagorean triples. The last case gives an idea of how any evolution (in space, time or some other variable) that obeys a kind of complex Riccati/Bernoulli equation can be quantized.

Chapter 8 summarizes the results, points out related aspects that were not mentioned before or discussed in detail and cites some future perspectives. The appendices comprise some explicit calculations that are not essential for understanding the general idea but are still useful for completing the approach.

References

1. W. Heisenberg, *Der Teil und das Ganze* (Deutscher Taschenbuch Verlag, München, 1973), p. 17
2. G. Reinisch, Macroscopic Schroedinger quantization of the early chaotic solar system. Astron. Astrophys. **337**, 299–310 (1998)
3. G. Reinisch, Nonlinear quantum mechanics. Physica A **206**, 229–252 (1994)

4. E. Schrödinger, Quantisierung als Eigenwertproblem (Erste Mitteilung). Ann. d. Phys. **79**, 361–376 (1926)
5. W. Heisenberg, Über quantentheoretische Umdeutung kinematischer und mechanischer Beziehungen. Z. Phys. **33**, 879–893 (1925)
6. P.A.M. Dirac, *The Principles of Quantum Mechanics* (Oxford University Press, Oxford, 1935), p. 34
7. C.N. Yang, Square Root of Minus One, Complex Phases and Erwin Schrödinger, in *Schrödinger – Centenary Celebration of a Polymath*, ed. by C.W. Kilmister (Cambridge University Press, Cambridge, 1987), p. 53
8. E. Wigner, The unreasonable effectiveness of mathematics in the natural sciences. Richard courant lecture in mathematical sciences delivered at New York University, May 11, 1959. Commun. Pure Appl. Math. **13**, 1–14 (1960)
9. S. Weinberg, Testing quantum mechanics. Ann. Phys. (N.Y.) **194**, 336–386 (1989)
10. I. Bialynicki-Birula, J. Mycielski, Nonlinear wave mechanics. Ann. Phys. (N.Y.) **100**, 62–93 (1976)
11. N. Gisin, Weinberg's non-linear quantum mechanics and supraluminal communications. Phys. Lett. A **143**, 1–2 (1990)
12. N. Gisin, Relevant and Irrelevant Nonlinear Schrödinger Equations, in *Nonlinear, Deformed and Irreversible Quantum Systems*, ed. by H.-D. Doebner, V.K. Dobrev, P. Nattermann (World Scientific, Singapore, 1995), pp. 109–119
13. G.A. Goldin, Diffeomorphism Group Representation and Quantum Nonlinearity: Gauge Transformations and Measurement, in *Nonlinear, Deformed and Irreversible Quantum Systems*, ed. by H.-D. Doebner, V.K. Dobrev, P. Nattermann (World Scientific, Singapore, 1995), pp. 125–139
14. W. Lücke, Nonlinear Schrödinger Dynamics and Nonlinear Observables, in *Nonlinear, Deformed and Irreversible Quantum Systems*, ed. by H.-D. Doebner, V.K. Dobrev, P. Nattermann (World Scientific, Singapore, 1995), pp. 140–154

Chapter 2
Time-Dependent Schrödinger Equation and Gaussian Wave Packets

2.1 Dynamics of Mean Values and Uncertainties

In the following, the one-dimensional time-dependent Schrödinger equation (TDSE) for problems with analytic solutions in the form of Gaussian wave packets (WPs) is considered. This applies to potentials that are at most quadratic (or bilinear) in position and momentum variables. In particular, the discussion focuses on the harmonic oscillator (HO) ($V = \frac{m}{2}\omega^2 x^2$) with constant frequency, $\omega = \omega_0$, or the parametric oscillator with TD frequency, $\omega = \omega(t)$, where the corresponding expressions for the free motion ($V = 0$) are obtained in the limit $\omega \to 0$. The Gaussian function (see Fig. 2.1) is completely determined by its maximum and width.

In our case, both parameters can be TD. The evolution equations for these parameters can be obtained by inserting a general Gaussian WP ansatz,

$$\Psi(x,t) = N(t) \exp\left\{ \mathrm{i} \left[y(t)\, \tilde{x}^2 + \frac{\langle p \rangle}{\hbar}\, \tilde{x} + K(t) \right] \right\} \tag{2.1}$$

into the TDSE

$$\mathrm{i}\hbar \frac{\partial}{\partial t} \Psi(x,t) = \left\{ -\frac{\hbar^2}{2m} \frac{\partial^2}{\partial x^2} + \frac{m}{2}\omega^2 x^2 \right\} \Psi(x,t) \tag{2.2}$$

where $\hbar = \frac{h}{2\pi}$ with h = Planck's constant.

The variable $\tilde{x}$ in WP (2.1) is a shifted coordinate, $\tilde{x} = x - \langle x \rangle = x - \eta(t)$, where the mean value $\langle x \rangle = \int_{-\infty}^{+\infty} dx\ \Psi^* x \Psi = \eta(t)$ corresponds to the classical trajectory and defines the maximum of the WP, $\langle p \rangle = m\dot{\eta}$ represents the classical momentum and the coefficient of the quadratic term in the exponent, $y(t) = y_{\mathrm{R}}(t) + \mathrm{i}y_{\mathrm{I}}(t)$, is a complex function of time and related to the WP width. The (possibly TD) normalization factor $N(t)$ and the purely TD function $K(t)$ in the exponent are not relevant for the dynamics of the WP maximum and width and will be specified later.

D. Schuch, *Quantum Theory from a Nonlinear Perspective*,
Fundamental Theories of Physics 191, https://doi.org/10.1007/978-3-319-65594-9_2

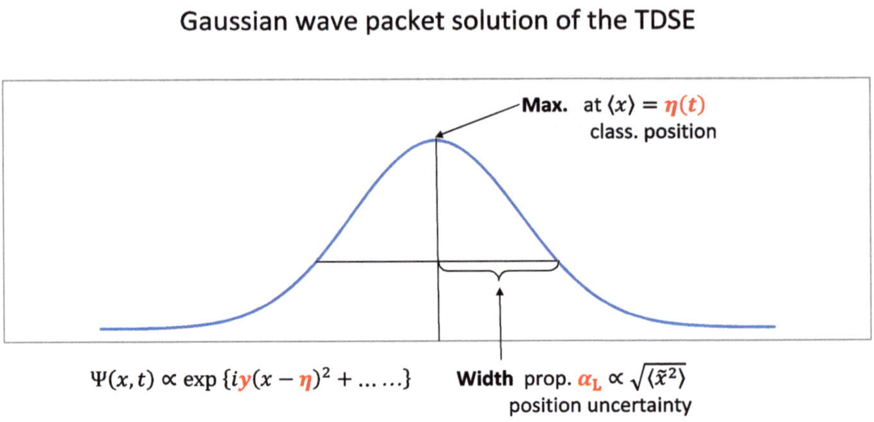

Fig. 2.1 Gaussian function, uniquely determined by its maximum and width

Inserting WP ansatz (2.1) into the TDSE (2.2) provides terms proportional to $\tilde{x}^2, \tilde{x}$ and independent of $\tilde{x}$. Equating the terms proportional to $\tilde{x}$ to zero provides the equation of motion for $\eta(t)$

$$\ddot{\eta} + \omega^2 \eta = 0 \,. \tag{2.3}$$

Similarly from the terms proportional to $\tilde{x}^2$, one obtains for $y(t)$, or $\frac{2\hbar}{m} y = \mathcal{C}$, the equation

$$\dot{\mathcal{C}} + \mathcal{C}^2 + \omega^2 \;=\; 0 \,. \tag{2.4}$$

Overdots denote derivatives with respect to time.

The Newtonian equation (2.3) simply states that the maximum of the WP, located at $x = \langle x \rangle = \eta(t)$, follows the classical trajectory. The equation for the quantity $\frac{2\hbar}{m} y = \mathcal{C}$ has the form of a *complex* NL Riccati equation and describes the time-dependence of the WP width that is related to the position uncertainty via $y_{\text{I}} = \frac{1}{4\langle \tilde{x}^2 \rangle}$ with $\langle \tilde{x}^2 \rangle = \langle x^2 \rangle - \langle x \rangle^2$ being the mean square deviation of position. Now it will be shown that the complex Riccati equation (2.4) not only provides the information about the evolution of the quantum uncertainties (and thus characteristic quantum mechanical properties like tunnelling currents) but all the dynamical information (and possibly more) that is supplied by the TDSE. For this purpose, different treatments of the Riccati equation (2.4) are discussed.

2.2 Direct Solution of the Riccati Equation

There are different ways of treating the (inhomogeneous) Riccati equation, illustrating different aspects of this equation [1, 2]. First, it can be solved directly by transforming it into a (homogeneous) NL (complex) Bernoulli equation providing a particular solution $\tilde{\mathcal{C}}$ of the Riccati equation is known. The general solution of Eq. (2.4) is then given by $\mathcal{C} = \tilde{\mathcal{C}} + \mathcal{V}(t)$ where $\mathcal{V}(t)$ fulfils the Bernoulli equation

$$\dot{\mathcal{V}} + 2\,\tilde{\mathcal{C}}\mathcal{V} + \mathcal{V}^2 = 0\,. \tag{2.5}$$

The coefficient of the linear term, occurring now instead of the inhomogeneity ω^2, depends on the particular solution $\tilde{\mathcal{C}}$. An advantage of Eq. (2.5) is its exact linearizability via $\mathcal{V} = \frac{1}{\kappa(t)}$ to

$$\dot{\kappa} - 2\,\tilde{\mathcal{C}}\,\kappa \;=\; 1 \tag{2.6}$$

which can be solved straightforwardly.

For constant $\tilde{\mathcal{C}}$, $\kappa(t)$ can be expressed in terms of exponential or hyperbolic functions (for real $\tilde{\mathcal{C}}$) or trigonometric functions (for imaginary $\tilde{\mathcal{C}}$). In this case, $\mathcal{C}$ can be written as

$$\mathcal{C}(t) \;=\; \tilde{\mathcal{C}} \;+\; \frac{\mathrm{e}^{-2\tilde{\mathcal{C}}t}}{\frac{1}{2\tilde{\mathcal{C}}}\left(1 - \mathrm{e}^{-2\tilde{\mathcal{C}}t}\right) \;+\; \kappa_0} \tag{2.7}$$

where the enumerator is obviously the derivative of the denominator (with constant initial condition κ_0), i.e., the second term on the rhs is just a (scale-invariant) logarithmic derivative.

For $\tilde{\mathcal{C}}$ being TD, $\kappa(t)$ and hence $\mathcal{V}$ can be expressed in terms of the integral $\mathcal{I}(t) = \int^t dt' \exp\{-\int^{t'} dt'' 2\tilde{\mathcal{C}}(t'')\}$. Then the general solution of Eq. (2.4) can be written as

$$\mathcal{C}(t) \;=\; \tilde{\mathcal{C}} \;+\; \frac{d}{dt}\ln\left[\mathcal{I}(t) + \kappa_0\right] \tag{2.8}$$

with the logarithmic derivative representing the solution of the Bernoulli equation. It defines a one-parameter family of solutions depending on the (complex) initial value $\kappa_0 = \mathcal{V}_0^{-1}$ as parameter.

The choice of this parameter can have enormous qualitative effects on the solution of the Riccati equation and thus the behaviour of the WP width (and tunnelling currents). This can be illustrated already using the HO with constant frequency ω, as an example. Choosing the particular solution $\tilde{\mathcal{C}}$ to be constant, from Eq. (2.4) it follows that $\tilde{\mathcal{C}} = \pm\,\mathrm{i}\,\omega_0 = \mathrm{i}\,\tilde{\mathcal{C}}_{\mathrm{I}}$, where only the plus-sign is physically reasonable because the minus-sign would lead to a positive sign for the quadratic term in the exponent of the WP, prohibiting normalizability. The parameter κ_0 then takes the form

$$\kappa_0 = \frac{1}{\mathcal{C}(0) - \tilde{\mathcal{C}}(0)} = \frac{\mathcal{C}_{\mathrm{R}}(0)}{\mathcal{C}_{\mathrm{R}}^2(0) + \left[\mathcal{C}_{\mathrm{I}}(0) - \tilde{\mathcal{C}}_{\mathrm{I}}\right]^2} - \mathrm{i}\frac{\mathcal{C}_{\mathrm{I}}(0) - \tilde{\mathcal{C}}_{\mathrm{I}}}{\mathcal{C}_{\mathrm{R}}^2(0) + \left[\mathcal{C}_{\mathrm{I}}(0) - \tilde{\mathcal{C}}_{\mathrm{I}}\right]^2}\,. \tag{2.9}$$

The imaginary part of $\mathcal{C}$ (or $\tilde{\mathcal{C}}$) is related to the WP width or position uncertainty via

$$C_I = \frac{\hbar}{2m\langle\tilde{x}^2\rangle}\,. \tag{2.10}$$

For the particular solution $\tilde{C}_I = \omega_0$ this leads to $\langle\tilde{x}^2\rangle = \frac{\hbar}{2m\omega_0}$ which is the well-known expression for the ground state wave function of the HO. Therefore, if the width of the initial WP is chosen to be that of the ground state, then C_I and $\tilde{C}_I$ are identical and, for[1] $C_R(0) = 0$, κ_0 diverges, so in Eq. (2.7) the second term on the rhs vanishes, simply leaving the particular solution corresponding to a WP with constant width. For any other choice of $\langle\tilde{x}^2\rangle(t = 0)$, κ_0 remains finite, leading to a WP with oscillating width (more details will be given in Sect. 2.3.1). This oscillating WP corresponds to the general solution of the Riccati equation (2.4) and only this one leads, in the limit $\omega_0 \to 0$, to the WP solution of the free particle problem, spreading quadratically with time, whereas in this limit the WP with constant width only turns into a plane wave-type solution.

Expressions with exactly the same form as (2.7) or (2.8) can be found in supersymmetric (SUSY) quantum mechanics if time t is replaced by a spatial variable in the context of isospectral potentials, also derived from Riccati equations. There too, the choice of the free parameter leads to drastic qualitative effects. More about this is mentioned in Sect. 2.1.

2.3 Alternative Treatment via the Ermakov Equation and Its Corresponding Dynamical Invariant

Alternative treatments of the complex Riccati equation (2.4) are also possible. It can be separated into real and imaginary parts,

$$\text{Re}: \quad \dot{C}_R + C_R^2 - C_I^2 + \omega^2 = 0\,, \tag{2.11}$$

$$\text{Im}: \quad \dot{C}_I + 2C_I C_R = 0\,. \tag{2.12}$$

Using (2.12), the real part C_R can be expressed in terms of C_I and its derivative. It turns out to be useful to introduce a new variable, $\alpha(t)$, that is connected with $C_I(t)$ via

$$C_I(t) = \frac{1}{\alpha^2(t)}, \tag{2.13}$$

where $\alpha(t)$ is directly proportional to the WP width or position uncertainty,

$$\alpha = \sqrt{\frac{2m\langle\tilde{x}^2\rangle}{\hbar}}\,. \tag{2.14}$$

[1] From the meaning of C_R that will become evident in the next subsection, it is obvious that in this case $C_R(0) \neq 0$ would not agree with the assumption of $C_I(0)$ being constant.

Inserting definition (2.13) into Eq. (2.12) shows that the real part of C just describes the relative change in time of the WP width,

$$C_R = \frac{\dot{\alpha}}{\alpha} = \frac{1}{2} \frac{\frac{d}{dt}\langle \tilde{x}^2 \rangle}{\langle \tilde{x}^2 \rangle} , \tag{2.15}$$

i.e., again a logarithmic derivate but now independent of the initial width of the WP.

Together with definition (2.13), this turns Eq. (2.11) into the so-called Ermakov equation[2] for $\alpha(t)$,

$$\ddot{\alpha} + \omega^2 \alpha = \frac{1}{\alpha^3} . \tag{2.16}$$

It had been shown by Ermakov [6] in 1880, 45 years before quantum mechanics was formulated by Schrödinger and Heisenberg, that from the pair of equations (2.3) and (2.16), coupled via ω^2, by eliminating ω^2 from the equations, a dynamical invariant can be obtained. Following Ermakov's method (see also Ray and Reid [7]) this leads to

$$\ddot{\alpha} - \frac{\ddot{\eta}}{\eta} \alpha = \frac{1}{\alpha^3} . \tag{2.17}$$

Multiplying this equation first by η,

$$\eta\, \ddot{\alpha} - \ddot{\eta}\, \alpha = \frac{d}{dt}(\eta\, \dot{\alpha} - \dot{\eta}\, \alpha) = \frac{\eta}{\alpha^3} , \tag{2.18}$$

and then by $\eta\, \dot{\alpha} - \dot{\eta}\, \alpha$,

$$(\eta\, \dot{\alpha} - \dot{\eta}\, \alpha) \frac{d}{dt}(\eta\, \dot{\alpha} - \dot{\eta}\, \alpha) = \frac{(\eta\, \dot{\alpha} - \dot{\eta}\, \alpha)\, \eta}{\alpha^3} , \tag{2.19}$$

allows one to express it as

$$\frac{1}{2}\frac{d}{dt}(\eta\, \dot{\alpha} - \dot{\eta}\, \alpha)^2 = -\frac{1}{2}\frac{d}{dt}\left(\frac{\eta}{\alpha}\right)^2 , \tag{2.20}$$

thus yielding the Ermakov invariant

$$I_L = \frac{1}{2}\left[(\dot{\eta}\, \alpha - \eta\, \dot{\alpha})^2 + \left(\frac{\eta}{\alpha}\right)^2\right] = \text{const.} \tag{2.21}$$

---

[2]This equation had been studied already in 1874, six years before Ermakov, by Steen [3]. However, Steen's work was ignored by mathematicians and physicists for more than a century because it was published in Danish in a journal not usually containing many articles on mathematics. An English translation of the original paper [4] and generalizations can be found in [5].

This invariant[3] was rediscovered by several authors, also in a quantum mechanical context; see, e.g. [8–10].

There are some remarkable properties of this invariant [11, 12]: (i) it is also a constant of motion for $\omega = \omega(t)$ (as the frequency gets eliminated in the course of the above derivation), in the case where the corresponding Hamiltonian does not have this property; (ii) apart from a missing constant m, i.e. mass of the system, it has the dimension of *action*, not of energy. The missing factor m is due to the fact that Ermakov used the *mathematical* Eq. (2.3) whereas, in a *physical* context, Newtons equation, i.e. Eq. (2.3) multiplied by m, is relevant.

Furthermore, as will be shown in Chap. 4, an invariant of this type also exists for certain dissipative systems, i.e. systems for which a conventional physical Hamiltonian (for the system alone without involving environmental degrees of freedom) does not even exist.

Also, factorization of the operator corresponding to the Ermakov invariant leads to generalized creation and annihilation operators (see Sect. 2.11). In this context the complex Riccati equation (2.4) again plays the central role.

2.3.1 Position and Momentum Uncertainties in Terms of Ermakov and Riccati Variables

The quantum uncertainties of position, momentum and their correlation can be determined directly by calculating the corresponding mean values using the Gaussian WP solution. They can be expressed in terms of $\alpha(t)$ and $\dot{\alpha}(t)$ or in terms of real and imaginary parts of $C(t)$, respectively, as

$$\langle \tilde{x}^2 \rangle(t) = \frac{\hbar}{2m}\alpha^2(t) = \frac{\hbar}{2m}\frac{1}{C_{\mathrm{I}}(t)}, \tag{2.22}$$

$$\langle \tilde{p}^2 \rangle(t) = \frac{m\hbar}{2}\left[\dot{\alpha}^2(t) + \frac{1}{\alpha^2(t)}\right] = \frac{m\hbar}{2}\frac{C_{\mathrm{R}}^2(t) + C_{\mathrm{I}}^2(t)}{C_{\mathrm{I}}(t)}, \tag{2.23}$$

$$\langle [\tilde{x}, \tilde{p}]_+ \rangle(t) = \langle \tilde{x}\tilde{p} + \tilde{p}\tilde{x} \rangle(t) = \hbar\,\alpha(t)\,\dot{\alpha}(t) = \hbar\frac{C_{\mathrm{R}}(t)}{C_{\mathrm{I}}(t)}, \tag{2.24}$$

where $[\ ,\ ]_+$ denotes the anti-commutator and $C(t)$ and $\alpha(t)$ are related by

$$C(t) = \frac{\dot{\alpha}}{\alpha} + \mathrm{i}\,\frac{1}{\alpha^2}\,. \tag{2.25}$$

[3]The index L indicates that this invariant corresponds to the conventional linear SE for non-dissipative systems. There is a similar invariant also for certain dissipative systems. Other formulations aside, these systems can also be described by NL modifications of the SE. Therefore, this invariant will be distinguished from the above one by the subscript NL.

It can be shown straitforwardly that also the Schrödinger–Robertson uncertainty relation [13, 14]

$$\langle \tilde{x}^2 \rangle \langle \tilde{p}^2 \rangle - \left(\frac{1}{2} \langle [\tilde{x}, \tilde{p}]_+ \rangle \right)^2 = \frac{\hbar^2}{4} \tag{2.26}$$

is fulfilled.

In order to obtain explicit expressions for the time-dependence of the uncertainties, essentially, the Ermakov equation (2.16) must be solved for given initial conditions $\alpha(t_0) \equiv \alpha_0$ and $\dot{\alpha}(t_0) \equiv \dot{\alpha}_0$, or the Riccati equation (2.4) for given κ_0.

It is interesting to note that the solution of the Ermakov equation (2.16) can also be obtained knowing two linear independent solutions $f_1(t)$ and $f_2(t)$ of the (linear) Newtonian equation (2.3). This can be achieved using the method of linear invariant operators and their relation with quadratic invariant operators, introduced by Man'ko et al. [15, 16] and outlined in Appendix A.[4]

The solution of the Ermakov equation (2.16) can then be given in the form

$$\alpha(t) = \left[\left(\dot{\alpha}_0^2 + \frac{1}{\alpha_0^2} \right) f_1^2(t) + \alpha_0^2 f_2^2(t) \mp 2\dot{\alpha}_0 \alpha_0 f_1(t) f_2(t) \right]^{\frac{1}{2}} \tag{2.27}$$

where the two solutions of the Newtonian equation and their time-derivatives have the initial conditions

$$f_1(t_0) = 0, \ \dot{f}_1(t_0) = -1, \quad f_2(t_0) = 1, \ \dot{f}_2(t_0) = 0 \tag{2.28}$$

with $f_1(t) = -\frac{1}{v_0}\eta(t)$ (for details, see Appendix A).

The initial conditions for α and $\dot{\alpha}$ can also be given in terms of the initial uncertainties and their correlation function as

$$\alpha_0 = \left(\frac{2m}{\hbar} \langle \tilde{x}^2 \rangle_0 \right)^{\frac{1}{2}}, \quad \dot{\alpha}_0 = \left(\frac{1}{2m\hbar \langle \tilde{x}^2 \rangle_0} \right)^{\frac{1}{2}} \langle [\tilde{x}, \tilde{p}]_+ \rangle_0 \, . \tag{2.29}$$

The relations between the initial condition κ_0 of the Riccati solution and the initial WP uncertainties can be expressed (using Eqs. (2.22–2.24)) as

$$\mathcal{C}(t_0) \equiv \mathcal{C}_0 = \frac{\dot{\alpha}_0}{\alpha_0} + \mathrm{i}\frac{1}{\alpha_0^2} \tag{2.30}$$

with $\mathcal{V}_0 = \frac{1}{\kappa_0} = \mathcal{C}_0 - \tilde{\mathcal{C}}_0$. Note that all previous results are valid for TD and well as TI frequency ω.

As an example, the WP solution of the HO with constant frequency ω_0 is now considered. For the particular solution $\tilde{\mathcal{C}}_+ = i\omega_0$ we have

---

[4] A different way to establish relations between linear and quadratic invariants and how the quadratic invariant can be related to the Ermakov invariant is given in [17].

$$\mathcal{V}_0 = \frac{\dot{\alpha}_0}{\alpha_0} + \mathrm{i}\left(\frac{1}{\alpha_0^2} - \omega_0\right), \quad \kappa_0 = \frac{\frac{\dot{\alpha}_0}{\alpha_0} - \mathrm{i}\left(\frac{1}{\alpha_0^2} - \omega_0\right)}{\frac{\dot{\alpha}_0^2}{\alpha_0^2} + \left(\frac{1}{\alpha_0^2} - \omega_0\right)^2} . \tag{2.31}$$

The particular solution itself corresponds to the WP with constant width, as the real part vanishes, $\tilde{C}_{+,\mathrm{R}} = \frac{\dot{\alpha}}{\alpha} = 0$, leading to

$$\langle \tilde{x}^2 \rangle = \frac{\hbar}{2m\omega_0}, \quad \langle \tilde{p}^2 \rangle = \frac{\hbar m \omega_0}{2}, \quad \langle [\tilde{x}, \tilde{p}]_+ \rangle = 0 , \tag{2.32}$$

which is also valid for the ground state of the HO.

The analytic expression for the solution of the Ermakov equation, corresponding to the WP with TD width, is

$$\alpha(t) = \alpha_0 \left[\left(\frac{\dot{\alpha}_0^2}{\alpha_0^2} + \frac{1}{\alpha_0^4} \right) \frac{1}{\omega_0^2} \sin^2 \omega_0 t + \cos^2 \omega_0 t \mp \frac{2}{\omega_0} \frac{\dot{\alpha}_0}{\alpha_0} \sin \omega_0 t \cos \omega_0 t \right]^{\frac{1}{2}} . \tag{2.33}$$

Even in the case where the initial spreading vanishes, $\dot{\alpha}_0 = 0$, the WP width is still oscillating. With the abbreviation $\beta_0 = \frac{1}{\alpha_0^2} = \frac{\hbar}{2m\langle \tilde{x}^2 \rangle_0}$, $\alpha(t)$ then takes the form

$$\alpha(t) = \alpha_0 \left[\cos^2 \omega_0 t + \left(\frac{\beta_0}{\omega_0} \sin \omega_0 t \right)^2 \right]^{\frac{1}{2}} . \tag{2.34}$$

So, whenever $\beta_0 \neq \omega_0$, i.e., $\langle \tilde{x}^2 \rangle_0$ is different from the ground state initial position uncertainty (as given in (2.32)), the width of the WP solution – even for constant ω_0 – is oscillating.

Knowing (2.33), the uncertainties and their correlation can be written explicitly as

$$\langle \tilde{x}^2 \rangle(t) = \frac{\hbar}{2m} \left[\left(\dot{\alpha}_0^2 + \frac{1}{\alpha_0^2} \right) \frac{1}{\omega_0^2} \sin^2 \omega_0 t + \alpha_0^2 \cos^2 \omega_0 t + 2\dot{\alpha}_0 \alpha_0 \frac{1}{\omega_0} \sin \omega_0 t \cos \omega_0 t \right] , \tag{2.35}$$

$$\langle \tilde{p}^2 \rangle(t) = \frac{m\hbar}{2} \left[\left(\dot{\alpha}_0^2 + \frac{1}{\alpha_0^2} \right) \cos^2 \omega_0 t + \alpha_0^2 \omega_0^2 \sin^2 \omega_0 t - 2\dot{\alpha}_0 \alpha_0 \omega_0 \sin \omega_0 t \cos \omega_0 t \right] , \tag{2.36}$$

$$\langle [\tilde{x}, \tilde{p}]_+ \rangle(t) = \frac{\hbar}{4} \left[\left\{ \left(\dot{\alpha}_0^2 + \frac{1}{\alpha_0^2} \right) \frac{1}{\omega_0} - \alpha_0^2 \omega_0 \right\} \sin 2\omega_0 t + 2\dot{\alpha}_0 \alpha_0 \cos 2\omega_0 t \right] . \tag{2.37}$$

The corresponding results for the free motion are obtained easily using

$$\lim_{\omega_0 \to 0} \frac{\sin \omega_0 t}{\omega_0 t} = t, \quad \lim_{\omega_0 \to 0} \cos \omega_0 t = 1 . \tag{2.38}$$

The solution of the Ermakov equation turns into

$$\alpha(t) = \alpha_0 \left[\left(\frac{\dot{\alpha}_0^2}{\alpha_0^2}\, t \mp 1 \right)^2 + \frac{t^2}{\alpha_0^4} \right]^{\frac{1}{2}} \tag{2.39}$$

and the uncertainties and their correlation into

$$\langle \tilde{x}^2 \rangle(t) = \frac{\hbar}{2m} \left[\left(\dot{\alpha}_0^2 + \frac{1}{\alpha_0^2} \right) t^2 + \alpha_0^2 \mp 2\dot{\alpha}_0 \alpha_0\, t \right] , \tag{2.40}$$

$$\langle \tilde{p}^2 \rangle(t) = \frac{m\hbar}{2} \left(\dot{\alpha}_0^2 + \frac{1}{\alpha_0^2} \right) , \tag{2.41}$$

$$\langle [\tilde{x}, \tilde{p}]_+ \rangle(t) = \frac{\hbar}{2} \left[\left(\dot{\alpha}_0^2 + \frac{1}{\alpha_0^2} \right) t + \dot{\alpha}_0 \alpha_0 \right] . \tag{2.42}$$

For $\dot{\alpha}_0 = 0$ the well-known textbook results are regained.

2.3.2 *Consequences of the Wave Packet Spreading for the Probability Current*

The non-classical aspect of the quantum mechanical WP solutions is expressed by the fact that the probability of finding the system somewhere in (position) space is not only restricted to a point, in this case the maximum of the WP, as in the classical situation, but also has non-vanishing positive values at all other positions in space. This probability distribution, characterized by the function $\varrho(x,t) = \Psi^*(x,t)\Psi(x,t)$ is not necessarily fixed in space and time but can evolve according to the continuity equation

$$\frac{\partial}{\partial t}\varrho + \frac{\partial}{\partial x}(\varrho\, v_-) = \frac{\partial}{\partial t}\varrho + \frac{\partial}{\partial x} j = 0 \tag{2.43}$$

with the probability current $j = \varrho v_-$, where the velocity field $v_-(x,t)$ is defined as

$$v_- = \frac{\hbar}{2mi} \left(\frac{\frac{\partial}{\partial x}\Psi}{\Psi} - \frac{\frac{\partial}{\partial x}\Psi^*}{\Psi^*} \right) = \frac{\hbar}{2mi} \frac{\partial}{\partial x} \ln \frac{\Psi}{\Psi^*} . \tag{2.44}$$

For a Gaussian WP, this velocity field is given by

$$v_- = \dot{\eta} + \frac{\dot{\alpha}}{\alpha}\tilde{x} , \tag{2.45}$$

i.e., the probability of finding the system at a particular point in space cannot only change due to the motion of the WP maximum, but also due to the (relative) change of its width, characterized by $\frac{\dot{\alpha}}{\alpha}$, being connected with a tunnelling current. The explicit form of this term determines the tunnelling dynamics.

Already for the free motion ($V = 0$) this term does not vanish and has the form of a Lorentzian curve,

$$\frac{\dot{\alpha}}{\alpha} = \beta_0 \frac{(\beta_0 t)}{1 + (\beta_0 t)^2} , \tag{2.46}$$

again with the abbreviation $\beta_0 = \frac{1}{\alpha_0^2}$, having the dimension of a frequency.

A formal difference compared to the continuity equation in classical statistical mechanics shall be mentioned. For Hamiltonian systems, in classical *phase space* (Γ-space) the divergence of the velocity field always vanishes, $\nabla_\Gamma \mathbf{v}_\Gamma = 0$, leading from the continuity equation to the Liouville equation. In our quantum mechanical situation (in *position* space only!) the divergence (here in one dimension) has the form

$$\frac{\partial}{\partial x} v_- = \frac{\dot{\alpha}}{\alpha} \tag{2.47}$$

and vanishes only for $\dot{\alpha} = 0$, i.e., WPs with constant width.

2.4 Linearization of the Complex Riccati Equation

Another property of the Riccati equation, particularly interesting in a quantum mechanical context, is the existence of a superposition principle for this NL differential equation [18, 19]. This is related to the fact that the Riccati equation can be linearized. This *linearization* is *not* an *approximation* of a NL equation by a linear one but an *exact* transformation. In our case, this can be achieved using the ansatz [20]

$$\left(\frac{2\hbar}{m} y\right) = C = \frac{\dot{\lambda}}{\lambda} \tag{2.48}$$

with complex $\lambda(t)$, leading to

$$\ddot{\lambda} + \omega^2(t)\lambda = 0 , \tag{2.49}$$

which has the form of the Newton-type equation (2.3) of the corresponding problem, but now for a complex variable.

First, a kind of geometric interpretation of the motion of $\lambda(t)$ in the complex plane shall be given. Expressed in Cartesian coordinates, λ can be written as $\lambda = u + \mathrm{i}z$, or in polar coordinates as $\lambda = \alpha \mathrm{e}^{\mathrm{i}\varphi}$. Inserting the polar form into Eq. (2.48) leads to

$$C = \frac{\dot{\alpha}}{\alpha} + \mathrm{i}\,\dot{\varphi} \tag{2.50}$$

where the real part is already identical to $\mathcal{C}_R$, as defined above.

The quantity α defined in $\mathcal{C}_I$ as being proportional to the position uncertainty is identical to the absolute value of λ if it can be shown that

$$\dot{\varphi} = \frac{1}{\alpha^2} \, . \tag{2.51}$$

This, however, can be proven by simply inserting real and imaginary parts of (2.50) into the imaginary part of the Riccati equation (2.4). Comparing relation (2.51), that can also be written in the form

$$\dot{z}u - \dot{u}z = \alpha^2 \dot{\varphi} = 1 \, , \tag{2.52}$$

with the motion of a particle under the influence of a central force in two-dimensional physical space, shows that this relation corresponds to the "conservation of angular momentum", but here for the motion in the *complex* plane!

Relation (2.52) also shows that real and imaginary parts, or phase and amplitude, respectively, of the complex quantity are not independent of each other but uniquely coupled. This coupling is due to the quadratic nonlinearity in the Riccati equation. We will find an analogous situation in the TI case, discussed in Chap. 3.

2.5 Time-Dependent Green Function or Feynman Kernel

After the physical meaning of the absolute value of $\lambda(t)$ in polar coordinates and its relation to the phase angle via (2.51), $\dot{\varphi} = \frac{1}{\alpha^2}$, have been clarified, the interpretation of the Cartesian coordinates u and z ($\lambda = u + \mathrm{i}z$) needs to be ascertained.

For this purpose, it can be utilized that the WP solution $\Psi_{\mathrm{WP}}(x,t)$ at *time t* can also be obtained with the help of an *initial* WP at time t' (e.g., $t' = 0$) and a *TD Green function*, also called time-propagator or Feynman kernel, via

$$\Psi_{\mathrm{WP}}(x,t) = \int dx' \, G(x,x',t,t'=0) \, \Psi_{\mathrm{WP}}\left(x',t'=0\right) \, . \tag{2.53}$$

For the considered Gaussian WP the initial distribution[5] is given by

$$\Psi_{\mathrm{WP}}\left(x',0\right) = \left(\frac{m}{\pi\hbar\alpha_0^2}\right)^{\frac{1}{4}} \exp\left\{\frac{\mathrm{i}m}{2\hbar}\left[\mathrm{i}\left(\frac{x'}{\alpha_0}\right)^2 + 2\frac{p_0}{m}x'\right]\right\} \, , \tag{2.54}$$

[5] If the initial change of the WP width, and thus $\dot{\alpha}_0$, is different from zero, the term $\mathrm{i}\left(\frac{x'}{\alpha_0}\right)^2$ in Eq. (2.54) must be replaced by $\left(\frac{\dot{\alpha}_0}{\alpha_0} + \mathrm{i}\frac{1}{\alpha_0^2}\right)x'^2 = \mathcal{C}_0 x'^2$. However, in the examples discussed in this section, this is not the case.

with $p_0 = \langle p \rangle(t=0) = m\dot{\eta}_0$. The integral kernel $G(x, x', t, t')$ was determined by Feynman using his path integral method [21]. Particularly for Gaussian WPs, the Green function also has the form of a Gaussian function[6] and can be written as

$$G\left(x, x', t, 0\right) = \left(\frac{m}{2\pi \mathrm{i}\hbar\alpha_0 z}\right)^{\frac{1}{2}} \exp\left\{\frac{im}{2\hbar}\left[\frac{\dot{z}}{z}x^2 - 2\frac{x}{z}\left(\frac{x'}{\alpha_0}\right) + \frac{u}{z}\left(\frac{x'}{\alpha_0}\right)^2\right]\right\} . \tag{2.55}$$

So far, $z(t)$ and $u(t)$ are only arbitrary TD parameters; only later will they be identified with imaginary and real parts of $\lambda(t)$!

As in the definition of $\Psi_{\mathrm{WP}}(x,t)$ according to (2.53) only G actually depends on x and t, therefore the kernel G, as defined in (2.55), must also fulfil the TDSE. Inserting (2.55) into the TDSE (2.2) and sorting according to powers of x shows that $z(t)$ and $u(t)$ not only fulfil the same Newtonian equation as $\eta(t)$ and $\lambda(t)$ but, in addition, are also uniquely *coupled* via the relation

$$\dot{z}u - \dot{u}z = 1 , \tag{2.56}$$

identical to the conservation law $\alpha^2\dot{\varphi} = 1$, as shown in Eq. (2.52).

The last step necessary for the identification of $z(t)$ and $u(t)$ is to explicitly perform the integration in (2.53) using (2.54) and (2.55) to yield the WP solution in the form

$$\Psi_{\mathrm{WP}}(x,t) = \left(\frac{m}{\pi\hbar}\right)^{\frac{1}{4}} \left(\frac{1}{u+\mathrm{i}z}\right)^{\frac{1}{2}} \exp\left\{\frac{im}{2\hbar}\left[\frac{\dot{z}}{z}x^2 - \frac{(x - \frac{p_0\alpha_0}{m}z)^2}{z(u+\mathrm{i}z)}\right]\right\} . \tag{2.57}$$

Comparison with the WP solution written in the form (2.1) shows that the following relations hold:

$$z(t) = \frac{m}{\alpha_0\, p_0}\, \eta(t) \tag{2.58}$$

and

$$\frac{\dot{z}}{z} - \frac{1}{z\,\lambda} = \frac{\dot{\lambda}}{\lambda} = C \tag{2.59}$$

where, in the latter case, $\lambda = u + \mathrm{i}z$ and Eq. (2.56) have been used. From Eq. (2.56), however, also $u(t)$ can be determined (up to an integration constant) once $z(t)$, i.e. the classical trajectory, is known, as

[6] The equivalence between deriving the TD Green function via a Gaussian ansatz or via Feynman's path integral method has been shown in [22, 23] where also the relation to the Ermakov invariant is considered. Starting from a more general Gaussian ansatz than (2.55), just using three TD parameters $a(t)$, $b(t)$ and $c(t)$, the Green function (2.55) was also obtained in [24]. An elegant method to derive the time propagator using the Ermakov system is shown in [25] and in [11] compared with the method described here.

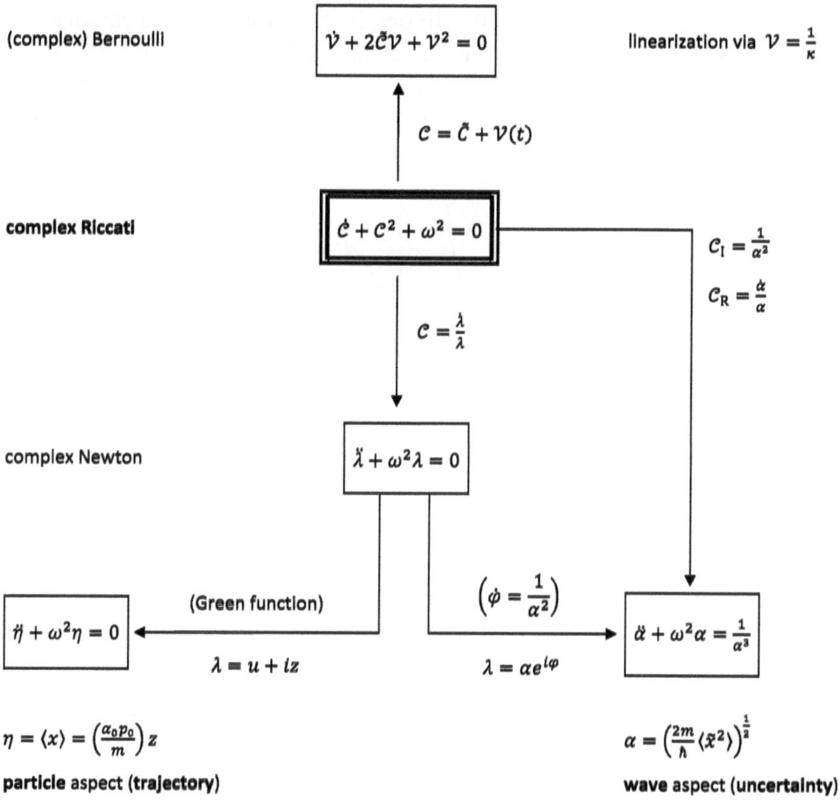

Fig. 2.2 Different treatments of the complex Riccati equation and relation to particle and wave aspects

$$u = -z \int^t \frac{1}{z^2(t')}\, dt' . \tag{2.60}$$

In Fig. 2.2 the different treatments of the complex Riccati equation (2.4) and their relation to particle and wave aspects of the system are summarized.

Knowing z and u, also $\alpha(t)$ can be determined via $\alpha = \left(u^2 + z^2\right)^{\frac{1}{2}}$. So, the knowledge of the solution of the classical Newtonian equation is sufficient to also obtain the solution of the Ermakov equation determining the dynamics of the quantum mechanical uncertainties (and thus the WP width). This is in agreement with the result in Sect. 2.3.1 where the solution of the Ermakov equation (following the method outlined in Appendix A) could also be written (see Eq. (2.27)) in terms of solutions of the corresponding Newtonian equation (there are two linear-independent solutions necessary, what is guaranteed here, as Eq. (2.56) shows that the Wronskian determinant of the two solutions z and u is different from zero).

On the other hand, it is also possible to obtain the solution of the classical Newtonian equation determining the dynamics of the WP maximum once the solution of the Ermakov equation is known. Knowing the amplitude $\alpha(t)$ of the complex quantity $\lambda = \alpha\, e^{i\varphi}$, the phase angle $\varphi(t)$ can be determined (up to an integration constant), using $\dot{\varphi} = \frac{1}{\alpha^2}$, via

$$\varphi = \int^{t} \frac{1}{\alpha^2(t')}\, dt' . \tag{2.61}$$

Knowing α and φ, also λ is known and thus, from its imaginary part, $z = \alpha \sin\varphi = \frac{m}{\alpha_0 p_0}\eta(t)$, the classical trajectory is obtained.

These interrelations between the dynamics of the classical and quantum mechanical aspects are summarized schematically in Fig. 2.3.

In conclusion, one can say that the complex quantity $\lambda(t)$ contains the particle as well as the wave aspects of the system. In polar coordinates, the absolute value α of λ is connected directly with the quantum mechanical position uncertainty. In Cartesian coordinates, the imaginary part of λ is directly proportional to the classical particle trajectory η. Absolute value and phase, or real and imaginary part, of λ

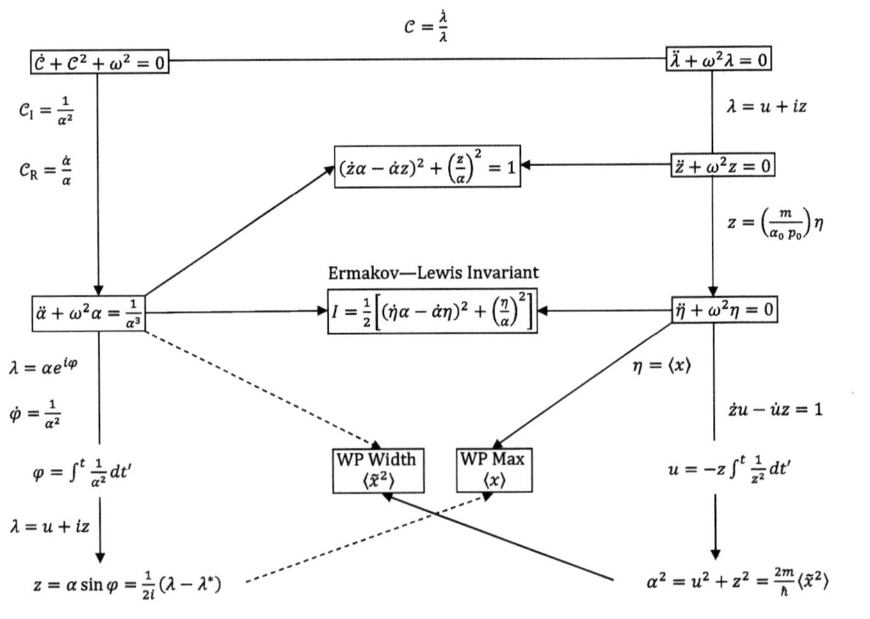

Fig. 2.3 Interrelations between the quantities determining the dynamics of maximum (classical aspect) and width (quantum mechanical aspect) of the wave packet solutions of the Schrödinger equation in position space

are not independent of each other but uniquely coupled via the conservation law (2.52) which has its origin in the quadratic nonlinearity of the corresponding Riccati equation (2.4).

2.5.1 Riccati Equations from the Green Function and Trigonometric Considerations

When inserting the TD Green function (2.55) into the SE (2.2) and sorting according to powers of x, from the terms proportional to x^2 one actually first obtains the Riccati equation

$$\frac{\partial}{\partial t}\left(\frac{\dot{z}}{z}\right)+\left(\frac{\dot{z}}{z}\right)^2+\omega^2=0\,, \tag{2.62}$$

which can then be linearized to the Newtonian equation for $z(t)$.

Then again, also the terms independent of x provide a Riccati equation although this is not obvious at first sight as they fulfil the equation

$$\frac{\partial}{\partial t}\left(\frac{u}{z}\right)+\frac{1}{z^2}=0\,. \tag{2.63}$$

Rewriting z and u in polar coordinates and using $\frac{\partial}{\partial t}=\left(\frac{d\varphi}{dt}\right)\frac{\partial}{\partial\varphi}=\frac{1}{\alpha^2}\frac{\partial}{\partial\varphi}$ leads to

$$\frac{\partial}{\partial t}\cot\varphi+\frac{1}{\alpha^2\sin^2\varphi}=\frac{1}{\alpha^2}\frac{\partial}{\partial\varphi}\cot\varphi+\frac{1}{\alpha^2\sin^2\varphi}=0\,, \tag{2.64}$$

or

$$\frac{\partial}{\partial\varphi}\cot\varphi=-\frac{1}{\sin^2\varphi}\,, \tag{2.65}$$

which is obviously correct. However, using the trigonometric relation $\cos^2\varphi+\sin^2\varphi=1$ to replace the 1 in the enumerator, Eq. (2.65) turns into the Riccati equation

$$\frac{\partial}{\partial\varphi}\cot\varphi+\cot^2\varphi+1=0\,. \tag{2.66}$$

Knowing that $\frac{\partial}{\partial\varphi}\tan\varphi=\frac{1}{\cos^2\varphi}$, it follows immediately that also the inverse function, $\tan\varphi$, fulfils a Riccati equation but now with a minus-sign for the derivative term,

$$-\frac{\partial}{\partial\varphi}\tan\varphi+\tan^2\varphi+1=0\,. \tag{2.67}$$

Changing to an imaginary variable, $\varphi \to i\varphi$ turns the trigonometric functions into hyperbolic ones. In this case, the derivatives are $\frac{\partial}{\partial\varphi}\coth\varphi = -\frac{1}{\sinh^2\varphi}$ and $\frac{\partial}{\partial\varphi}\tanh\varphi = \frac{1}{\cosh^2\varphi}$ and the relation $\cosh^2\varphi - \sinh^2\varphi = 1$ is valid, resulting in the Riccati equations

$$\frac{\partial}{\partial\varphi}\coth\varphi + \coth^2\varphi - 1 = 0 \tag{2.68}$$

and

$$\frac{\partial}{\partial\varphi}\tanh\varphi + \tanh^2\varphi - 1 = 0\,. \tag{2.69}$$

The differences compared with the trigonometric functions are that there is no change of sign of the derivative term and the sign of the inhomogeneity changes from plus to minus. The latter change would correspond in our TD quantum mechanical problem to a change from an attractive oscillator potential to a repulsive one, i.e. to $V = -\frac{m}{2}\omega^2 x^2$. This becomes even more obvious when changing the variables and their functions according to $\varphi \to \omega\varphi$, $f(\varphi) \to \omega f(\varphi)$, turning ± 1 into $\pm\omega^2$.

The hyperbolic functions and corresponding Riccati equations will become relevant in connection with dissipative systems, as discussed in Chap. 4, and when considering other fields of physics like statistical thermodynamics, NL dynamics or soliton theory, as will be done in Chap. 7.

The complex variable $\lambda = u + i\,z = \alpha\cos\varphi + i\,\alpha\sin\varphi$ also allows for a kind of trigonometric interpretation of the Ermakov invariant. For this purpose we take advantage of the imaginary part of λ being directly proportional to the classical trajectory, i.e., $z = \frac{m}{\alpha_0 p_0}\eta(t)$. Therefore, the second quadratic term of the invariant (2.21) is proportional to $\left(\frac{z}{\alpha}\right)^2 = \sin^2\varphi$. Consequently, the first term must be $\left(\frac{u}{\alpha}\right)^2 = \cos^2\varphi$ to yield a constant value for I_L. So the invariant can be written as

$$I_L = \frac{1}{2}\left(\frac{\alpha_0 p_0}{m}\right)^2\left[\left(\frac{u}{\alpha}\right)^2 + \left(\frac{z}{\alpha}\right)^2\right] = \text{const.} \tag{2.70}$$

Furthermore, the real part of λ can be expressed in terms of η, $\dot\eta$, α and $\dot\alpha$ as

$$u = \dot z\alpha^2 - z\dot\alpha\alpha = \left(\frac{m}{\alpha_0 p_0}\right)\left(\alpha^2\dot\eta - \dot\alpha\alpha\eta\right) = \left(\frac{m}{\alpha_0 p_0}\right)\alpha^2\left(\dot\eta - \frac{\dot\alpha}{\alpha}\eta\right)\,. \tag{2.71}$$

Defining a new variable $\mathcal{Y}(\varphi) = \frac{z}{\alpha} = \sin\varphi$ that depends on the angle φ instead of time t, the first term in the square brackets can be expressed as

$$\mathcal{Y}'(\varphi) = \frac{d}{d\varphi}\mathcal{Y} = \cos\varphi = \left(\frac{u}{\alpha}\right)\,. \tag{2.72}$$

In this form, the invariant (2.70) is formally equivalent to the Hamiltonian of a HO with angle-dependent variable $\mathcal{Y}(\varphi)$ (instead of TD variable $\eta(t)$) and unit frequency $\omega = 1$, leading to the corresponding equation of motion

$$\mathcal{Y}'' + 1^2 \mathcal{Y} = 0\,. \tag{2.73}$$

However, the time-dependence is implicitly contained in the time-dependence of the angle φ, i.e., $\varphi = \varphi(t)$. Therefore, expressing Eq. (2.73) as a differential equation with respect to time t instead of angle φ and using $\frac{d}{d\varphi} = \alpha^2 \frac{d}{dt}$ with $\mathcal{Y}(\varphi(t)) = \hat{\mathcal{Y}}(t)$ yields

$$\frac{d^2}{dt^2}\hat{\mathcal{Y}}(t) + 2\frac{\dot{\alpha}}{\alpha}\frac{d}{dt}\hat{\mathcal{Y}}(t) + \dot{\varphi}^2\hat{\mathcal{Y}}(t) = 0\,. \tag{2.74}$$

For $\dot{\alpha} = 0$, i.e. $\alpha =$ constant, with $z(t) \propto \eta(t)$ and $\dot{\varphi} =$ const. $= \omega_0$, Eq. (2.74) just turns into $\ddot{\eta} + \omega_0^2 \eta = 0$, i.e., Eq. (2.3) for TI frequency $\omega = \omega_0$ and a WP solution with constant width.

For $\dot{\alpha} \neq 0$, i.e., $\alpha = \alpha(t)$, an additional first-derivative term appears in (2.74) that looks like a linear velocity dependent friction force in the Langevin equation (see below, Chap. 4) with friction coefficient $2\frac{\dot{\alpha}}{\alpha}$. At first sight this looks contradictory as we are not dealing so far with dissipative systems with irreversible time-evolution. A closer look shows that Eq. (2.74) is actually still invariant under time-reversal as the coefficient of the second term also contains a time-derivative (in $\frac{\dot{\alpha}}{\alpha}$). So, together with $\frac{\partial}{\partial t}\mathcal{Y}$, this term also does not change its sign under time-reversal (unlike in the Langevin equation where the friction coefficient γ is usually assumed to be constant). Equation (2.74) takes into account that not only the angle $\varphi(t)$ of the complex quantity $\lambda(t)$ describing the system, but also its amplitude α may change in time.

2.6 Lagrange–Hamilton Formalism for Quantum Uncertainties

In classical mechanics, the Hamiltonian function is not only representing the energy of a (conservative) system but also supplies the equations of motion for the system. In quantum mechanics, the mean value of the Hamiltonian operator (H_{op}) (in the cases considered so far) does not only supply the classical energy (E_{cl}, equivalent to the classical Hamiltonian function) but also a quantum mechanical contribution ($\tilde{E}$) due to the position and momentum uncertainties, i.e.,

$$\begin{aligned}\langle H_{\rm op}\rangle &= \frac{1}{2m}\langle p^2\rangle + \frac{m}{2}\,\omega^2\langle x^2\rangle \\ &= \left(\frac{1}{2m}\langle p\rangle^2 + \frac{m}{2}\,\omega^2\langle x\rangle^2\right) + \left(\frac{1}{2m}\langle \tilde{p}^2\rangle + \frac{m}{2}\,\omega^2\langle \tilde{x}^2\rangle\right) \\ &= E_{\rm cl} + \tilde{E} = (T_{\rm cl} + V_{\rm cl}) + (\tilde{T} + \tilde{V})\end{aligned} \tag{2.75}$$

(with $\langle p^2\rangle = \langle p\rangle^2 + \langle \tilde{p}^2\rangle$ and $\langle x^2\rangle = \langle x\rangle^2 + \langle \tilde{x}^2\rangle$). For the WP solution of the HO, the quantum mechanical contribution $\tilde{E}$ just represents the ground state energy, i.e. $\tilde{E} = \frac{\hbar}{2}\omega_0$.

However, as we have seen that equations of motion exist also for the quantum uncertainties, could this quantum contribution to the energy also be formulated in a way that it provides a Lagrangian/Hamitonian formalism for the quantum uncertainties as in the classical situation? It is shown subsequently that this question can be answered positively.

For this purpose the difference between kinetic and potential energy uncertainties is written as Lagrangian function $\tilde{\mathcal{L}}$ depending on the variables α, φ and the corresponding velocities $\dot{\alpha}$ and $\dot{\varphi}$, i.e.,

$$\tilde{\mathcal{L}}(\alpha, \dot{\alpha}, \varphi, \dot{\varphi}) = \tilde{T} - \tilde{V} = \frac{\hbar}{4}\left(\dot{\alpha}^2 + \alpha^2\dot{\varphi}^2 - \omega^2\alpha^2\right) \tag{2.76}$$

where Eqs. (2.22), (2.23) and (2.51), as well as the analogy to the two-dimensional motion in a real plane, expressed in polar coordinates, have been used.

The corresponding Euler–Lagrange equations are then

$$\frac{d}{dt}\frac{\partial\tilde{\mathcal{L}}}{\partial\dot{\varphi}} - \frac{\partial\tilde{\mathcal{L}}}{\partial\varphi} = 0\,, \tag{2.77}$$

$$\frac{d}{dt}\frac{\partial\tilde{\mathcal{L}}}{\partial\dot{\alpha}} - \frac{\partial\tilde{\mathcal{L}}}{\partial\alpha} = 0\,. \tag{2.78}$$

From the first equation follows $\frac{d}{dt}\left(\frac{\hbar}{2}\alpha^2\dot{\varphi}\right) = 0$, or, $\alpha^2\dot{\varphi} = \text{const.}$, in agreement with Eq. (2.51) (for const. = 1); from the second equation follows $\ddot{\alpha} + \omega^2\alpha = \dot{\varphi}^2\alpha = \frac{(\text{const.})^2}{\alpha^3}$, equivalent to Eq. (2.16) (again for const. = 1; in general, the "constant" is proportional to an "angular momentum").

The corresponding canonical momenta are then given by

$$\frac{\partial\tilde{\mathcal{L}}}{\partial\dot{\varphi}} = \frac{\hbar}{2}\alpha^2\dot{\varphi} = p_\varphi\,, \tag{2.79}$$

$$\frac{\partial\tilde{\mathcal{L}}}{\partial\dot{\alpha}} = \frac{\hbar}{2}\dot{\alpha} = p_\alpha\,. \tag{2.80}$$

An interesting point is that, in the case of our Gaussian WP, we found particularly $\dot{\varphi} = \frac{1}{\alpha^2}$, therefore the "*angular momentum*" p_φ is not only constant but has the value

$$p_\varphi = \frac{\hbar}{2}\,, \tag{2.81}$$

a value that does not usually describe an orbital angular momentum in quantum mechanics, but the non-classical angular momentum-type quantity *spin*. So, is spin just an angular momentum for the motion, in this case of $\lambda(t)$, in the *complex plane*?

With the help of the canonical momenta, the quantum energy contribution $\tilde{E} = \tilde{T} + \tilde{V}$ can be written in a Hamiltonian form as

$$\tilde{\mathcal{H}}(\alpha, p_\alpha, \varphi, p_\varphi) = \frac{p_\alpha^2}{\hbar} + \frac{p_\varphi^2}{\hbar\alpha^2} + \frac{\hbar}{4}\omega^2\alpha^2 . \tag{2.82}$$

The Hamiltonian equations of motion then take the form

$$\frac{\partial\tilde{\mathcal{H}}}{\partial p_\varphi} = \frac{2}{\hbar}\frac{p_\varphi}{\alpha^2} = \frac{1}{\alpha^2} = \dot{\varphi} \ , \quad \frac{\partial\tilde{\mathcal{H}}}{\partial\varphi} = 0 = -\dot{p}_\varphi \tag{2.83}$$

$$\frac{\partial\tilde{\mathcal{H}}}{\partial p_\alpha} = \frac{2}{\hbar}p_\alpha = \dot{\alpha} \ , \quad \frac{\partial\tilde{\mathcal{H}}}{\partial\alpha} = \frac{\hbar}{2}\left(\omega^2\alpha - \frac{1}{\alpha^3}\right) = -\frac{\hbar}{2}\ddot{\alpha} = -\dot{p}_\alpha \ , \tag{2.84}$$

which is in agreement with the previous results.

With these variables, the uncertainty product can be written as

$$U = \langle\tilde{x}^2\rangle\langle\tilde{p}^2\rangle = p_\varphi^2 + (\alpha\ p_\alpha)^2 . \tag{2.85}$$

The second term on the rhs describes the deviation from the minimum uncertainty $p_\varphi^2 = \frac{\hbar^2}{4}$ and is given by the product of the "radial" variable α and the corresponding momentum p_α, depending on the time-dependence of the WP width according to $p_\alpha = \frac{\hbar}{2}\dot{\alpha}$.

Furthermore, the quantum uncertainties can also be expressed in terms of the complex quantities λ and $\dot{\lambda}$ and their complex conjugates as

$$\langle\tilde{x}^2\rangle = \frac{\hbar}{2m}\lambda\lambda^* \tag{2.86}$$

$$\langle\tilde{p}^2\rangle = \frac{\hbar m}{2}\dot{\lambda}\dot{\lambda}^* \tag{2.87}$$

$$\langle[\tilde{x}, \tilde{p}]_+\rangle = \frac{\hbar}{2}\frac{\partial}{\partial t}(\lambda\lambda^*) . \tag{2.88}$$

Expressed in these variables, the Lagrangian can be written as

$$\tilde{\mathcal{L}}(\lambda, \dot{\lambda}, \lambda^*, \dot{\lambda}^*) = \frac{\hbar}{4}\left(\dot{\lambda}\dot{\lambda}^* - \omega^2\lambda\lambda^*\right) , \tag{2.89}$$

leading to the Euler–Lagrange equations

$$\frac{d}{dt}\frac{\partial\tilde{\mathcal{L}}}{\partial\dot{\lambda}^*} - \frac{\partial\tilde{\mathcal{L}}}{\partial\lambda^*} = \frac{\hbar}{4}\left(\ddot{\lambda} + \omega^2\lambda\right) = 0 , \tag{2.90}$$

$$\frac{d}{dt}\frac{\partial\tilde{\mathcal{L}}}{\partial\dot{\lambda}} - \frac{\partial\tilde{\mathcal{L}}}{\partial\lambda} = \frac{\hbar}{4}\left(\ddot{\lambda}^* + \omega^2\lambda^*\right) = 0 , \tag{2.91}$$

i.e., the complex Newtonian equation (2.49) and its complex conjugate.

With the canonical momenta

$$p_\lambda = \frac{\partial \tilde{\mathcal{L}}}{\partial \dot{\lambda}} = \frac{\hbar}{4}\dot{\lambda}^* \quad , \quad p_{\lambda^*} = \frac{\partial \tilde{\mathcal{L}}}{\partial \dot{\lambda}^*} = \frac{\hbar}{4}\dot{\lambda} \, , \tag{2.92}$$

the Hamiltonian $\tilde{\mathcal{H}}$ can be obtained from $\tilde{\mathcal{H}} = \dot{\lambda}\; p_\lambda + \dot{\lambda}^*\; p_{\lambda^*} - \tilde{\mathcal{L}}$ as

$$\tilde{\mathcal{H}}(\lambda, \lambda^*, p_\lambda, p_{\lambda^*}) = \frac{4}{\hbar} p_\lambda p_{\lambda^*} + \frac{\hbar}{4}\lambda\lambda^* \, , \tag{2.93}$$

leading to the Hamiltonian equations of motion

$$\frac{\partial \tilde{\mathcal{H}}}{\partial p_\lambda} = \frac{4}{\hbar} p_{\lambda^*} = \dot{\lambda} \; , \quad \frac{\partial \tilde{\mathcal{H}}}{\partial \lambda} = \frac{\hbar}{4}\omega^2 \lambda^* = -\frac{\hbar}{4}\ddot{\lambda}^* = -\dot{p}_\lambda \tag{2.94}$$

$$\frac{\partial \tilde{\mathcal{H}}}{\partial p_{\lambda^*}} = \frac{4}{\hbar} p_\lambda = \dot{\lambda}^* \; , \quad \frac{\partial \tilde{\mathcal{H}}}{\partial \lambda^*} = \frac{\hbar}{4}\omega^2 \lambda = -\frac{\hbar}{4}\ddot{\lambda} = -\dot{p}_{\lambda^*} \, , \tag{2.95}$$

agreeing with the results in Sect. 2.4.

In conclusion it can be stated that the quantum mechanical energy contribution $\tilde{E} = \tilde{T} + \tilde{V}$, expressed in terms of the real variables $\alpha, \dot{\alpha}, \varphi$ and $\dot{\varphi}$ as well as in terms of the complex variables λ, $\dot{\lambda}$, λ^* and $\dot{\lambda}^*$ (and corresponding conjugate momenta) can be used as a Hamiltonian function (or $\tilde{\mathcal{L}} = \tilde{T} - \tilde{V}$ as a Lagrangian function) to obtain the correct equations of motion for the quantum uncertainties in a Hamiltonian (or Lagrangian) formalism.

2.7 Momentum Space Representation

The solution of the TDSE in the momentum space representation, where $p_{\mathrm{op}} = p$ and $x_{\mathrm{op}} = -\frac{\hbar}{\mathrm{i}}\frac{\partial}{\partial p}$, can be obtained via Fourier transformation of the solution in position space,

$$\Psi(p,t) = \frac{1}{\sqrt{2\pi}} \int_{-\infty}^{+\infty} dx \; \mathrm{e}^{\frac{\mathrm{i}}{\hbar} px} \; \Psi(x,t) \, . \tag{2.96}$$

In particular for the WP solution (2.1), the Fourier transformation leads to a Gaussian WP in momentum space,

$$\Psi_{\mathrm{WP}}(p,t) = \left(\frac{a}{\hbar}\right)^{\frac{1}{2}} N(t) \exp\left\{-\frac{a}{2\hbar^2}\tilde{p}^2 - \mathrm{i}\frac{\langle x \rangle}{\hbar}\tilde{p} + \mathrm{i}\left(K - \frac{\langle x \rangle \langle p \rangle}{\hbar}\right)\right\}, \tag{2.97}$$

with $\tilde{p} = p - \langle p \rangle$ and the complex quantity $a(t) = a_{\mathrm{R}}(t) + \mathrm{i}a_{\mathrm{I}}(t)$ that is related to $y(t)$ via

$$\left(\frac{am}{\mathrm{i}\hbar}\right) = \frac{\lambda}{\dot{\lambda}} = \left(\frac{2\hbar}{m}y\right)^{-1} = \mathcal{C}^{-1} \,. \tag{2.98}$$

Particularly, the real part of $a(t)$ is related to the momentum uncertainty via

$$\left(\frac{a_{\mathrm{R}}m}{\hbar}\right) = \frac{m\hbar}{2\langle\tilde{p}^2\rangle} \,. \tag{2.99}$$

The equation of motion for $a(t)$ can be obtained by inserting the WP $\Psi_{\mathrm{WP}}(p,t)$ into the TDSE that, in momentum space, has the form

$$i\hbar\frac{\partial}{\partial t}\Psi_{\mathrm{WP}}(p,t) = \left(\frac{p^2}{2m} - \frac{m}{2}\omega^2\hbar^2\frac{\partial^2}{\partial p^2}\right)\Psi_{\mathrm{WP}}(p,t) \,. \tag{2.100}$$

The terms proportional to $\tilde{p}^2$ again leads to a complex Riccati equation,

$$-\left(\frac{\dot{a}m}{\mathrm{i}\hbar}\right) + \omega^2\left(\frac{am}{\mathrm{i}\hbar}\right)^2 + 1 = 0 \,. \tag{2.101}$$

With the above-mentioned definition $\frac{am}{\mathrm{i}\hbar} = \frac{\lambda}{\dot{\lambda}}$, this equation can be linearized to provide the same complex Newtonian equation as in position space, i.e., $\ddot{\lambda} + \omega^2\lambda = 0$ (Eq. 2.49). So the equation of motion for the uncertainties in the respective spaces can be obtained simply by inverting the relevant quantities (without Fourier transformation).

In the case of a constant particular solution, Eq. (2.101) can be brought into a form that differs from the one in position space, i.e. Eq. (2.4), only by the sign of the derivative term. Defining a new variable

$$\mathcal{K}(t) = -\tilde{\mathcal{C}}^2\mathcal{C}^{-1} = \omega^2\mathcal{C}^{-1} \tag{2.102}$$

with $\tilde{\mathcal{C}} = \pm\mathrm{i}\omega$, Eq. (2.101) can be rewritten as

$$-\dot{\mathcal{K}} + \mathcal{K}^2 + \omega^2 = 0 \,, \tag{2.103}$$

in agreement with Eqs. (2.66) and (2.67), where also $\cot\varphi = \frac{\cos\varphi}{\sin\varphi}$ and $\tan\varphi = \frac{\sin\varphi}{\cos\varphi}$ are logarithmic derivatives and the inverse of each other, obeying Riccati equations that also only differ by the sign of the derivative term.

In particular for $V = 0$ (see Bernoulli equation in position space), one obtains from (2.98)

$$\frac{\lambda}{\dot{\lambda}} = \frac{am}{\mathrm{i}\hbar} = t - \mathrm{i}\,\alpha_0^2 \,, \tag{2.104}$$

i.e., $\alpha_0^2 = \frac{\hbar}{2m}\langle\tilde{x}^2\rangle_0$ acts as a kind of *imaginary time-variable* that is related to the position uncertainty. Attempts to complexify physical quantities (also in the context of dissipative systems (Dekker and $\mathcal{PT}$-symmetry [26–32]) have recently gained growing interest (some references can also be found in [33–35]). Also the transition from quantum mechanical descriptions to those in statistical thermodynamics by replacing t by $\mathrm{i}\,\frac{\hbar}{\mathsf{k}_\mathrm{B}\mathsf{T}}$ (with k_B = Boltzmann's constant and T = temperature) are familiar [36, 37]. A formal comparison of the imaginary part of (2.104) with this replacement would lead to $\alpha_0^2 = \frac{\hbar}{\mathsf{k}_\mathrm{B}\mathsf{T}}$, or $\frac{\hbar}{2}\frac{1}{\alpha_0^2} = \frac{\hbar}{2}\,\dot{\varphi}_0 = \frac{\hbar}{2}\,\omega_0 = \frac{1}{2}\mathsf{k}_\mathrm{B}\mathsf{T}$, i.e., relate the quantum mechanical ground state energy of the HO with $\frac{1}{2}\mathsf{k}_\mathrm{B}\mathsf{T}$, the energy attributed to each degree of freedom (that is quadratic in the canonical variables) in statistical thermodynamics. Further formal similarities will be mentioned in Sects. 5.4 and 7.1.

The connections between the NL Riccati equations in position and momentum space (for arbitrary particular solutions and ω) and the linear complex Newtonian equation for $\lambda(t)$ are schematically summarized in Fig. 2.4.

In order to obtain the quantities related to the respective uncertainties (position/momentum), it is not necessary to involve a Fourier transformation, only the variable that fulfils the complex Riccati equation must be inverted.

Also in momentum space, the complex Riccati equation is equivalent to a real NL Ermakov equation. For this purpose the imaginary part of $\left(\frac{am}{\mathrm{i}\hbar}\right) = C^{-1}$ that is proportional to $a_\mathrm{R}(t)$, is expressed in terms of a new real variable $\epsilon_\mathrm{L}(t)$ according to

$$\frac{a_\mathrm{R} m}{\hbar} = \frac{m\hbar}{2\langle\tilde{p}^2\rangle_\mathrm{L}} = \frac{1}{\epsilon_\mathrm{L}^2}\,, \tag{2.105}$$

where ϵ_L, given as $\epsilon_\mathrm{L} = \sqrt{\frac{2}{m\hbar}\langle\tilde{p}^2\rangle_\mathrm{L}}$, is proportional to the WP width in momentum space, in analogy with the relation between α_L and the WP width in position space.

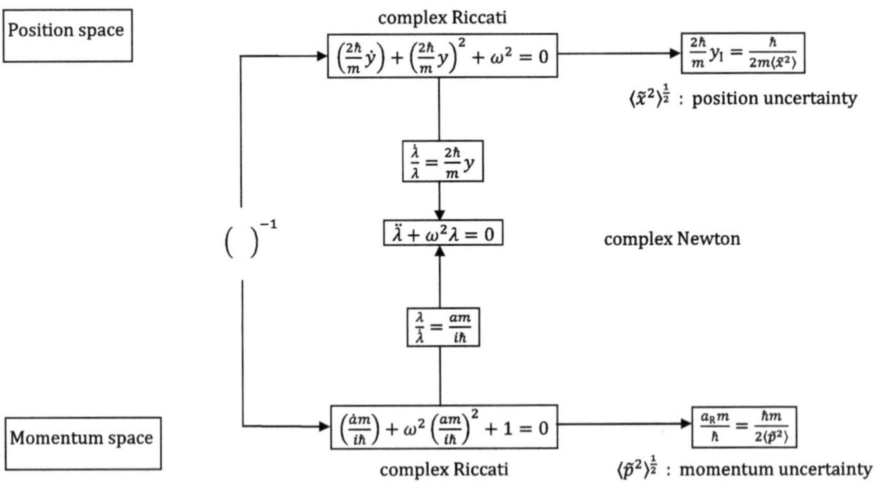

Fig. 2.4 Riccati equations in position and momentum space

Inserting this into Eq. (2.101) yields

$$\frac{a_{\text{I}}m}{\hbar} = \frac{1}{2\omega^2}\frac{\dot{a}_{\text{R}}}{a_{\text{R}}} = -\frac{1}{\omega^2}\frac{\dot{\epsilon}_{\text{L}}}{\epsilon_{\text{L}}} \tag{2.106}$$

and further

$$\ddot{\epsilon}_{\text{L}} + \omega^2 \epsilon_{\text{L}} = \frac{\omega^4}{\epsilon_{\text{L}}^3}\,, \tag{2.107}$$

i.e., an Ermakov equation for ϵ_{L}. This quantity is related to α_{L} via

$$\epsilon_{\text{L}}^2 = \left(\frac{4}{\hbar^2}U_{\text{L}}\right)\frac{1}{\alpha_{\text{L}}^2}\,, \tag{2.108}$$

where $U_{\text{L}} = \langle\tilde{x}^2\rangle_{\text{L}}\langle\tilde{p}^2\rangle_{\text{L}}$ is the uncertainty product. For a minimum uncertainty WP, i.e. $U_{\text{L}} = \frac{\hbar^2}{4}$, it follows $\epsilon_{\text{L}}^2 = \frac{1}{\alpha_{\text{L}}^2}$. If this is not the case, as for WPs with TD width, at least $\epsilon_{\text{L},0}^2 = \frac{1}{\alpha_{\text{L},0}^2}$ is valid, providing the WP fulfils at the initial time $t_0 = 0$ the minimum uncertainty requirement, i.e., $U_{\text{L},0} = \langle\tilde{x}^2\rangle_{\text{L},0}\langle\tilde{p}^2\rangle_{\text{L},0} = \frac{\hbar^2}{4}$. This is usually guaranteed in the conservative case if $\dot{\alpha}_0 = 0$ (but not necessarily in the dissipative one, as shown in Sect. 5.6.2).

One can achieve an even closer formal similarity to the position-space version by replacing ϵ_L according to

$$R_{\text{L}} = \frac{\epsilon_{\text{L}}}{\omega} \tag{2.109}$$

and rewriting the complex variable $\mathcal{K}(t)$ fulfilling the Riccati equation (2.103) in terms of R_{L} and its time-derivative. With

$$\mathcal{K}(t) = \omega^2\left(\frac{am}{i\hbar}\right) = \omega^2\left(\frac{a_{\text{I}}m}{\hbar} - \text{i}\frac{a_{\text{R}}m}{\hbar}\right) = \mathcal{K}_{\text{R}}(t) + \text{i}\,\mathcal{K}_{\text{I}}(t)\,, \tag{2.110}$$

imaginary and real parts can be rewritten as

$$\mathcal{K}_{\text{I}} = -\frac{\omega^2}{\epsilon_{\text{L}}^2} = -\frac{1}{R_{\text{L}}^2} \tag{2.111}$$

and

$$\mathcal{K}_{\text{R}} = -\frac{\dot{\epsilon}_{\text{L}}}{\epsilon_{\text{L}}} = -\frac{\dot{R}_{\text{L}}}{R_{\text{L}}} \tag{2.112}$$

for constant frequency $\omega = \omega_0$ which will be considered in the following. (Otherwise an additional term $\frac{\dot{\omega}}{\omega}$ on the rhs of (2.112) must be taken into account.)

This turns the complex Riccati equation (2.103) into the real Ermakov equation

$$\ddot{R}_{\text{L}} + \omega^2 R_{\text{L}} = \frac{1}{R_{\text{L}}^3}\,. \tag{2.113}$$

The equation of motion for the WP maximum is simply Newtons equation of motion that provides the time dependence of $\langle p\rangle = m\dot{\eta}$ via $m\ddot{\eta} + m\omega^2\eta = \frac{d}{dt}\langle p\rangle + m\omega^2\langle x\rangle = 0$. With $\Pi = \langle p\rangle = m\dot{\eta}$ this leads to the second equation of the Ermakov pair in the form

$$\ddot{\Pi} + \omega^2\Pi = 0\,. \tag{2.114}$$

Together with (2.113) this leads, via elimination of ω^2, to the invariant

$$I_{\mathrm{L,p}} = \frac{1}{2}\left[\left(\dot{\Pi}R_{\mathrm{L}} - \Pi\dot{R}_{\mathrm{L}}\right)^2 + \left(\frac{\Pi}{R_{\mathrm{L}}}\right)^2\right] = \text{const.} \tag{2.115}$$

For $\dot{\Pi}(0) = 0$ (i.e., $\dot{p}_0 = 0$) and $\dot{R}_{\mathrm{L}}(0) = 0$ (i.e., $\dot{\epsilon}_0 = 0$), this invariant takes the value

$$I_{\mathrm{L,p}} = \frac{1}{2}\left(\frac{p_0\,\omega_0}{\epsilon_0}\right)^2 = \frac{1}{2}\left(\frac{p_0\,\alpha_0}{m}\right)^2 m^2\omega_0^2\,, \tag{2.116}$$

where an initial minimum uncertainty WP has been assumed, i.e., $\epsilon_0 = \frac{1}{\alpha_0}$. So, apart from the constant factor $m^2\omega_0^2$, this invariant is identical to the one in position space.

Like in position space, also in momentum space is it possible to obtain the WP solution $\Psi_{\mathrm{WP}}(p,t)$ from an initial WP $\Psi_{\mathrm{WP}}(p',t'=0)$ via a corresponding propagator $G(p,p',t,t'=0)$ according to

$$\Psi_{\mathrm{WP}}(p,t) = \int dp'\; G(p,p',t,t'=0)\;\Psi_{\mathrm{WP}}\left(p',0\right)\,. \tag{2.117}$$

To obtain the explicit quantities, particularly for Gaussian WPs, one can start with a Fourier transformation of the initial state in position space, $\Psi_{\mathrm{WP}}(x',t'=0)$ and then use again a Gaussian ansatz for $G(p,p',t,t')$ like in position space.

Fourier transformation of the initial state (2.54) leads to[7]

$$\begin{aligned}
\Psi_{\mathrm{WP}}\left(p',0\right) &= \left(\frac{\alpha_0^2}{\pi\hbar m}\right)^{\frac{1}{4}} \exp\left\{-\frac{m\alpha_0^2}{2\hbar}\left(\frac{p'-p_0}{m}\right)^2\right\} \\
&= \left(\frac{1}{\pi\hbar m\epsilon_0^2}\right)^{\frac{1}{4}} \exp\left\{-\frac{m}{2\hbar}\left(\frac{p'-p_0}{m\epsilon_0}\right)^2\right\} \\
&= \left(\frac{\alpha_0^2}{\pi\hbar m}\right)^{\frac{1}{4}} \exp\left\{-\frac{m}{2\hbar}\left(\frac{\alpha_0 p_0}{m}\right)^2\right\} \exp\left\{\frac{\mathrm{i}m}{2\hbar}\left[\mathrm{i}\left(\frac{\alpha_0 p'}{m}\right)^2 - 2\mathrm{i}\alpha_0^2\frac{p_0}{m^2}\,p'\right]\right\} \\
&= \left(\frac{1}{\pi\hbar m\epsilon_0^2}\right)^{\frac{1}{4}} \exp\left\{-\frac{m}{2\hbar}\left(\frac{p_0}{m\epsilon_0}\right)^2\right\} \exp\left\{\frac{\mathrm{i}m}{2\hbar}\left[\mathrm{i}\left(\frac{p'}{m\epsilon_0}\right)^2 - 2\mathrm{i}\frac{p_0}{m^2\epsilon_0^2}\,p'\right]\right\}.
\end{aligned} \tag{2.118}$$

[7]The replacement of α_0 by $\frac{1}{\epsilon_0}$ implies that the initial state is a minimum uncertainty WP, i.e., $\langle\tilde{x}^2\rangle_0\langle\tilde{p}^2\rangle_0 = \frac{\hbar^2}{4}$.

The first exponential term on the rhs of line three and four is independent of p' and just a constant factor that can be taken care of after the integration performed in (2.117) via an appropriate normalization (note that the term in the exponent is, apart from a constant factor, simply the Ermakov invariant in position space). Without this first exponential term the expression for $\Psi_{WP}(p', 0)$ looks very much like the one for $\Psi_{WP}(x', 0)$, essentially replacing $\frac{x'}{\alpha_0}$ by $\frac{p'}{m\epsilon_0} = \frac{\alpha_0\, p'}{m}$.

The Gaussian propagator attains the form

$$\begin{aligned} G\left(p, p', t, 0\right) &= \left(\frac{-\alpha_0}{2\pi\hbar m\dot{z}}\right)^{\frac{1}{2}} \exp\left\{-\frac{im}{2\hbar}\left[\frac{z}{\dot{z}}\frac{p^2}{m^2} - 2i\frac{p}{m\dot{z}}\left(\frac{\alpha_0 p'}{m}\right) + \left(\frac{\dot{u}}{\dot{z}} + 2i\right)\left(\frac{\alpha_0 p'}{m}\right)^2\right]\right\} \\ &= \left(\frac{-1}{2\pi\hbar m\epsilon_0\dot{z}}\right)^{\frac{1}{2}} \exp\left\{-\frac{im}{2\hbar}\left[\frac{z}{\dot{z}}\frac{p^2}{m^2} - 2i\frac{p}{m\dot{z}}\left(\frac{p'}{m\epsilon_0}\right) + \left(\frac{\dot{u}}{\dot{z}} + 2i\right)\left(\frac{p'}{m\epsilon_0}\right)^2\right]\right\} . \end{aligned} \tag{2.119}$$

Performing the integration according to (2.117) yields the WP solution in the form

$$\Psi_{WP}\left(p, t\right) = \left(\frac{1}{\pi\hbar m}\right)^{\frac{1}{4}} \left(\frac{i}{\dot{\lambda}}\right)^{\frac{1}{2}} \exp\left\{-\frac{im}{2\hbar}\left[\frac{z}{\dot{z}}\frac{p^2}{m^2} + \frac{1}{\dot{z}\dot{\lambda}}\frac{\tilde{p}^2}{m^2}\right]\right\} . \tag{2.120}$$

Comparison with the corresponding WP in position space,

$$\Psi_{WP}\left(x, t\right) = \left(\frac{m}{\pi\hbar}\right)^{\frac{1}{4}} \left(\frac{1}{\lambda}\right)^{\frac{1}{2}} \exp\left\{\frac{im}{2\hbar}\left[\frac{\dot{z}}{z}x^2 - \frac{1}{z\lambda}\tilde{x}^2\right]\right\} , \tag{2.121}$$

shows that for the transition from position to momentum space or vice versa (in this form), essentially the following substitutions are required: $x \leftrightarrow \frac{p}{m}$, $+ \leftrightarrow -$ and z, u or $\lambda \leftrightarrow \dot{z}$, $\dot{u}$ or $\dot{\lambda}$.

A similar symmetry is also found when the WPs are finally written in the form

$$\Psi_{WP}(x, t) = \left(\frac{m}{\pi\hbar}\right)^{\frac{1}{4}} \left(\frac{1}{\lambda}\right)^{\frac{1}{2}} \exp\left\{\frac{im}{2\hbar}\left(\frac{\dot{\lambda}}{\lambda}\right)\tilde{x}^2 + \frac{i}{\hbar}\langle p\rangle\tilde{x} + \frac{i}{\hbar}\frac{\langle p\rangle\langle x\rangle}{2}\right\} \tag{2.122}$$

and

$$\Psi_{WP}(p, t) = \left(\frac{1}{\pi\hbar m}\right)^{\frac{1}{4}} \left(\frac{i}{\dot{\lambda}}\right)^{\frac{1}{2}} \exp\left\{-\frac{im}{2\hbar}\left(\frac{\lambda}{\dot{\lambda}}\right)\frac{\tilde{p}^2}{m^2} - \frac{i}{\hbar}\langle x\rangle\tilde{p} - \frac{i}{\hbar}\frac{\langle p\rangle\langle x\rangle}{2}\right\} . \tag{2.123}$$

2.8 Wigner Function and Ermakov Invariant

Recently, with the development of nanotechnology, etc., the transition between (microscopic) quantum mechanics and (macroscopic) classical mechanics has been an object of intensive theoretical as well as experimental studies. From an experimental point of view, an interesting approach is the study of single atoms caught in a trap like the Paul trap. The motion in such a trap can be represented by an oscillator with TD frequency $\omega(t)$. For such a system, the Hamiltonian is no longer a constant of motion whereas the Ermakov invariant still is.

From a theoretical viewpoint, the quantum mechanical object that comes closest to the classical phase space description is the so-called Wigner function. In the following, the relation between the Ermakov invariant and the Wigner function is investigated.

The Wigner function (for a pure state) can be obtained from the wave function in position space via the transformation

$$W(x, p) = \frac{1}{2\pi\hbar} \int_{-\infty}^{+\infty} dq \; e^{\frac{i}{\hbar}pq} \; \Psi^*(x + \frac{q}{2}) \; \Psi(x - \frac{q}{2}) \tag{2.124}$$

which has some similarities with a Fourier transformation and was introduced by Eugene P. Wigner in the context of quantum mechanical corrections to thermodynamic equilibrium [36, 37] without any real explanation.

The aim is to describe the motion of a system from position x' to x'' (corresponding to the transformation from x' to x in the Green function method mentioned above).

Therefore, a quantum-jump from x' to x'' shall be considered, i.e. a jump over the distance $q = x'' - x'$. One can define a centre of the jump via $x = \frac{x'+x''}{2}$ and introduce, instead of the coordinates x' and x'', the centre x and the distance q of the jump via

$$x' = x - \frac{q}{2} \tag{2.125}$$

$$x'' = x + \frac{q}{2} \, . \tag{2.126}$$

The momentum p of the particle is associated with the jump from x' to x'', i.e., q.

As the momentum distribution follows from the position distribution via Fourier transformation, a Fourier transformation with respect to the quantum jump q is performed, i.e.,

$$W(x, p) \;=\; \frac{1}{2\pi\hbar} \int_{-\infty}^{+\infty} dq \; e^{\frac{i}{\hbar}pq} \; \tilde{\varrho}(x, q) \tag{2.127}$$

with $\tilde{\varrho}(x, q) = \varrho(x'', x') = \Psi^*(x + \frac{q}{2}) \; \Psi(x - \frac{q}{2})$, i.e., the Wigner transformation defined above is applied. Indeed, this density ϱ depends on *two* position variables

(x' and x'' or q and x). Only one of them (q) is transformed and the other (x) plus the momentum variable (p) is still maintained after Fourier transformation. So, there are still *two variables*, the momentum of the jump p and the centre of the jump x, both being c-numbers, not operators! (For further details see, e.g. [38].)

In the following, the Wigner transformation is first applied to a TI Gaussian WP (corresponding to an initial state) and, afterwards, to a TD Gaussian WP (corresponding to a solution of the TDSE).

(a) TI Gaussian WP:

The Wigner transformation is applied to the initial state WP

$$\Psi(x) = N \exp\left\{-\frac{(x-x_0)^2}{2\delta^2} + \frac{\mathrm{i}}{\hbar} p_0 x\right\} \tag{2.128}$$

corresponding to the probability density in position space

$$\varrho(x) = \Psi^*(x)\,\Psi(x) = NN^* \exp\left\{-\frac{(x-x_0)^2}{\delta^2}\right\} \tag{2.129}$$

with $NN^* = \sqrt{\frac{1}{\pi\delta^2}}$.

Inserting $\Psi(x)$ into the Wigner transformation leads to

$$W(x,p) = \frac{1}{\pi\hbar} \exp\left\{-\frac{(x-x_0)^2}{\delta^2} - \frac{\delta^2(p-p_0)^2}{\hbar^2}\right\} \tag{2.130}$$

where $\delta^2 = 2\langle\tilde{x}^2\rangle$ and $\frac{\hbar^2}{\delta^2} = 2\langle\tilde{p}^2\rangle$ for $\langle\tilde{x}^2\rangle\langle\tilde{p}^2\rangle = \frac{\hbar^2}{4}$.

(b) TD Gaussian WP:

More interesting is the case where, for Ψ, the TD Gaussian WP

$$\Psi_{\mathrm{WP}}(x,t) = N(t) \exp\left\{\mathrm{i}\left[y(t)\,\tilde{x}^2 + \frac{\langle p\rangle}{\hbar}\,\tilde{x} + K(t)\right]\right\} \tag{2.131}$$

is inserted into the Wigner transformation [39]. This leads to the Wigner function in the form

$$W(x,p,t) = \frac{1}{\pi\hbar} \exp\left\{-2y_{\mathrm{I}}\left(1+\frac{y_{\mathrm{R}}^2}{y_{\mathrm{I}}^2}\right)\tilde{x}^2 - \frac{\tilde{p}^2}{2\hbar^2 y_{\mathrm{I}}} + \frac{2}{\hbar}\left(\frac{y_{\mathrm{R}}}{y_{\mathrm{I}}}\right)\tilde{x}\tilde{p}\right\}. \tag{2.132}$$

Using the relations between y_{I}, y_{R} and $\langle\tilde{x}^2\rangle$, $\langle\tilde{p}^2\rangle$, $\langle[\tilde{x},\tilde{p}]_+\rangle$ (where the anti-commutator equals $\hbar\left(\frac{y_{\mathrm{R}}}{y_{\mathrm{I}}}\right)$), $W(x,p,t)$ can be written as

$$W(x,p,t) = \frac{1}{\pi\hbar} \exp\left\{-\frac{2}{\hbar^2}\left[\langle\tilde{p}^2\rangle\tilde{x}^2 - \langle[\tilde{x},\tilde{p}]_+\rangle\tilde{x}\tilde{p} + \langle\tilde{x}^2\rangle\tilde{p}^2\right]\right\}. \tag{2.133}$$

Expressed with the help of α and $\dot{\alpha}$ (see above) this yields

$$W(x,p,t) = \frac{1}{\pi\hbar}\exp\left\{-\frac{m}{\hbar}\left[\left(\dot{\alpha}^2+\frac{1}{\alpha^2}\right)\tilde{x}^2 - 2\dot{\alpha}\alpha\tilde{x}\frac{\tilde{p}}{m} + \alpha^2\frac{\tilde{p}^2}{m^2}\right]\right\} . \quad (2.134)$$

The exponent can then be rewritten in the form

$$W(x,p,t) = \frac{1}{\pi\hbar}\exp\left\{-\frac{m}{\hbar}\left[\left(\dot{\alpha}\tilde{x} - \alpha\frac{\tilde{p}}{m}\right)^2 + \left(\frac{\tilde{x}}{\alpha}\right)^2\right]\right\} . \quad (2.135)$$

The expression in the square brackets has already much similarity with the Ermakov invariant I_{L} and, particularly at the origin of phase space, i.e. for $x = 0$ and $p = 0$, it is up to a constant factor identical to it:

$$\begin{aligned} W(0,0,t) &= \frac{1}{\pi\hbar}\exp\left\{-\frac{m}{\hbar}\left[(\dot{\eta}\alpha - \eta\dot{\alpha})^2 + \left(\frac{\eta}{\alpha}\right)^2\right]\right\} \\ &= \frac{1}{\pi\hbar}\exp\left\{-\frac{2m}{\hbar}I_L\right\} = \text{const.} \end{aligned} \quad (2.136)$$

It can be shown (for the cases considered here) that the Wigner function fulfils the same Liouville equation as the classical phase space probability distribution function, i.e.

$$\frac{\partial}{\partial t}W(x,p,t) = -\frac{p}{m}\frac{\partial W}{\partial x} + \frac{\partial V}{\partial x}\frac{\partial W}{\partial p} . \quad (2.137)$$

Inserting the Wigner function in the form (2.135) into this equation yields again the pair of equations (2.3) and (2.16) corresponding to the Ermakov invariant, providing the uncertainties are expressed in terms of α and $\dot{\alpha}$.

Sorting the results from inserting the Wigner function in the form (2.133) into the Liouville equation (2.137) according to terms proportional to $\tilde{x}^2$, $\tilde{p}^2$ or $\tilde{x}\tilde{p}$, one obtains the following set of coupled differential equations determining the time-evolution of the uncertainties:

$$\frac{\partial}{\partial t}\langle\tilde{x}^2\rangle = \frac{1}{m}\langle[\tilde{x},\tilde{p}]_+\rangle \quad (2.138)$$

$$\frac{1}{4}\frac{\partial}{\partial t}\langle[\tilde{x},\tilde{p}]_+\rangle = \frac{1}{2m}\langle\tilde{p}^2\rangle - \frac{m}{2}\omega^2\langle\tilde{x}^2\rangle = \tilde{\mathcal{L}} \quad (2.139)$$

$$\frac{\partial}{\partial t}\langle\tilde{p}^2\rangle = -m\omega^2\langle[\tilde{x},\tilde{p}]_+\rangle \quad (2.140)$$

where both Eqs. (2.138) and (2.140) are equivalent to the Ermakov equation (2.16).

From Eq. (2.139), the action function $\tilde{S}$ for the uncertainties can be obtained via

$$\tilde{S} = \int_0^t dt' \, \tilde{\mathcal{L}}(t') = \frac{1}{4} \int_0^t dt' \frac{\partial}{\partial t'} \langle [\tilde{x}, \tilde{p}]_+ \rangle$$
$$= \frac{1}{4} \langle [\tilde{x}, \tilde{p}]_+ \rangle \, |_{t_0}^t = \frac{\hbar}{4} \left(\dot{\alpha}\alpha - \dot{\alpha}_0 \alpha_0 \right), \tag{2.141}$$

which is for $\dot{\alpha}_0 = 0$ just $\tilde{S} = \frac{\hbar}{4} \, \dot{\alpha}\alpha = \frac{1}{2} \, \alpha p_\alpha$.

Obviously, the Ermakov invariant appears in different functions and different contexts when TD quantum mechanical problems are considered. These properties are investigated in more detail in the final part of this subsection and the rest of this chapter.

The classical energy for the HO with constant frequency ω_0, $E_{cl} = \frac{m}{2} \dot{\eta}^2(t) + \frac{m}{2} \omega^2 \eta^2(t)$ has the constant value $E_{cl} = \frac{p_0^2}{2m}$ for the initial conditions $\eta(0) = \eta_0 = 0$ and $\dot{\eta}(0) = \frac{p_0}{m}$ with the (maximum) initial momentum p_0. In this case, the invariant can be written as

$$I_{\text{L}} = \frac{1}{2} \left(\frac{\alpha_0 p_0}{m} \right)^2 = \frac{\alpha_0^2}{m} \, E_{cl} \tag{2.142}$$

and the WP width and the frequency of the oscillator are related via $\frac{\hbar}{2m\langle \tilde{x}^2 \rangle_0} = \beta_0 = \frac{1}{\alpha_0^2} = \omega_0$. As $\dot{\alpha} = 0$, the quantum mechanical contribution to the WP energy is just the ground state energy of the oscillator and can be written as

$$\tilde{E} = \tilde{T} + \tilde{V} = \frac{1}{2m} \langle \tilde{p}^2 \rangle + \frac{m}{2} \, \omega_0^2 \langle \tilde{x}^2 \rangle = \frac{\hbar}{2} \, \omega_0 = \frac{\hbar}{2\alpha_0^2} \, . \tag{2.143}$$

In (2.133) the exponent of the Wigner function (particularly for $x = p = 0$) which, apart from a constant factor, is identical to I_{L} is written as a sum of terms each being a product of a classical dynamical variable ($(\eta(t), \dot{\eta}(t) = \frac{p(t)}{m}$ in $\tilde{x}$ and $\tilde{p}$) and the conjugate quantum mechanical uncertainty $\left(\langle \tilde{x}^2 \rangle(t), \langle \tilde{p}^2 \rangle(t) \right)$. Now, however, the same invariant can be rewritten as a ratio of only classical to only quantum mechanical energy contributions, i.e.

$$I_{\text{L}} = \frac{\hbar}{2m} \frac{E_{cl}}{\tilde{E}} \, . \tag{2.144}$$

For the WP with constant width this might appear trivial since both $\langle \tilde{x}^2 \rangle$ and $\langle \tilde{p}^2 \rangle$ are constants. However, for the HO with TD width, these uncertainties also become functions of time, namely

$$\langle \tilde{x}^2 \rangle(t) = \frac{\hbar}{2m} \, \alpha_0^2 \left(\cos^2 \omega_0 t + \frac{\beta_0^2}{\omega_0^2} \sin^2 \omega_0 t \right) \tag{2.145}$$

$$\langle \tilde{p}^2 \rangle(t) = \frac{\hbar m}{2} \frac{1}{\alpha_0^2} \left(\cos^2 \omega_0 t + \frac{\omega_0^2}{\beta_0^2} \sin^2 \omega_0 t \right) \tag{2.146}$$

thus also $\tilde{T} = \tilde{T}(t)$ and $\tilde{V} = \tilde{V}(t)$ are now time-dependent for $\beta_0 = \frac{1}{\alpha_0^2} \neq \omega_0$. The quantum mechanical energy contribution now takes the form

$$\tilde{E} = \tilde{T} + \tilde{V} = \frac{\hbar}{4}\,\omega_0 \left[\frac{\omega_0}{\beta_0} + \frac{\beta_0}{\omega_0}\right] = \frac{\hbar}{2}\omega_0 \left[\frac{\omega_0^2 + \beta_0^2}{2\beta_0\omega_0}\right] = \frac{\hbar}{2}\omega_0 \left[1 + \frac{(\omega_0 - \beta_0)^2}{2\beta_0\omega_0}\right] > \frac{\hbar}{2}\omega_0 \,, \tag{2.147}$$

i.e., it is larger than the ground state energy $\frac{\hbar}{2}\omega_0$ of the WP with constant width, independent if ω_0 or β_0 is larger.

Again, α_0^2 in (2.142) can be expressed in terms of $\tilde{E}$, now as $\alpha_0^2 = \frac{\hbar}{4\tilde{E}}\left(\frac{\beta_0^2+\omega_0^2}{\beta_0^2}\right)$, turning Eq. (2.144) into

$$I_{\rm L} = \frac{\hbar}{4m}\left(\frac{\omega_0^2 + \beta_0^2}{\beta_0^2}\right)\frac{E_{cl}}{\tilde{E}} \tag{2.148}$$

(which turns into (2.144) for $\beta_0 = \omega_0$). Apart from a different constant prefactor, again we obtain a ratio of only classical TD variables to, this time also TD, purely quantum mechanical contributions.

There is yet another interpretation of $I_{\rm L}$ possible. For this purpose it should be noted that the WP solution of the TDSE for the HO can be expanded in terms of the stationary wave functions of the problem with constant superposition coefficients and TD phase factors. For an initial Gaussian WP the coefficients can be determined in closed form (see, e.g. [40]). The quantum number n_0, for which the coefficient becomes a maximum, can be determined in good approximation to be $n_0 \approx \frac{1}{2\hbar}m\omega_0\eta_{max}^2$ where η_{max} is the maximum amplitude of the corresponding classical oscillator. The energy of the respective quantum state n_0 is given by $E_{n_0} = (n_0 + \frac{1}{2})\hbar\omega_0 \approx \frac{m}{2}\omega_0^2\eta_{max}^2 + \frac{\hbar}{2}\omega_0$ and is the energy level from whose neighbourhood most of the contribution to the WP comes. This energy is, for $n_0 \gg 1$, approximately equal to the energy $E_{cl} = \frac{m}{2}\omega_0^2\eta_{max}^2$ of a classical oscillator that has the same amplitude.

Therefore, the relation

$$\frac{E_{cl}}{\tilde{E}} \approx \frac{n_0\,\hbar\omega_0}{\frac{\hbar}{2}\,\omega_0} = 2n_0 \tag{2.149}$$

or

$$I_{\rm L} = \frac{\hbar}{m}n_0 \,, \tag{2.150}$$

is valid, i.e., apart from the constant factors $\hbar$ and m, the invariant $I_{\rm L}$ is identical to the quantum number n_0.

This, however, resembles very much the interpretation of the Ermakov invariant in a classical cosmological context discussed by Ray [41]. There, the classical trajectory $\eta(t)$ is replaced by a field amplitude ϕ and α corresponds to his quantity ρ fulfilling the Ermakov equation (2.16) in Sect. 2.2. The invariant $I = \frac{1}{2}\left[\left(\rho\dot{\phi} - \phi\dot{\rho}\right)^2 + \left(\frac{\phi}{\rho}\right)^2\right] = N$ (in the adiabatic regime) defines in his case the particle number of the model and "provides an interesting alternative for calculating particle production in cosmological models". We will return to Ermakov systems in a cosmological context in Sect. 7.7.

So far, the discussion of the TD case included only systems where the potential is at most quadratic in its variables. This might not be as restrictive as it seems at first sight as one may sometimes perform canonical transformations to reduce a given Hamiltonian to a quadratic form [42] which has been shown explicitly by Sarlet [43] for some polynomial Hamiltonians. To what extent this method can also be applied in the case discussed here requires further studies.

Generalized Ermakov systems where the rhs of Eqs. (2.3) and (2.16) are functions of $\frac{\alpha}{\eta}$ or $\frac{\eta}{\alpha}$ are discussed, e.g., in Refs. [44–49].

2.9 Representation of Canonical Transformations in Quantum Mechanics

In classical Hamiltonian mechanics the time-evolution of a physical system is described by canonical transformations in phase space that keep the Poisson brackets of the transformed coordinates and momenta with respect to the initial ones unchanged. This transformation in phase space can be described (for a one-dimensional problem in physical space and, therefore, a two-dimensional one in phase space, to which the following discussion will be restricted again) by the so-called two-dimensional real symplectic group $Sp(2, R)$, represented by 2×2 matrices with a determinant equal to 1. (In order to compare the TD results with the ones for the TI case, only the homogeneous symplectic group without translations is considered, not the inhomogeneous symplectic group $ISp(2, R)$.) It has been shown in [27] how it is possible to obtain the representation of the group of linear canonical transformations in TI quantum mechanics via the determination of the configuration space representation of the unitary operator that connects quantum mechanically the transformed variables x and p with the initial ones, x' and p'. A subsequent Wigner transformation shows explicitly that, for the TI problems considered by this method, essentially the classical results are reproduced.

In the TD case however, there are, at least formal, differences between the classical and the quantum mechanical descriptions of the system even already for such simple ones like the free motion (see, e.g. Sect. 2.3.2). These differences are intimately connected with the time-dependence of the typical quantum mechanical aspect of the system, namely the uncertainties of position and momentum. Therefore, in the following, the influence of the time-dependence on the representation of the group of linear canonical transformations in quantum mechanics is investigated.

After a short summary of the main results of the TI situation, the TD case and the characteristic differences compared with the TI case are considered; in particular the role of the time-dependence of the uncertainties is discussed. Again, this will be restricted to systems with analytic solutions,i.e. at most quadratic Hamiltonians.

Time-independent case:

The time-evolution in classical Hamiltonian mechanics, described by canonical transformations in phase space, can be represented by

$$\begin{pmatrix} x \\ p \end{pmatrix} = \begin{pmatrix} a & b \\ c & d \end{pmatrix} \begin{pmatrix} x' \\ p' \end{pmatrix} \tag{2.151}$$

where a, b, c and d are real and the determinant of the 2×2 matrix is 1, i.e., $ad - bc = 1$. The group of transformations represented by the 2×2 matrices is the so-called two-dimensional real symplectic group $Sp(2, R)$.

Following [50] (see Chap. 35 and reference cited therein), it is possible to obtain the representation of the group of linear canonical transformations (2.151) in quantum mechanics. Referring to [50, 51], the main objective is to determine the configuration space representation

$$\langle x|U|x'\rangle = G(x, x') \tag{2.152}$$

of the unitary operator U that provides the quantum mechanical relation between x, p and x', p', according to

$$x = Ux'U^{-1} \quad , \quad p = Up'U^{-1} \,. \tag{2.153}$$

With the help of the kernel $G(x, x')$, the effect of any canonical transformation (2.151) can be described as

$$\Psi(x) = \int_{-\infty}^{+\infty} dx' \; G(x, x') \; \Psi(x') \,. \tag{2.154}$$

The integral kernel $G(x, x')$ has been derived taking into consideration that it must satisfy the following two differential equations[8] [52, 53]

$$\left(a\,x + b\,\frac{\hbar}{\mathrm{i}}\frac{\partial}{\partial x}\right) G(x, x') = x'\, G(x, x') \tag{2.155}$$

$$\left(c\,x + d\,\frac{\hbar}{\mathrm{i}}\frac{\partial}{\partial x}\right) G(x, x') = -\frac{\hbar}{\mathrm{i}}\frac{\partial}{\partial x'}\, G(x, x') \,. \tag{2.156}$$

An exponential ansatz, bilinear in x and x', finally leads to $G(x, x')$ in the form

$$G(x, x') = \left(\frac{1}{2\pi b\hbar}\right)^{\frac{1}{2}} \exp\left\{-\frac{\mathrm{i}}{2b\hbar}\left[ax^2 - 2xx' + dx'^2\right]\right\} . \tag{2.157}$$

[8] The minus sign on the rhs of Eq. (2.156) originates from $\langle x|p|x'\rangle = \frac{\hbar}{\mathrm{i}}\frac{\partial}{\partial x}\,\delta(x - x') = -\frac{\hbar}{\mathrm{i}}\frac{\partial}{\partial x'}\,\delta(x - x')$.

This kernel $G(x, x')$, related to the specific canonical transformation, is formulated in configuration space whereas the corresponding classical canonical transformation is formulated in phase space. Therefore, it is interesting to discuss the representation of this canonical transformation in the phase space version of quantum mechanics that was developed by Wigner [36, 37]. Applying the Wigner transformation specified in Eq. (2.124) to the kernel (2.157) leads to the phase space kernel in the form [54]

$$
\begin{aligned}
G(x, x', p, p') &= \frac{1}{2\pi\hbar^2} = \int_{-\infty}^{+\infty}\int_{-\infty}^{+\infty} dz\, dz'\, e^{\frac{i}{\hbar}(pz-p'z')}\, G^*\left(x - \frac{z}{2}, x' - \frac{z'}{2}\right) G\left(x + \frac{z}{2}, x' + \frac{z'}{2}\right) \\
&= \delta[x' - (ax + bp)]\,\delta[p' - (cx + dp)]\,,
\end{aligned} \tag{2.158}
$$

showing that, for this linear canonical transformation, the kernel coincides with its classical limit. So, the quantum mechanical problem mainly reproduces the classical situation without any additional quantum mechanical aspect.

Time-dependent case:

Now it is investigated how far this is still true in the case of specific quantum dynamical aspects entering the problem. Considering TD problems in quantum mechanics in terms of the TDSE or equivalent formulations, one finds that not only classical position and momentum can change in time (in a way that can be described by canonical transformations) but also the typical quantum mechanical degrees of freedom, like position and momentum uncertainties, may be TD (corresponding, e.g., to WPs with TD width). In Sect. 2.5 it has been shown, according to Eq. (2.53), how to achieve the transition of a Gaussian WP from initial position x' and time t' (in configuration space) to any later position x and time t with the help of a TD kernel (or propagator) $G(x, x', t, t')$. This kernel had the explicit form

$$
G\left(x, x', t, 0\right) = \left(\frac{m}{2\pi i\hbar\alpha_0 z}\right)^{1/2} \exp\left\{\frac{im}{2\hbar}\left[\frac{\dot{z}}{z}x^2 - 2\frac{x}{z}\left(\frac{x'}{\alpha_0}\right) + \frac{u}{z}\left(\frac{x'}{\alpha_0}\right)^2\right]\right\} \tag{2.159}
$$

where the time dependence enters via the parameters $z(t)$ and $u(t)$. In the limit $t \to 0$, the kernel turns into a delta function.

In order to compare the TD kernel (2.159) with the kernel $G(x, x')$ in (2.157), one must take into account that $G(x, x')$ has been obtained via Eqs. (2.155) and (2.156) which describe the transformation of x and p into the initial values x' and p' whereas $G(x, x', t, t')$ in (2.159) describes the inverse transformation from x' to x. This is expressed, e.g., by the different signs in the exponents of (2.157) and (2.159). For a direct comparison one must therefore take the inverse transformation of (2.159), obtained by changing the sign in the exponent and interchanging $\dot{z}$ and u. Inserting this kernel into Eqs. (2.155) and (2.156), one obtains the corresponding equations for the TD problem,

$$
\dot{z}x - z\frac{p}{m} = \frac{x'}{\alpha_0}\,, \tag{2.160}
$$

$$-\dot{u}x + u\frac{p}{m} = -\frac{\alpha_0 p'}{m} , \tag{2.161}$$

or, in matrix notation,

$$\begin{pmatrix} \frac{x'}{\alpha_0} \\ -\frac{\alpha_0 p'}{m} \end{pmatrix} = \begin{pmatrix} \dot{z} & -z \\ -\dot{u} & u \end{pmatrix} \begin{pmatrix} x \\ \frac{p}{m} \end{pmatrix} = \mathbf{M} \begin{pmatrix} x \\ \frac{p}{m} \end{pmatrix} . \tag{2.162}$$

The transformation matrix $\mathbf{M}$ has again a determinant equal to 1, $\dot{z}u - \dot{u}z = 1$. As shown in Sect. 2.4, this is equal to $\dot{\varphi} = \frac{1}{\alpha^2}$, corresponding to a kind of conservation of angular momentum (in the complex plane). However, different from the TI case, the initial state is not only characterized by the initial position x' and momentum p' but also by the corresponding initial uncertainties as $\alpha_0 = \left(\frac{2m}{\hbar}\langle\tilde{x}^2\rangle_0\right)^{\frac{1}{2}}$ is proportional to the initial position uncertainty and (for a WP with initial minimum uncertainty $\langle\tilde{x}^2\rangle_0\langle\tilde{p}^2\rangle_0 = \frac{\hbar^2}{4}$, i.e. $\dot{\alpha}_0 = 0$) the inverse $\frac{1}{\alpha_0} = \left(\frac{2}{m\hbar}\langle\tilde{p}^2\rangle_0\right)^{\frac{1}{2}}$ is proportional to the initial momentum uncertainty, i.e. $\frac{x'}{\alpha_0} \propto \frac{x'}{\sqrt{\langle\tilde{x}^2\rangle_0}}$ and $\frac{\alpha_0 p'}{m} \propto \frac{p'}{\sqrt{\langle\tilde{p}^2\rangle_0}}$.

Also in this case we are interested in the corresponding (TD) Wigner function in phase space (the one already derived differently in Sect. 2.8 via the Wigner transformation of the Gaussian WP $\Psi_{\text{WP}}(x,t)$, see Eqs. (2.133–2.135)), now using the initial Wigner function $W(x', p', t' = 0)$ and the TD kernel $G(x, x', p, p', t, t')$ in phase space,

$$W(x, p, t) = \int_{-\infty}^{+\infty}\int_{-\infty}^{+\infty} dx'dp'\, G(x, x', p, p', t, t')\, W(x', p', t' = 0) \tag{2.163}$$

where the initial state can be written in the form

$$\begin{aligned} W(x', p', t' = 0) &= \frac{1}{\pi\hbar} \exp\left\{-\frac{x'^2}{2\langle\tilde{x}^2\rangle_0} - \frac{p'^2}{2\langle\tilde{p}^2\rangle_0}\right\} \\ &= \frac{1}{\pi\hbar} \exp\left\{-\frac{m}{\hbar}\left[\left(\frac{x'}{\alpha_0}\right)^2 + \left(\frac{\alpha_0 p'}{m}\right)^2\right]\right\} . \end{aligned} \tag{2.164}$$

The TD kernel in phase space, obtained in the same way as in the TI case by Wigner transform of the kernel in configuration space, is again given by the product of two delta functions where now, however, x' is replaced by $\frac{x'}{\alpha_0}$ and p' is replaced by $\frac{\alpha_0 p'}{m}$ and the transformed variables in the delta functions are determined by (2.160) and (2.161), i.e.

$$G(x, x', p, p', t, t' = 0) = \delta\left[\left(\frac{x'}{\alpha_0}\right) - \left(\dot{z}x - z\frac{p}{m}\right)\right] \delta\left[\left(\frac{\alpha_0 p'}{m}\right) - \left(u\frac{p}{m} - \dot{u}x\right)\right] . \tag{2.165}$$

Applying this kernel to the initial Wigner distribution function (2.164) yields $W(x, p, t)$ as

$$W(x, p, t) = \frac{1}{\pi\hbar} \exp\left\{-\frac{m}{\hbar}\left[\left(\dot{z}x - z\frac{p}{m}\right)^2 + \left(u\frac{p}{m} - \dot{u}x\right)^2\right]\right\} . \quad (2.166)$$

Using the definition of $\lambda(t)$ in Cartesian coordtinates, $\lambda = u+iz$, the determinant of $\mathbf{M}$, $\dot{z}u - \dot{u}z = 1$, and the relations (2.86–2.88) between λ, $\dot{\lambda}$ and the position and momentum uncertainties as given in Sect. 2.6, finally allows one to write

$$W(x, p, t) = \frac{1}{\pi\hbar} \exp\left\{-\frac{2}{\hbar^2}\left[\langle\tilde{p}^2\rangle x^2 - \langle[\tilde{x}, \tilde{p}]_+\rangle xp + \langle\tilde{x}^2\rangle p^2\right]\right\} , \quad (2.167)$$

where the time-dependence of the uncertainties is determined totally by the time-dependence of $z(t)$ and $u(t)$. In the case of TD Gaussian WPs, the classical time-dependence is expressed by the fact that the maximum of the WP follows the classical trajectory $\langle x\rangle = \eta(t)$. This is taken into account by shifting the variables of position and momentum from x to $\tilde{x} = x - \langle x\rangle = x - \eta$ and p to $\tilde{p} = p - \langle p\rangle = p - m\dot{\eta}$. As $\langle x\rangle$ and $\langle p\rangle$ are purely TD quantities, $\tilde{x}$ and $\tilde{p}$ replace x and p in Eqs. (2.155) and (2.156) because these equations only contain derivatives with respect to space, not time. So x and p in (2.167) would be replaced by $\tilde{x}$ and $\tilde{p}$ which would lead to the result (2.133) obtained in Sect. 2.8 showing the connection between the exponent of the TD Wigner function and the dynamical Ermakov invariant that can also be expressed using the parameters $z(t)$ and $u(t)$ of the TD kernel $G(x, x', t, t')$ and has been defined, e.g., in (2.21) and (2.70).

In the quantum mechanical phase space picture according to Wigner, this results not only in changing initial position and momentum uncertainties into their values at time t, but also an additional contribution occurs from the time-change of $\langle\tilde{x}^2\rangle$, or α^2, respectively, expressed by the term proportional $\langle[\tilde{x}, \tilde{p}]_+\rangle$, or $\dot{\alpha}\alpha$, respectively, in the exponent of $W(x, p, t)$.

All these quantum dynamical aspects are contained in the TD parameters $z(t)$ and $u(t)$, entering the transformation matrix in (2.162). In particular, the change of the position uncertainty (expressed by $\dot{\alpha}$) is taken into account by the parameter u, which can be expressed as shown in Sect. 2.4, as

$$u = \dot{z}\alpha^2 - z\dot{\alpha}\alpha = \left(\frac{m}{\alpha_0 p_0}\right)\left[\dot{\eta}\alpha^2 - \eta\dot{\alpha}\alpha\right] . \quad (2.168)$$

As long as the *position* uncertainty is constant, $\alpha = \alpha_0$, u is simply proportional to the classical velocity $\dot{\eta}(t)$; for $\dot{\alpha} \neq 0$, however, the situation can become quite different. As examples, the HO with constant width and the free motion with $\alpha = \alpha(t)$ (expressed by the spreading of the corresponding WP solution) are discussed briefly.

For this purpose, also $\dot{u}$ is now given in term of η and $\dot{\eta}$ where the equations of motion (2.3) (for $\eta(t)$) and (2.16) (for $\alpha(t)$) are applied. So it follows from (2.168) that

$$\dot{u} = \left(\frac{m}{\alpha_0 p_0}\right)\left[\dot{\eta}\dot{\alpha}\alpha - \eta\left(\dot{\alpha}^2 + \frac{1}{\alpha^2}\right)\right] . \tag{2.169}$$

For constant $\alpha = \alpha_0$, all terms proportional to $\dot{\alpha}$ vanish and the transformation maxtrix (2.162) can be written as

$$\mathbf{M} = \begin{pmatrix} \dot{z} & -z \\ -\dot{u} & u \end{pmatrix} = \left(\frac{m}{\alpha_0 p_0}\right)\begin{pmatrix} \dot{\eta} & -\eta \\ \frac{\eta}{\alpha_0^2} & \alpha_0^2\dot{\eta} \end{pmatrix} . \tag{2.170}$$

This matrix $\mathbf{M}$, for the HO with *constant* width (in agreement with $\dot{\alpha} = 0$), i.e., $\alpha = \alpha_0 = \frac{1}{\sqrt{\omega}}$, and $\eta(t) = \frac{v_0}{\omega}\sin\omega t$ (for $\eta(0) = \eta_0 = 0$) and $\dot{\eta}(t) = v_0\cos\omega t$ (for $\dot{\eta}(0) = \dot{\eta}_0 = v_0 = \frac{p_0}{m}$), turns into

$$\mathbf{M}_{\mathrm{HO}} = \left(\frac{m}{\alpha_0 p_0}\right)\begin{pmatrix} v_0\cos\omega t & -\frac{v_0}{\omega}\sin\omega t \\ v_0\sin\omega t & \frac{v_0}{\omega}\cos\omega t \end{pmatrix} = \begin{pmatrix} \frac{1}{\alpha_0}\cos\omega t & -\alpha_0\sin\omega t \\ \frac{1}{\alpha_0}\sin\omega t & \alpha_0\cos\omega t \end{pmatrix} , \tag{2.171}$$

i.e., (up to the constant α_0 that also occurs in the column vector) just the classical result is reproduced.

However, for the free motion with $\eta(t) = v_0 t$, $\dot{\eta}(t) = v_0$, for constant $\alpha = \alpha_0$, one would obtain

$$\tilde{\mathbf{M}}_{\mathrm{fr}} = \left(\frac{m}{\alpha_0 p_0}\right)\begin{pmatrix} v_0 & -v_0 t \\ \frac{1}{\alpha_0^2}v_0 t & \alpha_0^2 v_0 \end{pmatrix} , \tag{2.172}$$

that is different from the classical situation where the matrix element in the first column and second row $\left(\frac{1}{\alpha_0^2}v_0 t\right)$ would be zero. The consequence for a free motion WP with constant width would be that the transformation matrix would no longer describe a canonical transformation as its determinant would no longer be equal to 1 but

$$\det(\tilde{\mathbf{M}}_{\mathrm{fr}}) = \left[1 + \left(\frac{t}{\alpha_0^2}\right)^2\right] , \tag{2.173}$$

which just describes the time-dependence of the WP spreading (see Eq. (2.40) for $\dot{\alpha}_0 = 0$).

For $\dot{\alpha} \neq 0$, the well-known time-dependence of the WP width given by $\alpha^2(t) = \alpha_0^2\left[1 + \left(\frac{t}{\alpha_0^2}\right)^2\right]$ (as can be obtained as a solution of Eq. (2.16) for $\omega = 0$) leads to the correct transformation matrix

$$\mathbf{M}_{fr} = \left(\frac{m}{\alpha_0 p_0}\right)\begin{pmatrix} v_0 & -v_0 t \\ 0 & \alpha_0^2 v_0 \end{pmatrix} , \tag{2.174}$$

with $\det(\mathbf{M}_{\mathrm{fr}}) = 1$. This shows explicitly the influence of the time-dependence of the uncertainties of $\alpha(t)$ on the transformation describing the dynamics of the system.

It should also be mentioned that the determinant of **M**, written in terms of η, $\dot{\eta}$, α and $\dot{\alpha}$ takes just the form of the Ermakov invariant, i.e.,

$$\mathbf{M} = \left(\frac{m}{\alpha_0 p_0}\right) \begin{pmatrix} \dot{\eta} & -\eta \\ -\dot{\eta}\dot{\alpha}\alpha + \eta\left(\dot{\alpha}^2 + \frac{1}{\alpha^2}\right) & \dot{\eta}\alpha^2 - \eta\dot{\alpha}\alpha \end{pmatrix} \tag{2.175}$$

yields

$$\begin{aligned} \det(\mathbf{M}) &= \left(\frac{m}{\alpha_0 p_0}\right)^2 \left[\dot{\eta}^2\alpha^2 - 2\eta\dot{\eta}\alpha\dot{\alpha} + \eta^2\left(\dot{\alpha}^2 + \frac{1}{\alpha^2}\right)\right] \\ &= \left(\frac{m}{\alpha_0 p_0}\right)^2 \left[(\dot{\eta}\alpha - \dot{\alpha}\eta)^2 + \left(\frac{\eta}{\alpha}\right)^2\right] = \left(\frac{m}{\alpha_0 p_0}\right)^2 2\, I_{\mathrm{L}} \\ &= 1 = \dot{z}u - \dot{u}z = \dot{\varphi}\,\alpha^2 \,. \end{aligned} \tag{2.176}$$

This leads to the interesting result that the Ermakov invariant is actually equivalent to the conservation law $\dot{z}u - \dot{u}z = \dot{\varphi}\,\alpha^2 = 1$, describing the "conservation of angular momentum in the complex plane", or the Wronskian determinant showing the linear independence of the two solutions of the linear Newtonian equation of motion, repectively. This result has been confirmed in [55].

The second result of this subsection is that, in the TD quantum mechanical problem, the transformation (2.162) corresponding to the classical linear canonical transformation (2.151) and its TI quantum mechanical analogue ((2.155), (2.156)), not only transform the *initial position* and *momentum* into its values at a later time but, also, does the same *simultaneously* with the *corresponding uncertainties*! How far this is connected with the existence of a Lagrangian/Hamiltonian formulation of the dynamics of the quantum uncertainties, as presented in Sect. 2.6, needs further investigation.

2.10 Algebraic Derivation of the Ermakov Invariant

In cases where the (classical) Hamiltonian is no longer a constant of motion, e.g., for the parametric oscillator with TD frequency $\omega = \omega(t)$, another constant of motion, the Ermakov invariant, may still exist that does not have the dimension of an energy but essentially that of an action.

In Sect. 2.3 it has been shown how this invariant can be obtained by eliminating $\omega^2(t)$ from the Newtonian equation (2.3)[9] and the Ermakov equation (2.16), leading to

$$I = \frac{1}{2}\left[(\dot{q}\,\alpha - q\,\dot{\alpha})^2 + \left(\frac{q}{\alpha}\right)^2\right] = \text{const.} \tag{2.177}$$

[9]In this subsection, η is replaced by q to be in agreement with a more conventional notation in classical phase space; $\dot{\eta}$ consequently turns into $\dot{q} = \frac{p}{m}$.

The same invariant can also be obtained in a different way via an *algebraic* approach (see e.g. [56]). In this case, the Hamiltonian is rewritten in terms of (not explicitly TD) phase space functions Γ_n as

$$H = \sum_n h_n(t)\, \Gamma_n \tag{2.178}$$

where the dynamical algebra is the Lie algebra of the functions Γ_n which is closed with relation to the Poisson brackets $\{\ ,\ \}_-$,

$$\{\Gamma_n, \Gamma_m\}_- = \sum_r C^r_{nm}\, \Gamma_r \tag{2.179}$$

with the C^r_{nm} being the structure constants of the algebra.

The time-evolution of any phase space function $F(q, p, t)$ is given by

$$\frac{d}{dt}F = \{F, H\}_- + \frac{\partial}{\partial t}F\,. \tag{2.180}$$

In particular, a *dynamical invariant* is characterized by $\frac{d}{dt}I = 0$, i.e.,

$$\frac{\partial}{\partial t}I = \{H, I\}_-\,. \tag{2.181}$$

Looking for an invariant that is also a member of the dynamical algebra, i.e.,

$$I = \sum_n \kappa_n(t)\, \Gamma_n\,, \tag{2.182}$$

Eq. (2.181) leads to a coupled set of evolution equations for the expansion coefficients κ_n,

$$\dot{\kappa}_r + \sum_n \left(\sum_m C^r_{nm}\, h_m(t) \right) \kappa_n = 0\,. \tag{2.183}$$

In the following, we consider the dynamical algebra of

$$\Gamma_1 = \frac{p^2}{2m}\,, \quad \Gamma_2 = p\, q\,, \quad \Gamma_3 = \frac{mq^2}{2} \tag{2.184}$$

with Poisson brackets

$$\{\Gamma_1, \Gamma_3\}_- = -\,\Gamma_2, \qquad \{\Gamma_1, \Gamma_2\}_- = -2\,\Gamma_1, \qquad \{\Gamma_3, \Gamma_2\}_- = 2\,\Gamma_3\,. \tag{2.185}$$

For the Hamiltonian of the oscillator with TD frequency,

$$H = \frac{1}{2m}\, p^2 + \frac{m}{2}\, \omega^2(t)\, q^2 \,, \tag{2.186}$$

with $h_1 = 1$, $h_2 = 0$ and $h_3 = \omega^2(t)$, the set of equations for κ_n can be written as

$$\dot{\kappa}_1 = -\,2\,\kappa_2 \,, \tag{2.187}$$
$$\dot{\kappa}_2 = \omega^2 \kappa_1 \;-\; \kappa_3 \,, \tag{2.188}$$
$$\dot{\kappa}_3 = 2\,\omega^2\,\kappa_2 \,. \tag{2.189}$$

This coupled set of equations can be reduced to a single second-order differential equation by introducing a new variable $\alpha(t)$ via $\kappa_1 = \frac{1}{m}\alpha^2$, leading to the Ermakov equation (2.16).

Assuming the same relations hold between α and $\dot{\alpha}$ on the one side and the position and momentum uncertainties on the other, as in Sects. 2.3 and 2.3.1, the κ_n can be written as

$$\kappa_1 = \frac{1}{m}\alpha^2 = \frac{2}{\hbar}\,\langle \tilde{x}^2 \rangle \,, \tag{2.190}$$
$$\kappa_2 = -\frac{1}{m}\dot{\alpha}\alpha = -\frac{1}{\hbar m}\langle [\tilde{x}, \tilde{p}]_+ \rangle \,, \tag{2.191}$$
$$\kappa_3 = \frac{1}{m}\left(\dot{\alpha}^2 + \frac{1}{\alpha^2}\right) = \frac{2}{\hbar m^2}\langle \tilde{p}^2 \rangle \,. \tag{2.192}$$

Inserting (2.190–2.192) into (2.187–2.189), the latter equations, expressed in terms of the uncertainties, simply turn into Eqs. (2.138–2.140) obtained from the Liouville equation of the Wigner function in Sect. 2.8. With κ_n, expressed in terms of α and $\dot{\alpha}$, the invariant (2.182) can be written as

$$I = \frac{1}{2}\left[\alpha^2 \frac{p^2}{m^2} - 2\dot{\alpha}\alpha \frac{p}{m}\, q + \left(\dot{\alpha}^2 + \frac{k^2}{\alpha^2}\right) q^2\right] , \tag{2.193}$$

which is identical to (2.21) for $\eta = q$, $p = m\dot{q}$ and $k = 1$.

2.11 Generalized Creation and Annihilation Operators and Coherent States

The Ermakov invariant cannot only be expressed in terms of α and $\dot{\alpha}$ but also in terms of real and imaginary parts of the complex Riccati variable $\mathcal{C} = \mathcal{C}_{\mathrm{R}} + i\mathcal{C}_{\mathrm{I}} = \frac{\dot{\alpha}}{\alpha} + i\frac{1}{\alpha^2}$ as the sum of two quadratic terms or as a product of two linear terms, complex conjugate to each other, as[10]

[10] As this subsection again deals with TD Gaussian WPs and their dynamics, the notation using η and $\dot{\eta}$ instead of q and $\frac{p}{m}$ is applied again.

$$
\begin{aligned}
I &= \frac{1}{2}\,\alpha^2\left[(\dot{\eta} - C_{\mathrm{R}}\eta)^2 \;+\; (C_{\mathrm{I}}\eta)^2\right] \\
&= \frac{1}{2}\,\alpha^2\left[(\dot{\eta} - C\eta)\,(\dot{\eta} - C^*\eta)\right] .
\end{aligned} \tag{2.194}
$$

This brings back to mind the rewriting of the HO Hamiltonian (in quantum mechanics) as a product of two (adjoint) operators as

$$
H_{\mathrm{op}} = \frac{1}{2m}\,p_{\mathrm{op}}^2 \;+\; \frac{m}{2}\,\omega_0^2\,x^2 = \hbar\omega_0\left(a^+a \;+\; \frac{1}{2}\right) \tag{2.195}
$$

with $p_{\mathrm{op}} = \frac{\hbar}{\mathrm{i}}\frac{\partial}{\partial x}$, or divided by $\hbar\omega_0$, as

$$
\tilde{H} = \frac{H}{\hbar\omega_0} = \left(a^+a + \frac{1}{2}\right) \tag{2.196}
$$

where a^+a is the so-called *number* operator as it counts the number of *quanta of action* because $\frac{H}{\omega_0}$ has the dimension of an *action*!

The *creation* and *annihilation* operators a^+ and a are defined as

$$
a = \mathrm{i}\sqrt{\frac{m}{2\hbar\omega_0}}\,\left(\frac{p_{\mathrm{op}}}{m} - \mathrm{i}\,\omega_0\,x\right) , \tag{2.197}
$$

$$
a^+ = -\mathrm{i}\sqrt{\frac{m}{2\hbar\omega_0}}\,\left(\frac{p_{\mathrm{op}}}{m} + \mathrm{i}\,\omega_0\,x\right) . \tag{2.198}
$$

This *factorization* method had been applied by Schrödinger himself [57] but was already mentioned earlier by Dirac in his book *"The Principles of Quantum Mechanics"* [58].

With the help of the annihilation operator a, the ground state wave function of the HO can be obtained and, by successive application of the creation operator a^+, all the excited states can be "*created*".

By superimposing all these states (with appropriate weight-factors), a Gaussian WP with *constant* width as solution for the SE of the HO, whose maximum follows the classical trajectory, was already obtained by Schrödinger [59]. Generalizations of Schrödinger's approach were achieved in the description of coherent light beams emitted by lasers and are connected with the names Glauber, Sudarshan and Klauder [60–62], considering what is now known as *coherent states* (CS).[11]

There are at least three different definitions of these states in the literature:

1. *minimum uncertainty CS*, meaning Gaussian WPs that minimize Heisenberg's uncertainty relation: $U = \langle\tilde{x}^2\rangle\langle\tilde{p}^2\rangle = \frac{\hbar^2}{4}$;

2. *annihilation operator CS*, meaning eigenfunctions of the operator a with complex eigenvalue w;

[11] For more details see also [63] and literature cited therein.

3. *displacement operator CS*, meaning displaced vacuum states obtained from the ground state $\Psi_0 = |0\rangle$ by applying the displacement operator $D(w) = \exp\{wa^+ - w^*a\}$.

In the following, it is shown how the creation and annihilation operators can be generalized to provide also CSs (WPs) with TD width, i.e., cases with position-momentum correlations ($\langle[\tilde{x}, \tilde{p}]_+\rangle = \hbar\dot{\alpha}\alpha \neq 0$ or $\dot{\alpha} \neq 0$).

The CS of the HO with constant width is associated with the particular solution of the corresponding Riccati equation

$$\tilde{C} = \mathrm{i}\,\tilde{C}_{\mathrm{I}} = \mathrm{i}\,\omega_0 = \mathrm{i}\frac{1}{\alpha_0^2}\,. \tag{2.199}$$

Replacing ω_0 accordingly in a and a^+, they can be rewritten as

$$a = \mathrm{i}\sqrt{\frac{m}{2\hbar}}\,\alpha_0\,\left(\frac{p_{\mathrm{op}}}{m} - \mathrm{i}\,C_{\mathrm{I}}\,x\right)\,, \tag{2.200}$$

$$a^+ = -\mathrm{i}\sqrt{\frac{m}{2\hbar}}\,\alpha_0\,\left(\frac{p_{\mathrm{op}}}{m} + \mathrm{i}\,C_{\mathrm{I}}x\right)\,. \tag{2.201}$$

As shown above, already for $\omega = \omega_0 =$ constant, solutions exist with TD (oscillating) WP width, i.e., $\dot{\alpha} \neq 0$ and $\alpha = \alpha(t)$; hence, also $C_{\mathrm{R}} = \frac{\dot{\alpha}}{\alpha}$ must be taken into account. Obviously, the same also applies for the oscillator with TD frequency $\omega = \omega(t)$. Therefore, in the definitions of a and a^+, α_0 must be replaced by $\alpha(t)$ and $\mathrm{i}C_{\mathrm{I}}$ by C thus leading to [63, 64][12]

$$a(t) = \mathrm{i}\sqrt{\frac{m}{2\hbar}}\,\alpha\,\left(\frac{p_{\mathrm{op}}}{m} - C\,x\right)\,, \tag{2.202}$$

$$a^+(t) = -\mathrm{i}\sqrt{\frac{m}{2\hbar}}\,\alpha\,\left(\frac{p_{\mathrm{op}}}{m} - C^*\,x\right)\,. \tag{2.203}$$

It is easy to check that the commutator relation $[a(t), a^+(t)]_- = 1$ is fulfilled.

At least for TD frequency $\omega(t)$ the corresponding Hamiltonian is no longer a constant of motion. So one might ask whether $a(t)$ and $a^+(t)$ are constants of motion or not; in other words, do they fulfil

$$\frac{\partial}{\partial t}a + \frac{1}{\mathrm{i}\hbar}\,[a, H]_- = 0\,. \tag{2.204}$$

To answer this, x, p and hence H must be expressed in terms of $a(t)$ and $a^+(t)$, leading to

[12] The notation used in the following is not consistent with the one used in [63] as $a(t)$ and $\tilde{a}(t)$ are interchanged!

$$x = \frac{\alpha}{2} \sqrt{\frac{2\hbar}{m}} \, (a + a^+) \, , \tag{2.205}$$

$$p = m \, \frac{\alpha}{2} \sqrt{\frac{2\hbar}{m}} \, \left(\mathcal{C}^* \, a + \mathcal{C} \, a^+ \right) \tag{2.206}$$

with

$$H = \frac{\hbar}{4} \alpha^2 \, \left[\mathcal{C}_{\rm R}(a + a^+) + {\rm i} \, \mathcal{C}_{\rm I}(a^+ - a) \right]^2 + \frac{\hbar}{4} \, \omega^2 \alpha^2 \, \left(a^+ + a \right)^2 \, . \tag{2.207}$$

This finally leads to

$$\frac{\partial}{\partial t} a + \frac{1}{{\rm i}\hbar} \, [a, H]_- = -{\rm i} \, \frac{1}{\alpha^2} \, a \neq 0 \, . \tag{2.208}$$

So, $a(t)$ and $a^+(t)$ are no constants of motion but can be turned into such by simply introducing a phase factor according to

$$\tilde{a}(t) = a(t) \, {\rm e}^{{\rm i} \int^t dt' \frac{1}{\alpha^2}} = a(t) \, {\rm e}^{{\rm i}\varphi} \, , \tag{2.209}$$

$$\tilde{a}^+(t) = a^+(t) \, {\rm e}^{-{\rm i} \int^t dt' \frac{1}{\alpha^2}} = a^+(t) \, {\rm e}^{-{\rm i}\varphi} \tag{2.210}$$

where we can use our knowledge that $\lambda = \alpha {\rm e}^{{\rm i}\varphi}$ to show that, for the transition from a, a^+ to $\tilde{a}, \tilde{a}^+$, just the parameter $\alpha(t)$ in $a(t)$ must be replaced by $\lambda(t)$ and by $\lambda^*(t)$ in $a^+(t)$.

The phase factor shall be omitted in the following as it can be absorbed in the purely TD function $K(t)$ in the exponent of the WP/CS or into $N(t)$ (see Eq. (2.1)). The difference between a, a^+ and $\tilde{a}, \tilde{a}^+$ will also be mentioned again in connection with the displacement operator later in this subsection.

Next, it is shown that one can use the generalized annihilation operator $a(t)$ to obtain a CS $|w\rangle$ according to definition (2) of CSs, i.e. a CS as *eigenstate* of $a(t)$ with *complex eigenvalue* w,

$$a(t) \, |w\rangle = w \, |w\rangle \, . \tag{2.211}$$

First, the complex eigenvalue w is determined in terms of $\langle x \rangle_w = \eta$ and $\langle p \rangle_w = m\dot{\eta}$, i.e., mean values calculated with $|w\rangle$ where Eq. (2.211) is assumed to be valid.

From

$$\langle x \rangle_w = \sqrt{\frac{\hbar}{2m}} \, \alpha \, (w^* + w) = \sqrt{\frac{2\hbar}{m}} \, \alpha \, w_{\rm R} = \eta \tag{2.212}$$

$$\langle p \rangle_w = \sqrt{\frac{\hbar}{2m}} \, \alpha \, \left[\mathcal{C} \, w^* + \mathcal{C}^* \, w \right] = m\dot{\eta} \tag{2.213}$$

it follows that

$$w_{\mathrm{R}} = \sqrt{\frac{m}{2\hbar}} \left(\frac{\eta}{\alpha}\right) = \frac{1}{\sqrt{2}}\sqrt{\frac{m}{\hbar}}\, \alpha\, C_{\mathrm{I}}\, \eta \tag{2.214}$$

$$w_{\mathrm{I}} = \sqrt{\frac{m}{2\hbar}}\, (\dot{\eta}\alpha - \eta\dot{\alpha}) = \frac{1}{\sqrt{2}}\sqrt{\frac{m}{\hbar}}\, \alpha\, [\dot{\eta} - C_{\mathrm{R}}\, \eta] \tag{2.215}$$

or

$$w = \sqrt{\frac{m}{2\hbar}} \left[\left(\frac{\eta}{\alpha}\right) + \mathrm{i}\,(\dot{\eta}\alpha - \eta\dot{\alpha})\right] = \frac{1}{\sqrt{2}}\sqrt{\frac{m}{\hbar}}\, \alpha\, [C_{\mathrm{I}}\, \eta\, +\, \mathrm{i}\,(\dot{\eta} - C_{\mathrm{R}}\, \eta)] \, = \mathrm{i}\sqrt{\frac{m}{2\hbar}}\, \alpha\, [\dot{\eta}\, -\, C\eta]\, . \tag{2.216}$$

This shows the connection between the eigenvalues w (or w^*) and the Ermakov invariant as

$$I = \frac{\hbar}{m}\left(w_{\mathrm{R}}^2 + w_{\mathrm{I}}^2\right) = \frac{\hbar}{m} w w^* = \frac{\hbar}{m}|w|^2\, . \tag{2.217}$$

From this it immediately follows that the *operator* corresponding to the *Ermakov invariant*, when $p = m\dot{x}$ is replaced by the operator $p_{\mathrm{op}} = \frac{\hbar}{\mathrm{i}}\frac{\partial}{\partial x}$ (and taking into account that $[p_{\mathrm{op}}, x]_- = \frac{\hbar}{\mathrm{i}}$), can be written in terms of the generalized creation and annihilation operators as[13]

$$I_{\mathrm{op}} = \frac{\hbar}{m}\left[a^+(t)a(t) + \frac{1}{2}\right]\, . \tag{2.218}$$

In the position-space representation, the CS that is eigenstate of $a(t)$,

$$\langle x|a(t)|w\rangle = w\langle x|w\rangle \text{ or } \mathrm{i}\sqrt{\frac{m}{2\hbar}}\alpha \left\{\frac{\hbar}{m\mathrm{i}}\frac{\partial}{\partial x} - Cx\right\} \Psi_w(x) = w\, \Psi_w(x)\, , \tag{2.219}$$

can be given as

$$\Psi_w(x,t) = M(t)\exp\left\{\frac{\mathrm{i}m}{2\hbar}C(x - \langle x\rangle)^2 + \frac{\mathrm{i}}{\hbar}\langle p\rangle x - \frac{\mathrm{i}}{2\hbar}\langle p\rangle\langle x\rangle + \frac{1}{2}\left(w^2 + |w|^2\right)\right\} \tag{2.220}$$

which is, for $M(t) = N(t)\, \mathrm{e}^{-\frac{1}{2}(w^2+|w|^2)}$ and $N(t) = \left(\frac{m}{\pi\hbar}\right)^{\frac{1}{4}}\left(\frac{1}{\lambda}\right)^{\frac{1}{2}}$, identical to the normalized Gaussian WP (2.1) (written in the form (2.122)), i.e.,

$$\Psi_w(x,t) \;=\; \left(\frac{m}{\pi\hbar}\right)^{\frac{1}{4}}\left(\frac{1}{\lambda}\right)^{\frac{1}{2}} \exp\left\{\frac{\mathrm{i}m}{2\hbar}C\tilde{x}^2 + \frac{\mathrm{i}}{\hbar}\langle p\rangle\tilde{x} + \frac{\mathrm{i}}{2\hbar}\langle p\rangle\langle x\rangle\right\} \;=\; \Psi_{\mathrm{WP}}(x,t)\, . \tag{2.221}$$

---

[13]In any case, the phase factors cancel in the product; so $a^+a = \tilde{a}^+\tilde{a}$.

A comment on the normalization factor: In order to obtain the simple form $K(t) = \frac{1}{2\hbar}\langle p\rangle\langle x\rangle$ for the purely TD function $K(t)$ in the exponent, the normalization factor $N(t)$ becomes complex, $N(t) = \left(\frac{m}{\pi\hbar}\right)^{\frac{1}{4}}\left(\frac{1}{\lambda}\right)^{\frac{1}{2}}$ due to the complex $\lambda(t)$. For this quantity, using the polar form $\lambda = \alpha e^{i\varphi}$ with $\varphi = \int^t dt'\frac{1}{\alpha^2}$, $N(t)$ can be written as $N(t) = \left(\frac{m}{\pi\hbar\alpha^2}\right)^{\frac{1}{4}} e^{-i\frac{\varphi}{2}} = \left(\frac{1}{2\pi\langle\tilde{x}^2\rangle(t)}\right)^{\frac{1}{4}} e^{-\frac{i}{2}\int^t dt'\frac{1}{\alpha^2}}$. For $\alpha^2 = \alpha_0^2 = \omega_0^{-1}$, $N(t)$ turns into $N(t) = \left(\frac{m\omega_0}{\pi\hbar}\right)^{\frac{1}{4}} e^{-\frac{i}{2}\omega_0 t}$ and contributes the ground state energy of the HO that corresponds to $\tilde{E} = \frac{1}{2m}\langle\tilde{p}^2\rangle + \frac{m}{2}\omega_0^2\langle\tilde{x}^2\rangle = \frac{\hbar}{2}\omega_0$ in the case of the WP with constant width. Here we obtain a generalization for $\alpha = \alpha(t)$. The phase factors $e^{\mp i\int^t dt'\frac{1}{\alpha^2}}$ occurring in the creation/annihilation operators as defined in (2.209, 2.210) can be absorbed into $N(t)$ in a similar way.

As mentioned before, the CS can also be defined as a displaced vacuum state $|0\rangle$, i.e.

$$|w\rangle = \exp\left\{wa^+(t) - w^*a(t)\right\}|0\rangle = D(w)\,|0\rangle \tag{2.222}$$

with the displacement operator $D(w)$.

In the position-space representation, the vacuum state ϕ_0 can be obtained via

$$\langle x|a(t)|0\rangle = i\sqrt{\frac{m}{2\hbar}}\alpha\left\{\frac{\hbar}{mi}\frac{\partial}{\partial x} - \mathcal{C}x\right\}\phi_0(x) = 0 \tag{2.223}$$

as

$$\langle x|0\rangle = \phi_0(x,t) = N(t)\,e^{i\frac{m}{2\hbar}\mathcal{C}(t)x^2} \tag{2.224}$$

with the same $N(t)$ as above.

Note that due to $\mathcal{C}(t)$ the exponent is now *complex*, in particular via $i\mathcal{C}_R x^2$, the term $\frac{im}{2\hbar}\frac{\dot{\alpha}}{\alpha}x^2$ already occurs here naturally, whereas it must be introduced in the approach of Hartley and Ray [65] via a unitary transformation.

Using the Baker–Campbell–Hausdorff formula, the CS can now be written in the usual form

$$|w\rangle = D(w)\,|0\rangle = e^{-\frac{1}{2}|w|}\exp(wa^+(t))\,|0\rangle = e^{-\frac{1}{2}|w|^2}\sum_{n=0}^{\infty}\frac{w^n(a^+)^n}{n!}\,|0\rangle\,, \tag{2.225}$$

now with $a^+(t)$ as defined in (2.203). Evaluating e^{wa^+} with the help of w and $|w|^2$, as given in (2.216) and (2.217), leads again in the position-space representation to $\Psi_w(x,t)$ in the form as given in (2.122).

At this point, a comment concerning the displacement operator $D(w)$ should be added. In the case of the HO with constant width, the creation and annihilation operators are given by (2.202) and (2.203) with $\mathcal{C}_R = 0$ and $\mathcal{C}_I = \frac{1}{\alpha}^2$ with $\alpha = \alpha_0 = \left(\frac{1}{\omega_0}\right)^{\frac{1}{2}}$ and, according to (2.205), the sum of both operators is proportional to the position variable

$$a + a^+ = \left(\frac{2m}{\hbar}\right)^{\frac{1}{2}} \alpha\, C_{\mathrm{I}}\, x = \left(\frac{2m\alpha^2}{\hbar}\right)^{\frac{1}{2}} \frac{1}{\alpha^2}\, x = \left(\frac{2m\omega_0}{\hbar}\right)^{\frac{1}{2}} x\,. \tag{2.226}$$

Consequently, the corresponding eigenvalues w and w^* fulfil

$$w + w^* = \left(\frac{2m}{\hbar}\right)^{\frac{1}{2}} \alpha\, C_{\mathrm{I}}\, \eta = \left(\frac{2m\alpha^2}{\hbar}\right)^{\frac{1}{2}} \frac{1}{\alpha^2}\, \eta = \left(\frac{2m\omega_0}{\hbar}\right)^{\frac{1}{2}} \eta\,, \tag{2.227}$$

so the "shifted" creation/annihilation operators

$$d = a - w\,, \qquad d^+ = a^+ - w^* \tag{2.228}$$

fulfil

$$d + d^+ = (a + a^+) - (w + w^*) = \left(\frac{2m\omega_0}{\hbar}\right)^{\frac{1}{2}} (x - \eta) = \left(\frac{2m}{\hbar}\right)^{\frac{1}{2}} \frac{1}{\alpha_0}\, \tilde{x}\,, \tag{2.229}$$

i.e., the same as (2.226), only for a position variable displaced by the classical trajectory $\eta(t) = \langle x \rangle$ (in a negative direction). The shifted operators d and d^+ can be obtained from the original ones by a similarity transformation with the help of a unitary operator $D(w)$ according to [66]

$$D_-^+\, a\, D_- = a - w = d\,, \tag{2.230}$$

$$D_-^+\, a^+\, D_- = a^+ - w^* = d^+\,. \tag{2.231}$$

To find the explicit form of the operator $D_-(w)$ one can apply an exponential ansatz of the form

$$D_-(w) = \mathrm{e}^{\Lambda(w)} \tag{2.232}$$

and make use of $D_-(w)$ being unitary, i.e., $D_-^+ = \mathrm{e}^{\Lambda^+} = D_-^{-1} = \mathrm{e}^{-\Lambda}$, hence Λ must be anti-Hermitian, $\Lambda^+(z) = -\Lambda(z)$.

With this ansatz, it follows that (2.230) is fulfilled for $[\Lambda, a]_- = w$. Due to $[a, a^+]_- = 1$, Λ can be written as $\Lambda = -wa^+$. This form of Λ, however, is not anti-Hermitian. Knowing that $[a, a]_- = 0$, any linear combination of a can be added to Λ without changing $[\Lambda, a]_- = w$. So, an extension to

$$\Lambda(w) = -(wa^+ - w^* a) \tag{2.233}$$

also fulfils Eq. (2.231) and leads to the unitary displacement operator

$$D_-(w) = \exp\{-(wa^+ - w^* a)\}\,, \tag{2.234}$$

causing a displacement by $-\eta(t)$ (in agreement with (2.222) where $D(w) \mathrel{\hat{=}} D_+(w) = D_-^{-1}(w) = D_-^+(w)$ causes a displacement by the same positive amount).

How is this affected by the above treatment? As mentioned earlier, the operators $a(t)$ and $a^+(t)$ are not constants of motion but can be turned into such by multiplying them by a phase factor, leading to $\tilde{a}$ and $\tilde{a}^+$ according to Eqs. (2.209) and (2.210). The same would also apply to the corresponding eigenvalues of the CSs, $\tilde{w}$ and $\tilde{w}^*$. The shift of the position variable x would then, in analogy with Eq. (2.227), be given by $\tilde{w}+\tilde{w}^*$, where α (in the first term on the rhs) must be replaced by $\alpha e^{i\varphi} = \lambda(t)$ and C_I by C in $\tilde{w}$ and by $\lambda^*(t)$ and C^* in $\tilde{w}^*$. With $\eta = \frac{\alpha_0 p_0}{m} z$ and $C = \frac{\dot{\lambda}}{\lambda}$ (with $\lambda = u + iz$), $\tilde{w}$ and $\tilde{w}^*$ can then be written as

$$\begin{aligned}\tilde{w} &= i\sqrt{\frac{m}{2\hbar}}\lambda\left(\dot{\eta} - \frac{\dot{\lambda}}{\lambda}\eta\right) = i\sqrt{\frac{m}{2\hbar}}\frac{\alpha_0 p_0}{m}\left(\lambda\dot{z} - \dot{\lambda}z\right)\\ &= i\frac{\alpha_0 p_0}{m}\sqrt{\frac{m}{2\hbar}}\left(u\dot{z} - \dot{u}z\right) = i\frac{\alpha_0 p_0}{m}\sqrt{\frac{m}{2\hbar}}\end{aligned} \tag{2.235}$$

(where Eq. (2.56) has been used in the last step) and

$$\begin{aligned}\tilde{w}^* &= -i\sqrt{\frac{m}{2\hbar}}\lambda^*\left(\dot{\eta} - \frac{\dot{\lambda}^*}{\lambda^*}\eta\right) = -i\sqrt{\frac{m}{2\hbar}}\frac{\alpha_0 p_0}{m}\left(\lambda^*\dot{z} - \dot{\lambda}^* z\right)\\ &= -i\frac{\alpha_0 p_0}{m}\sqrt{\frac{m}{2\hbar}}\left(u\dot{z} - \dot{u}z\right) = -i\frac{\alpha_0 p_0}{m}\sqrt{\frac{m}{2\hbar}}\end{aligned} \tag{2.236}$$

leading to the displacement $\tilde{w} + \tilde{w}^* = 0$, i.e. *NO* displacement.

As the same would also apply if in the conventional creation and annihilation operators α_0 would be replaced by $\alpha_0 e^{i\omega_0 t}$, i.e., by the complex quantity λ, it follows that in order to obtain the shift by the amount $\eta(t)$, not the invariant operators $\tilde{a}$ and $\tilde{a}^+$ (including the phase factor), but a and a^+ which are *NOT* constants of motion must be applied in the definition of the displacement operator $D(w)$. Therefore, one can conclude that the phase factor $e^{\mp i\int^{t} dt' \frac{1}{\alpha^2}} = e^{i\varphi}$ not only turns the creation/annihilation operators into dynamical invariants but, to the contrary, its absence introduces the dynamics that moves the position variable by an amount $\eta(t)$ that corresponds to the classical trajectory.

Finally, it is shown that our CS also fulfils the Schrödinger–Robertson uncertainty relation and how the uncertainties can be expressed in terms of w, w^*, α, C_R and C_I.

In terms of these quantities the mean value of $\langle x^2\rangle$ can be written as

$$\langle x^2\rangle_w = \frac{\hbar}{2m}\,\alpha^2\,\left(w^{*2} + w^2 + 2|w|^2 + 1\right) \tag{2.237}$$

leading (together with Eq. (2.212)) to the mean square deviation of position

$$\langle\tilde{x}^2\rangle_w = \langle x^2\rangle_w - \langle x\rangle_w^2 = \frac{\hbar}{2m}\,\alpha^2\,. \tag{2.238}$$

In the same way, from

$$\langle p^2 \rangle_w = \frac{\hbar m}{2}\, \alpha^2 \left[\mathcal{C}^2 w^{*2} + \mathcal{C}^{*2} w^2 + |\mathcal{C}|^2 \left(2|w|^2 + 1\right) \right] \tag{2.239}$$

and using Eq. (2.213), one obtains

$$\langle \tilde{p}^2 \rangle_w = \langle p^2 \rangle_w - \langle p \rangle_w^2 = \frac{\hbar m}{2}\, \alpha^2\, |\mathcal{C}|^2 = \frac{\hbar m}{2}\, \alpha^2 \left[\mathcal{C}_{\mathrm{R}}^2 + \mathcal{C}_{\mathrm{I}}^2 \right] \tag{2.240}$$

and from

$$\langle [x, p]_+ \rangle_w = \langle xp + px \rangle_w = \hbar\, \alpha^2 \left[\mathcal{C} w^{*2} + \mathcal{C}^* w^2 + \mathcal{C}_{\mathrm{R}} \left(2|w|^2 + 1\right) \right] \tag{2.241}$$

with (2.212) and (2.213) the correlation uncertainty

$$\left\langle \frac{1}{2}[\tilde{x}, \tilde{p}]_+ \right\rangle_w = \left\langle \frac{1}{2}[x, p]_+ \right\rangle_w - \langle x \rangle_w \langle p \rangle_w = \frac{\hbar}{2}\, \alpha^2 \mathcal{C}_{\mathrm{R}}\,. \tag{2.242}$$

From

$$\langle \tilde{x}^2 \rangle_w \langle \tilde{p}^2 \rangle_w - \left\langle \frac{1}{2}[\tilde{x}, \tilde{p}]_+ \right\rangle_w^2 = \frac{\hbar^2}{4}\, \alpha^4 \left[\mathcal{C}_{\mathrm{R}}^2 + \mathcal{C}_{\mathrm{I}}^2 \right] - \frac{\hbar^2}{4}\, \alpha^4\, \mathcal{C}_{\mathrm{R}}^2 = \frac{\hbar^2}{4} \tag{2.243}$$

it follows that our CS also fulfils the Schrödinger–Robertson uncertainty condition.

Like the annihilation operator of the HO can be used to obtain the corresponding ground state wave function and the creation operators to obtain the excited states, the generalized annihilation and creation operators (2.202) and (2.203) can be used to obtain solutions of the TDSE where the (at most quadratic) potential of the problem enters via the complex TD variable $\mathcal{C}$ of the corresponding Riccati equation or the real TD variable $\alpha(t)$ of the equivalent Ermakov equation.

So, e.g., for the free motion ($V = 0$), the application of $a(t)$ on the ground state wave function $|0\rangle$ leads (in position representation) via $a(t)|0 >= 0$ to the ground state as given in Eq. (2.224), where $\mathcal{C}(t)$ is obtained from the Riccati equation (2.4) for $\omega = 0$. Note that this ground state (in contrast to the TI HO) has a position dependent phase via $\frac{im}{2\hbar}\mathcal{C}_{\mathrm{R}} x^2$.

The corresponding excited states can be obtained by applying $a^+(t)$ on this ground state according to $\Psi_n(x,t) = (a^+)^n|0\rangle$ in position space representation as

$$\Psi_n(x,t) = \left[\left(\frac{m}{\pi\hbar}\right)^{\frac{1}{2}} \frac{1}{\alpha}\, \frac{1}{n!2^n} \right]^{\frac{1}{2}} \mathcal{H}_n\left(\frac{x}{x_0}\right)\, \exp\left\{ -\frac{1}{2}\left(\frac{x}{x_0}\right)^2 + \mathrm{i}\frac{m}{2\hbar}\frac{\dot{\alpha}}{\alpha} x^2 \right\} \tag{2.244}$$

with $x_0 = \sqrt{\frac{\hbar}{m}}\alpha(t)$ being TD due to $\alpha(t)$ and $\mathcal{C}_{\mathrm{R}} = \frac{\dot{\alpha}}{\alpha}$. Hence, the dimensionless variable $\xi = \frac{x}{x_0}$ that usually appears in this context in the Hermite polynomials $\mathcal{H}_n(\xi)$ is now TD [64]!

Similar solutions for the TD free motion SE have also been found by Guerrero et al. [67] in a different way in connection with the quantum Arnold transformation [68–70].

Note: the $\Psi_n(x,t)$ are exact solutions of the TDSE for $V = 0$ but they are *no eigenfunctions* of the operator $H_{\text{op}} = -\frac{\hbar^2}{2m}\frac{\partial^2}{\partial x^2}$, however, they are eigenfunctions of the Ermakov operator (2.218).

The possibility of expanding solutions of the TDSE in terms of eigenfunctions of the Ermakov operator (2.218) will be demonstrated in the next subsection in order to transform TDSEs into formally TI ones.

2.12 Application of the Ermakov Invariant to Transform Time-Dependent into Time-Independent Schrödinger Equations

In the previous subsection it was shown that the operator that can be obtained from the Ermakov invariant (2.21) by replacing $m\dot{\eta}$ with the momentum operator $p_{\text{op}} = \frac{\hbar}{\text{i}}\frac{\partial}{\partial x}$, can be expressed in terms of the generalized TD creation and annihilation operators $a^+(t)$ and $a(t)$, Eq. (2.218), exactly as known for the HO with constant frequency and TI operators a^+ and a. In particular, the ground state wave function and a complete set of eigenfunctions of I_{op} can be obtained in the usual way. These eigenfunctions of I_{op} and the corresponding eigenvalues will be used in the following to find the solutions of SEs with explicitly TD Hamiltonians.

For this purpose we consider a classical Ermakov system with Hamiltonian[14]

$$H(t) = \frac{1}{2}\,p^2 + \frac{1}{2}\omega^2(t)\,x^2 \tag{2.245}$$

and the corresponding set of Ermakov equations (equivalent to (2.3) and (2.16)) and the Ermakov invariant I_{L}.

In the quantized version, obtained via the above specified replacement of p, I_{L} turns into a Hermitian operator that, as a constant of motion, therefore fulfils the relation

$$\frac{d}{dt}I_{\text{op}} = \frac{1}{\text{i}\hbar}\left[I_{\text{op}}, H_{\text{op}}\right]_- + \frac{\partial}{\partial t}I_{\text{op}} = 0\,, \tag{2.246}$$

equivalent to the classical expression (2.181) in Sect. 2.10, naturally replacing the Poisson bracket { , } by the commutator divided by $\text{i}\hbar$, $\frac{1}{\text{i}\hbar}[\ ,\ ]_-$.

Important is that $H = H(t)$ varies in time, depending on the frequency $\omega(t)$. Thus, the eigenvalues of H will also change in time and cannot be determined until $\omega(t)$ is specified. On the other hand, I_{op} satisfies (2.246) and has *constant eigenvalues* λ_n which can be determined from

---

[14] In this subsection $m = 1$ will be set, and hence $\dot{x} = p$, to simplify comparison between $H(t)$ and I; the factor m is usually omitted in I; see also the comment below Eq. (2.21).

$$I_{\text{op}}\psi_n(x,t) = \lambda_n \psi_n(x,t)\,. \tag{2.247}$$

The action of I_{op} on a state $|\;\rangle$ that solves the TDSE with the Hamilton operator $H_{\text{op}}(t)$ corresponding to (2.245)

$$\mathrm{i}\hbar\frac{\partial}{\partial t}|\;\rangle = H_{\text{op}}(t)|\;\rangle \tag{2.248}$$

leads to *another state* ($I_{\text{op}}|\;\rangle$) that is also a solution of this TDSE which can be proven easily using (2.246) (for further details see [71]).

Let us assume there is a complete set of (orthonormal) *eigenstates* $\psi_n(x,t)$ of I_{op} with *eigenvalues* λ_n. Then the general solution of the TDSE with the above Hamiltonian

$$\mathrm{i}\hbar\frac{\partial}{\partial t}\Psi(x,t) = \left\{-\frac{\hbar^2}{2}\frac{\partial^2}{\partial x^2} + \frac{1}{2}\omega^2(t)x^2\right\}\Psi(x,t)\,, \tag{2.249}$$

can be expressed in terms of the eigenfunctions of I_{op} in the form

$$\Psi(x,t) = \sum_n c_n \mathrm{e}^{\mathrm{i}b_n(t)}\psi_n(x,t) \tag{2.250}$$

with constant superposition coefficients c_n and TD phase functions $b_n(t)$. It can be shown (see [71] and below) that $\Psi(x,t)$ constructed in this way always satisfies the SE (2.249), providing $b_n(t)$ fulfils

$$\hbar\frac{d}{dt}b_n = \left\langle \psi_n \left| \mathrm{i}\hbar\frac{\partial}{\partial t} - H \right| \psi_n \right\rangle\,. \tag{2.251}$$

In the following, a short outline of this method is given. The central equation is (2.247) where the operator I_{op} corresponds to the invariant $I_{\text{L}} = \frac{1}{2}\left[(\alpha p - \dot{\alpha}x)^2 + k\left(\frac{x}{\alpha}\right)^2\right]$ with constant[15] k and c-numbers $\alpha(t)$ and $\dot{\alpha}(t)$. This operator can be written as

$$I_{\text{op}} = \frac{1}{2}\left\{-\hbar^2\alpha^2\frac{\partial^2}{\partial x^2} - 2\alpha\dot{\alpha}\frac{\hbar}{\mathrm{i}}x\frac{\partial}{\partial x} + \dot{\alpha}^2x^2 + k\left(\frac{x}{\alpha}\right)^2 - \alpha\dot{\alpha}\frac{\hbar}{\mathrm{i}}\right\}\,. \tag{2.252}$$

In a next step, a unitary transformation is applied to ψ_n,

$$\psi_n'(x,t) \;=\; \exp\left(-\frac{\mathrm{i}}{2\hbar}\frac{\dot{\alpha}}{\alpha}x^2\right)\psi_n(x,t) \;=\; U\,\psi_n(x,t)\,. \tag{2.253}$$

---

[15]This is just a generalization of the case considered in our WP example. Therefore, $\frac{1}{\alpha^3}$ has to be replaced by $\frac{k}{\alpha^3}$ on the rhs of Ermakov equation (2.16).

Note that this transformation essentially compensates the phase factor proportional to x^2 in our Gaussian WP solutions that was necessary (via $\mathcal{C}_{\mathrm{R}} = \frac{\dot{\alpha}}{\alpha}$) in order to take into account the time-dependence of the WP width.

In Sect. 2.8, Eq. (2.145), it has been shown explicitly that the width of an oscillating WP is TD if $\beta_0 = \frac{\hbar}{2m\langle\tilde{x}^2\rangle_0} \neq \omega_0$ in the harmonic case and certainly always TD for TD frequency as $\beta_0 = \text{const.} \neq \omega(t)$. Therefore, the unitary transformation U simply eliminates the phase factor originating from the time-dependence of $\psi_n(x,t)$ due to $\omega(t)$ to turn it into $\psi_n'(x,t)$, compatible with a WP with constant width and thus constant frequency $\omega = \omega_0$, corresponding to a constant Hamiltonian instead of $H(t)$.

Under this transformation, the operator I_{op} changes into

$$I'_{\mathrm{op}} = U\, I_{\mathrm{op}}\, U^+ \tag{2.254}$$

with

$$I'_{\mathrm{op}}\, \psi_n'(x,t) = \lambda_n\, \psi_n'(x,t)\;, \tag{2.255}$$

i.e., the eigenvalue spectrum remains unchanged. Straightforward calculation yields

$$I'_{\mathrm{op}} = \frac{1}{2}\left\{-\hbar^2\alpha^2\frac{\partial^2}{\partial x^2} + k\left(\frac{x}{\alpha}\right)^2\right\}\,. \tag{2.256}$$

Defining a new independent variable $\sigma = \frac{x}{\alpha}$ allows one to rewrite Eq. (2.255) as

$$\left\{-\frac{\hbar^2}{2}\frac{d^2}{d\sigma^2} + \frac{1}{2}k\sigma^2\right\}\phi_n(\sigma) = \lambda_n\,\phi_n(\sigma)\;, \tag{2.257}$$

or

$$I'_{\mathrm{op}}\,\phi_n(\sigma) = \lambda_n\,\phi_n(\sigma)\;, \tag{2.258}$$

where $\psi_n'(x,t)$ has been replaced by the formally TI function $\phi(\sigma)$ according to

$$\psi_n'(x,t) = \frac{1}{\alpha^{\frac{1}{2}}}\,\phi_n(\sigma) = \frac{1}{\alpha^{\frac{1}{2}}}\,\phi_n\left(\frac{x}{\alpha}\right)\,, \tag{2.259}$$

removing the explicit time-dependence in ψ_n' and in the potential (replacing $\omega^2(t)$ with k). The factor $\frac{1}{\alpha^{\frac{1}{2}}}$ is introduced to fulfil the normalizability condition $\int \psi_n'^*(x,t)\, \psi_n'(x,t)\,dx = \int \phi_n^*(\sigma)\,\phi_n(\sigma)\,d\sigma = 1$. The important point is that the transformed eigenvalue problem (2.257) is now an ordinary one-dimensional TISE with a not explicitly TD potential $V(\sigma) = \frac{1}{2}k\sigma^2$, formally equivalent to a HO with frequency $\omega = \sqrt{k}$ and well-known eigenfunctions and eigenvalues.

Knowing $\phi_n(\sigma)$, it is easy to reconstruct the eigenfunctions $\psi_n(x,t)$ of the original eigenvalue problem (2.247). In order to obtain the solution $\Psi(x,t)$ of the TD problem with $H(t)$ as given in (2.245), it still remains to determine the phases $b_n(t)$ which satisfy Eq. (2.251).

Carrying out the unitary transformation $\psi'_n = U\psi_n$ on the rhs of this equation, it turns into

$$\hbar\frac{d}{dt}b_n(t) = \left\langle \psi'_n \left| \mathrm{i}\hbar\frac{\partial}{\partial t} - \frac{\dot{\alpha}}{\alpha}\frac{\hbar}{\mathrm{i}}\frac{\partial}{\partial x} + \frac{\dot{\alpha}}{\alpha}\,\mathrm{i}\frac{\hbar}{2} - \frac{1}{\alpha^2}\,I'_{\mathrm{op}} \right| \psi'_n \right\rangle . \qquad (2.260)$$

With the substitution (2.259) for ψ'_n and keeping in mind that via $\alpha(t)$ there is a contribution to the $\frac{\partial}{\partial t}$-term, this leads to

$$\hbar\frac{d}{dt}b_n(t) = \left\langle \phi_n \left| -\frac{1}{\alpha^2}\,I'_{\mathrm{op}} \right| \phi_n \right\rangle \qquad (2.261)$$

which provides with (2.258)

$$\hbar\frac{d}{dt}b_n(t) = -\frac{\lambda_n}{\alpha^2}\,, \qquad (2.262)$$

or

$$b_n(t) = -\frac{\lambda_n}{\hbar}\int dt'\frac{1}{\alpha^2(t')} \qquad (2.263)$$

where, because of $\dot{\varphi} = \frac{\sqrt{k}}{\alpha^2}$ the integral corresponds to the phase angle like in $\lambda = \mathrm{e}^{\mathrm{i}\varphi}$; so $b_n(t)$ is determined via this angle and the eigenvalues λ_n of the Ermakov invariant.

To summarize, the solution $\Psi(x,t)$ of the SE (2.249) can be obtained from the eigenfunctions and eigenvalues of the corresponding Ermakov operator via

$$\begin{aligned}
&\Psi(x,t) = \sum_n c_n \mathrm{e}^{\mathrm{i}b_n(t)}\psi_n(x,t)\\
\text{with}\quad &\psi_n(x,t) = U^+\,\psi'_n(x,t) = \exp\left(\frac{\mathrm{i}}{2\hbar}\frac{\dot{\alpha}}{\alpha}x^2\right)\frac{1}{\alpha^{\frac{1}{2}}}\,\phi_n\left(\frac{x}{\alpha}\right)\\
\text{and}\quad &b_n(t) = -\frac{\lambda_n}{\hbar}\int dt'\frac{1}{\alpha^2(t')}\\
&c_n = \langle\psi_n(x,t=0)|\Psi(x,t=0)\rangle \mathrm{e}^{-\mathrm{i}b_n(0)}\\
&\quad = \int dx\,\psi_n^*(x)\,\Psi(x)\mathrm{e}^{-\mathrm{i}b_n}\,.
\end{aligned} \qquad (2.264)$$

An advantage of this method transforming the original TD problem into a formally TISE is that all the knowledge about the solution of TI problems, like perturbation theory, that is not always available in the TD case, can be applied to obtain the solutions ϕ_n that, in the way described above, enables one to also get the solution of the TD problem (more details and references can be found in [72, 73]).

There is yet another aspect to the possibility of expressing solutions of the TDSE in terms of eigenfunctions of the Ermakov operator. As the expression for Ψ in

(2.264), via c_n, depends also on the initial function at $t' = 0$ and the corresponding position x' (because c_n is TI, it can be determined at any time t' and position x'), it can be rearranged so that it provides $\Psi(x,t)$ from a given initial $\Psi(x',t')$ via a transformation that is equivalent to the TD Green function or Feynman kernel discussed in Sect. 2.5. For this purpose, we write (2.264) in the form

$$\begin{aligned}\Psi(x,t) &= \sum_n \mathrm{e}^{-\mathrm{i}b_n(t')}\left[\int dx'\psi_n^*(x',t')\,\Psi(x',t')\right]\mathrm{e}^{\mathrm{i}b_n(t)}\psi_n(x,t)\\ &= \int dx'\sum_n \mathrm{e}^{\mathrm{i}(b_n(t)-b_n(t'))}\psi_n^*(x',t')\psi_n(x,t)\,\Psi(x',t')\\ &= \int dx' G(x,t,x',t')\,\Psi(x',t')\,,\end{aligned} \tag{2.265}$$

with

$$G\left(x,x',t,t'\right) = \sum_n \mathrm{e}^{\mathrm{i}(b_n(t)-b_n(t'))}\psi_n(x,t)\,\psi_n^*(x',t')\,. \tag{2.266}$$

From the expression for b_n in (2.264) and with $\dot{\varphi} = \frac{1}{\alpha^2}$ and $\lambda_n = \hbar(n+\frac{1}{2})$, one obtains $b_n(t) = -(n+\frac{1}{2})\varphi(t)$. The eigenfunctions of I_{op} have been given in (2.244) where $x_0 = \sqrt{\frac{\hbar}{m}}\alpha(t)$ and $x_0' = \sqrt{\frac{\hbar}{m}}\alpha_0$ have to be taken.

Inserting all of this into Eq. (2.265), the TD Green function, at first sight, takes the more complicated-looking form

$$\begin{aligned}G\left(x,x',t,t'\right) &= \left(\frac{m}{\pi\hbar\alpha_0\alpha}\right)^{\frac{1}{2}}\mathrm{e}^{-\frac{\mathrm{i}}{2}\varphi(t)}\exp\left\{\frac{\mathrm{i}m}{2\hbar}\left[\frac{\dot{\alpha}}{\alpha}x^2 - \frac{\dot{\alpha}_0}{\alpha_0}x'^2 + \mathrm{i}\left(\left(\frac{x}{\alpha}\right)^2 + \left(\frac{x'}{\alpha_0}\right)^2\right)\right]\right\}\\ &\quad\times\sum_n \mathrm{e}^{-\mathrm{i}n\varphi(t)}\left(\frac{1}{2^n n!}\right)\mathcal{H}_n\left(\sqrt{\frac{m}{\hbar}}\frac{x'}{\alpha_0}\right)\mathcal{H}_n\left(\sqrt{\frac{m}{\hbar}}\frac{x}{\alpha}\right)\end{aligned} \tag{2.267}$$

where $\varphi(t') = \varphi_0 = 0$ has been taken and $\mathcal{H}_n^* = \mathcal{H}_n$, as $\mathcal{H}_n$ is real. Howerver, this can be simplified using the relation [74]

$$\sum_{n=0}^{\infty}\frac{1}{2^n}\frac{t^n}{n!}\mathcal{H}_n(r)\mathcal{H}_n(s) = \left(1-t^2\right)^{-\frac{1}{2}}\exp\left\{\frac{2rst-(r^2+s^2)t}{1-t^2}\right\} \tag{2.268}$$

where $t = e^{-\mathrm{i}\varphi} = \cos\varphi - \mathrm{i}\sin\varphi$, $r = \sqrt{\frac{m}{\hbar}}\frac{x}{\alpha}$, $s = \sqrt{\frac{m}{\hbar}}\frac{x'}{\alpha_0}$ and $1-t^2 = \mathrm{i}2t\sin\varphi$ to yield the TD Green function in a form like (2.55),

$$G\left(x,x',t,t'\right) = \left(\frac{m}{2\pi\mathrm{i}\alpha_0 z(t)}\right)^{\frac{1}{2}}\exp\left\{\frac{\mathrm{i}m}{2\hbar}\left[\frac{\dot{\alpha}}{\alpha}x^2 - \frac{\dot{\alpha}_0}{\alpha_0}x'^2 + \frac{u(t)}{z(t)}\left(\frac{x}{\alpha(t)}\right)^2 - 2\frac{x}{z(t)}\left(\frac{x'}{\alpha_0}\right) + \frac{u(t)}{z(t)}\left(\frac{x'}{\alpha_0}\right)^2\right]\right\} \tag{2.269}$$

where $\frac{\dot{z}}{z} = \frac{\dot{\alpha}}{\alpha} + \frac{u}{z}\frac{1}{\alpha^2}$ has been used and the more general case, where initially $\frac{\dot{\alpha}_0}{\alpha_0}$ is different from zero, has also been taken into account.

This confirms the close connection and interrelations between the Ermakov system, the TD Green function and the linearized form of the Riccati equation in terms of a complex Newtonian equation.

Furthermore, the method can be generalized to Hamiltonians of the form

$$H(t) = \frac{1}{2}\,p^2 + \frac{1}{2}\omega^2(t)\,x^2 + \frac{1}{\alpha^2} f\left(\frac{x}{\alpha}\right) \tag{2.270}$$

with unchanged equation for $\alpha(t)$, $\ddot{\alpha} + \omega^2\alpha = \frac{k}{\alpha^3}$; but from (2.270) follows for $x(t)$:

$$\ddot{x} + \omega^2(t)x = \frac{1}{\alpha^3} f'\left(\frac{x}{\alpha}\right) \tag{2.271}$$

with $f' = \frac{d}{d\left(\frac{x}{\alpha}\right)} f$. Consequently, the Ermakov invariant changes into

$$I = \frac{1}{2}\left[(\alpha p - \dot{\alpha}x)^2 + k\left(\frac{x}{\alpha}\right)^2\right] + f\left(\frac{x}{\alpha}\right) . \tag{2.272}$$

In the quantized version, this leads to the TDSE

$$\mathrm{i}\hbar\frac{\partial}{\partial t}\Psi(x,t) = \left\{-\frac{\hbar^2}{2}\frac{\partial^2}{\partial x^2} + \frac{1}{2}\omega^2(t)x^2 + \frac{1}{\alpha^2} f\left(\frac{x}{\alpha}\right)\right\}\Psi(x,t) . \tag{2.273}$$

The operator I_{op} corresponding to this new invariant I has the same additional (c-number) function $f\left(\frac{x}{\alpha}\right)$. Therefore, after the unitary transformation to I' and finally in the "TISE" depending on the variable σ, also the same function turns up, i.e.

$$\left\{-\frac{\hbar^2}{2}\frac{\partial^2}{\partial\sigma^2} + \frac{1}{2}k\sigma^2 + f(\sigma)\right\}\phi_n(\sigma) = \lambda_n\,\phi_n(\sigma) , \tag{2.274}$$

looking like a TISE with potential[16]

$$V(\sigma) = \frac{1}{2}k\sigma^2 + f(\sigma) . \tag{2.275}$$

Apart from $f = 0$, only for $f = c\frac{\alpha^2}{x^2}$ the parameter $\alpha(t)$ does not occur explicitly in the potential $V(\sigma)$. In the other cases it can be interpreted as an external field the time-dependence of which must be determined from the $\ddot{\alpha}$-equation.

---

[16]For further details see also [72, 73] and references cited therein.

At the end of this section, it has been indicated that Ermakov systems (and equivalent complex Riccati equations) can also play a useful role in TI quantum mechanics and for systems going beyond the harmonic or parametric oscillator. In the next section, this is intensified to consider, firstly, real Riccati equations occurring in the context of supersymmetric (SUSY) quantum mechanics. Afterwards, a reformulation of the TISE in terms of an Ermakov or complex Riccati equation is presented that is valid for *any* potential.

Before this, a short summary of the results obtained from the TDSE is given.

2.13 Interrelations Between the Different Treatments and Properties of the Complex Riccati Equation for Time-Dependent Systems

The equation of major interest, the complex Riccati equation (2.4), was introduced in Sect. 2.1 in the context of the TDSE (in position space) with exact Gaussian-shaped WP solutions and different treatments of this equation have been discussed (see also Fig. 2.5).

(1) Direct solution of this equation can be achieved if a particular solution is found, allowing for the transformation of the inhomogeneous Riccati equation into a homogeneous Bernoulli equation that can be linearized and integrated. The solution, and thus also the solution of the NL Riccati equation, depends on the initial conditions

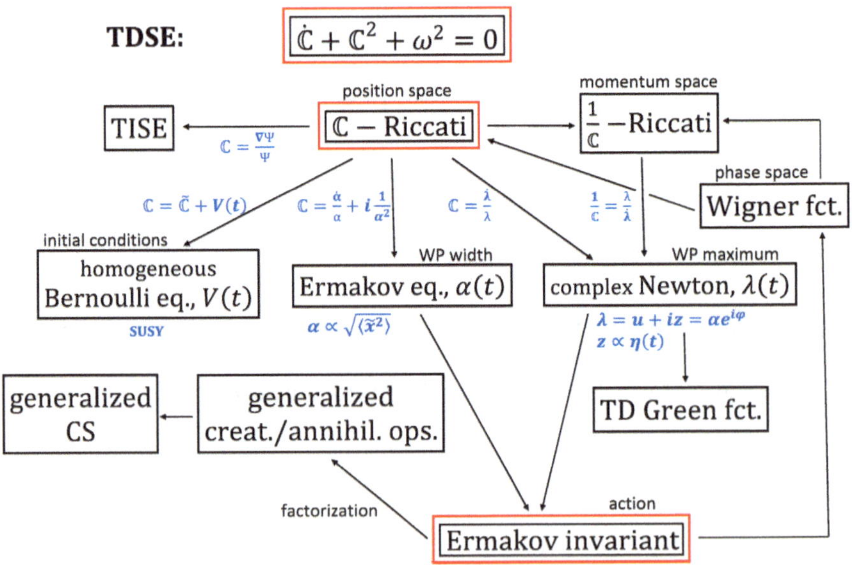

Fig. 2.5 Different treatments of the complex Riccati equation for the TDSE

(like initial WP width) as a parameter. The sensitivity of NL differential equations to changes of the initial conditions applies also in this case. Consider, for example, the HO where a constant WP width changes into an oscillating one when the initial WP width is varied. Further examples will follow for open systems in Chap. 5. Formal similarities with isospectral potentials in SUSY quantum mechanics are specified in Sect. 2.1.

(2) Introduction of a new real variable $\alpha(t)$, that is proportional to the WP width or position uncertainty, transforms the complex NL Riccati equation (2.4) into the real NL Ermakov equation (2.16). Together with the Newtonian equation (2.3) for the mean value of position, this Ermakov equation forms a coupled system of differential equations that, via elimination of the (possibly TD) coupling parameter $\omega^2(t)$, leads to the dynamical invariant (2.21) that is a constant of motion even if the Hamiltonian is not (e.g., for an oscillator with TD frequency $\omega(t)$). This Ermakov invariant has essentially the dimension of an action, the physical quantity that is actually quantized in portions of $\hbar$.

(3) A Newtonian equation like (2.3) can also be obtained from the Riccati equation (2.4) if its complex variable $\mathcal{C}$ is written as a logarithmic derivative of another complex quantity $\lambda(t)$, $\mathcal{C} = \frac{\dot{\lambda}}{\lambda}$, linearizing the Riccati equation to the complex Newtonian equation (2.49). This quantity $\lambda(t)$ can be written in terms of amplitude and phase as $\lambda = \alpha e^{i\varphi}$ or in terms of real and imaginary parts as $\lambda = u + iz$. Both components (in both cases) are not independent of each other but coupled via a conservation law that is obtained from the imaginary part of the Riccati equation when the logarithmic derivative of $\lambda(t)$ is inserted. In polar coordinates, it couples phase and amplitude via a kind of conservation of angular momentum in the complex plane, $\dot{\varphi} = \frac{1}{\alpha^2}$, showing that the amplitude α of λ is exactly the variable of the Ermakov equation (2.16) and thus proportional to the position uncertainty or WP width.

Written in Cartesian coordinates this conservation law couples real and imaginary parts of λ via $\dot{z}u - \dot{u}z = 1$. The physical meaning of $u(t)$ and $z(t)$ becomes more transparent knowing that both completely determine the time-dependence of the Green function or Feynman kernel that transforms the initial Gaussian WP at x' and $t' = 0$ into the WP solution of the TDSE at position x and time t. Expressing the TD WP with these parameters shows that the imaginary part $z(t)$, according to $z(t) = \frac{m}{\alpha_0 p_0}\eta(t)$, Eq. (2.58), is directly proportional to the classical position $\eta(t) = \langle x\rangle(t)$. With the help of the above-mentioned *"conservation of angular momentum"*, the real part $u(t)$ of λ can be determined (up to a constant) if the imaginary part $z(t)$ is known via $u = -z \int^t \frac{1}{z^2(t')}\,dt'$ (Eq. (2.60)) (and vice versa). The above-mentioned conservation law in Cartesian coordinates is then the Wronskian determinant of the two solutions u and z of the Newtonian equation, thus guaranteeing their linear independence (in this case, u and z are even orthogonal to each other - in the complex plane).

The quantum mechanical contribution to the energy of the Gaussian WP solution of the TDSE, depending on position and momentum uncertainties $\langle\tilde{x}^2\rangle$ and $\langle\tilde{p}^2\rangle$, can be expressed in terms of α, $\dot{\alpha}$, φ and $\dot{\varphi}$ or λ, $\dot{\lambda}$, λ^* and $\dot{\lambda}^*$ in a form that can be used to develop a Lagrangian formalism. With the corresponding conjugate momenta, a

Hamiltonian formalism is developed that allows one to obtain the equations of motion for these quantities and thus for the quantum uncertainties in exactly the same way as the equations of motion for the classical position and momentum are obtained from the corresponding classical Lagrangian or Hamiltonian function.

The inverse of the variable $\mathcal{C}$ fulfilling the complex Riccati equation (2.4) also fulfils a complex Riccati equation, now describing the time-evolution of the momentum uncertainties corresponding to the WP solutions of the TDSE in momentum space. This Riccati equation can be linearized to exactly the same complex Newtonian equation (2.49) for $\lambda(t)$.

The dynamical Ermakov invariant (2.21) depends on $\alpha(t)$, the amplitude of the complex quantity $\lambda(t)$ and on $\eta(t)$ which is essentially the imaginary part of $\lambda(t)$. As the equation for $\lambda(t)$ can be obtained via linearization in position as well as in momentum space, the information of the dynamics in both spaces should also be contained in $\lambda(t)$ and thus in the Ermakov invariant. Therefore, it is not too surprising that the Ermakov invariant is also intimately connected to the Wigner function of the system, giving the quantum mechanical description of the time-evolution in phase space; essentially in our case, the Ermakov invariant is (up to a constant) identical to the exponent of the Wigner function.

From the continuity equation in phase space, fulfilled by this Wigner function, it also follows a closed set of coupled linear differential equations showing the interrelations between the dynamics of the uncertainties of position and momentum and their correlation. Exactly the same result has also been obtained by the algebraic approach to find the Ermakov invariant in Sect. 2.10.

There is yet another aspect to the Ermakov invariant. Not only can it be written as a sum of two quadratic terms but also be factorized as a product of two linear terms, exactly as in the case of the HO when written in terms of creation and annihilation operators. Quantization of the two product terms of the Ermakov invariant leads to generalized creation and annihilation operators that allow for the construction of generalized coherent states corresponding to WPs with a TD width. Furthermore, they also provide (unusual) exact solutions of the TDSE like Hermitian polynomials (with TD variable) for the free motion (which are not eigenfunctions of the corresponding Hamiltonian but of the operator associated with the Ermakov invariant).

Finally, the Ermakov invariant also allows for the transformation of the TDSE formally into a TISE thus paving the way for application of techniques that are available for TI problems but not necessarily for TD ones. Moreover, a link between this transformation and the TD Green function discussed in Sect. 2.5 is also possible.

References

1. D. Schuch, Connections between Newton- and Schrödinger-type equations in the description of reversible and irreversible dynamics. Int. J. Quant. Chem., Quant. Chem. Symp. **36**(S23), 59–72 (1989)

2. D. Schuch, On the complex relations between equations describing the dynamics of wave and particle aspects. Int. J. Quant. Chem. **42**, 663–683 (1992)
3. A. Steen, Om Formen for Integralet af den lineaere Differentialligning af anden Orden. Overs. over d. K. Danske Vidensk. Selsk. Forh. 1–12 (1874)
4. R. Redheffer, Steen's equation and its generalizations. Aequationes Math. **58**, 60–72 (1999)
5. R. Redheffer, I. Redheffer, Steen's 1874 paper: historical survey and translation. Aequationes Math. **61**, 131–150 (2001)
6. V.P. Ermakov, Second-order differential equations: conditions of complete integrability. Univ. Jzv. Kiev **20**, No 9, 1–25 (1880); A.O. Harin (English translation), Appl. Anal. Discrete Math. **2**, 123–145 (2008)
7. J.R. Ray, J.L. Reid, More exact invariants for the time-dependent harmonic oscillator. Phys. Lett. A **71**, 317–318 (1979)
8. W.E. Milne, The numerical determination of characteristic numbers. Phys. Rev. **35**, 863–867 (1930)
9. E. Pinney, The nonlinear differential equation $y'' + p(x)y + cy^{-3} = 0$. Proc. Amer. Math. Soc. **1**, 681 (1950)
10. H.R. Lewis Jr., Classical and quantum systems with time-dependent harmonic-oscillator-type Hamiltonians. Phys. Rev. Lett. 18, 510 (1967); Erratum, Classical and quantum systems with time-dependent harmonic-oscillator-type Hamiltonians. Phys. Rev. Lett. **18**, 636 (1967)
11. D. Schuch, M. Moshinsky, Connection between quantum-mechanical and classical time evolution via a dynamical invariant. Phys. Rev. A **73**, 062111 (2006)
12. H. Cruz, D. Schuch, O. Castaños, O. Rosas-Ortiz, Time-evolution of quantum systems via a complex nonlinear Riccati equation II. Dissipative Syst. Ann. Phys. **360**, 44–60 (2015)
13. E. Schrödinger, Zum Heisenbergschen Unschrfeprinzip. *Sitzungsber. Preuss. Akad. Wiss. (Math. Phys. Kl.)* XIX, 296–303(Berlin, 1930); annotated by A. Angelow, M.-C. Batoni, About Heisenberg uncertainty relation. arXiv:quant-ph/9903100
14. H.P. Robertson, The uncertainty principle. Phys. Rev. **34**, 163–164 (1929); ibid **35** 667 (1930); ibid **46**, 794 (1934)
15. I.A. Malkin, V.I. Man'ko, D.A. Trifonov, Coherent states and transition probabilities in a time-dependent electromagnetic field. Phys. Rev. D **2**, 1371 (1970)
16. V.V. Dodonov, V.I. Man'ko, D.E. Nikonov, Exact propagators for time-dependent Coulomb, delta and other potentials. Phys. Lett. A **162**, 359–364 (1992)
17. M.C. Bertin, B.M. Pimentel, J.A. Ramirez, Construction of time-dependent dynamical invariants: a new approach. J. Math. Phys. **53**, 042104 (2012)
18. J.G. Hartley, J.R. Ray, Ermakov systems and quantum-mechanical superposition laws. Phys. Rev. A **24**, 2873 (1981)
19. P.G. Kevrekidis, Y. Drossinos, Nonlinearity from linearity: The Ermakov–Pinney equation revisited. Math. Comp. Sim. **74**, 196–202 (2007)
20. D. Schuch, Connections between Newton- and Schrödinger-type equations in the description of reversible and irreversible dynamics. Int. J. Quant. Chem., Quant. Chem. Symp. **36**(S23), 59–72 (1989)
21. R.P. Feynman, A.R. Hibbs, *Quantum Mechanics and Path Integrals* (McGraw-Hill, New York, 1965)
22. I.A. Malkin, V.I. Man'ko, D.A. Trifonov, Linear adiabatic invariants and coherent states. J. Math. Phys. **14**, 576 (1973)
23. I.A. Malkin, V.I. Man'ko and D.A. Trifonov, *Invariants and Evolution of Nonstationary Quantum Systems*, Vol. **183**, ed. by M.A. Markov (Nova Science, New York, 1989)
24. I.I. Goldman, W.D. Kriwtschenkow, *Aufgabensammlung zur Quantenmechanik*, Hochschulbücher für Physik **38**, ed. by O. Lucke and R. Rompe (VEB Deutscher Verlag der Wissenschaften, Berlin, 1963), p. 129
25. A.K. Dhara, S.V. Lawande, Time-dependent invariants and the Feynman propagator. Phys. Rev. A **30**, 560 (1984)
26. H. Dekker, Classical and quantum mechanics of the damped harmonic oscillator. Phys. Rep. **80**, 1–110 (1981)

27. H. Dekker, On the quantization of dissipative systems in the Lagrange–Hamilton formalism. Z. Physik B **21**, 295–300 (1975)
28. H. Dekker, On the noise operator approach to quantized dissipative systems. Z. Physik B **26**, 273–275 (1977)
29. H. Dekker, Quantization of the linearly damped harmonic oscillator. Phys. Rev. A **16**, 2126 (1977); Erratum. Phys. Rev. A **17**, 1240 (1978)
30. H. Dekker, On the master equation and pure state representations for the quantized damped oscillator. Phys. Lett. A **74**, 15–17 (1979)
31. H. Dekker, Damped oscillator pure state representations. Phys. Lett. A **80**, 369–371 (1980)
32. C.M. Bender, S. Boettcher, Real spectra in non-hermitian Hamiltonians having PT symmetry. Phys. Rev. Lett. **80**, 5243 (1998)
33. O. Rosas-Ortiz, O. Castaños, D. Schuch, New supersymmetry-generated complex potentials with real spectra. J. Phys. A: Math. Theor. **48**, 445302 (2015)
34. T.B. Pedersen, H. Koch, On the time-dependent Lagrangian approach in quantum chemistry. J. Chem. Phys. **108**, 5194 (1998)
35. D.V. Shalashilin, I. Burkhardt, Gaussian based techniques for quantum propagation from the time-dependent variational principle: formulation in terms of trajectories of coupled classical and quantum variables. J. Chem. Phys. **129**, 084104 (2008)
36. E.P. Wigner, On the quantum correction for thermodynamic equilibrium. Phys. Rev. **40**, 749 (1932)
37. M. Hillery, R.F. O'Connell, M.O. Scully, E.P. Wigner, Distribution functions in physics: fundamentals. Phys. Rep. **106**, 121–167 (1984)
38. W.P. Schleich, *Quantum Optics in Phase Space* (Wiley-VCh, Berlin, 2001), p. 92
39. D. Schuch, On the relation between the Wigner function and an exact dynamical invariant. Phys. Lett. A **338**, 225–231 (2005)
40. L.I. Schiff, *Quantum Mechanics*, 3rd ed. (McGraw Hill Kogakusha, Tokyo, 1968), Chap. 13
41. J.R. Ray, Cosmological particle creation. Phys. Rev. D **20**, 2632 (1979)
42. J.R. Ray, Time-dependent invariants with applications in physics. Lett. Nuovo Cimento **27**, 424–428 (1980)
43. W. Sarlet, Class of Hamiltonians with one degree-of-freedom allowing application of Kruskal's asymptotic theory in closed form. II. Ann. Phys. (N.Y.) **92**, 248–261 (1975)
44. J.L. Reid, J.R. Ray, Ermakov systems, nonlinear superposition, and solutions of nonlinear equations of motion. J. Math. Phys. **21**, 1583 (1980)
45. J.R. Ray, Nonlinear superposition law for generalized Ermakov systems. Phys. Lett. A **78**, 4–6 (1980)
46. W. Sarlet, F. Cantrijn, A generalization of the nonlinear superposition idea for Ermakov systems. Phys. Lett. A **88**, 383–387 (1982)
47. C. Athorne, C. Rogers, U. IURamgulam, A. Osbaldestin, On linearization of the Ermakov system. Phys. Lett. A **143**, 207–212 (1990)
48. P.G.L. Leach, Generalized Ermakov systems. Phys. Lett. A **158**, 102–106 (1991)
49. K.S. Govinder, P.G.L. Leach, Integrability of generalized Ermakov systems. J Phys. A: Math. Gen. **27**, 4153 (1994)
50. M. Moshinsky, Y.F. Smirnov, *The Harmonic Oscillator in Modern Physics* (Harwood Academic Publishers, Amsterdam, 1996)
51. D. Schuch, M. Moshinsky, Wigner distribution functions and the representation of canonical transformations in time-dependent quantum mechanics. SIGMA **4**, 054 (2008)
52. M. Moshinsky, C. Quesne, Canonical transformations and matrix elements. J. Math. Phys. **2**, 1772 (1971)
53. P.A. Mellow, M. Moshinsky, Nonlinear canonical transformations and their representations in quantum mechanics. J. Math. Phys. **16**, 2017 (1975)
54. G. García-Calderoń, M. Moshinsky, Wigner distribution functions and the representation of canonical transformations in quantum mechanics. J. Phys. A: Math. Gen. **13**, L185 (1990)
55. M.F. Guasti, The Wronskian and the Ermakov-Lewis invariant. Intern Math. Forum **4**, 795–804 (2009)

56. H.J. Korsch, Dynamical invariants and time-dependent harmonic systems. Phys. Lett. A **74**, 294–296 (1979)
57. E. Schrödinger, A method of determining quantum-mechanical eigenvalues and eigenfunctions. Proc. Roy. Irish Acad. **46**(A2), 9–16 (1940)
58. P.A.M. Dirac, *The Principles of Quantum Mechanics*, Chap. 34 (Oxford University Press, Oxford, 1935)
59. E. Schrödinger, Der stetige Übergang von der Mikro- zur Makromechanik. Naturwissenschaften **14**, 664–666 (1926)
60. R.J. Glauber, The quantum theory of optical coherence. Phys. Rev. **130**, 2529 (1963)
61. E.C.G. Sudarshan, Equivalence of semiclassical and quantum mechanical descriptions of statistical light beams. Phys. Rev. Lett. **10**, 277 (1963)
62. J.R. Klauder, Continuous-representation theory. I. Postulates of continuous-representation theory. J. Math. Phys. **4**, 1055 (1963)
63. O. Castaños, D. Schuch, O. Rosas-Ortiz, Generalized coherent states for time-dependent and nonlinear Hamiltonian operators via complex Riccati equations. J. Phys. A: Math. Theor. **46**, 075304 (2013)
64. D. Schuch, O. Castaños, O. Rosas-Ortiz, Generalized creation and annihilation operators via complex nonlinear Riccati equations. J. Phys. Conf. Series **442**, 012058 (2013)
65. J.G. Hartley, J.R. Ray, Coherent states for the time-dependent harmonic oscillator. Phys. Rev. D **25**, 382 (1982)
66. W. Greiner, J. Reinhardt, *Theoretische Physik*, vol. 7A (Verlag Harri Deutsch, Thun, 1993)
67. J. Guerrero, F.F. López-Ruiz, V. Aldaya, F. Cossio, Harmonic states for the free particle. J. Phys. A: Math. Theor. **44**, 445307 (2011)
68. V. Aldaya, F. Cossio, J. Guerrero, F.F. López-Ruiz, The quantum Arnold transformation. J. Phys. A: Math. Theor. **44**, 065302 (2011)
69. J. Guerrero, F.F. López-Ruiz, The quantum Arnold transformation and the Ermakov–Pinney equation. Phys. Scr. **87**, 038105 (2013)
70. F.F. López-Ruiz, J. Guerrero, Generalizations of the Ermakov system through the quantum Arnold transformation. J. Phys. Conf. Ser. **538**, 012015 (2014)
71. H.R. Lewis Jr., W.B. Riesenfeld, An exact quantum theory of the time-dependent harmonic oscillator and of a charged particle in a time-dependent electromagnetic field. J. Math. Phys. **10**, 1458 (1969)
72. J.G. Hartley, J.R. Ray, Solutions to the time-dependent Schrödinger equation. Phys. Rev. A **25**, 2388 (1982)
73. J.R. Ray, Minimum-uncertainty coherent states for certain time-dependent systems. Phys. Rev. D **25**, 3417 (1982)
74. A. Erdélyi, W. Magnus, F. Oberhettinger, F.G. Tricomi, *Higher Transcendental Functions*, vol. II (McGraw-Hill, New York, 1953), p. 194

Chapter 3
Time-Independent Schrödinger and Riccati Equations

3.1 On Supersymmetry and Riccati Equations

In Sect. 2.11 it has been shown how a generalization of the creation and annihilation operators, known from the algebraic treatment of the harmonic oscillator (HO) problem, can be generalized by replacing $\pm i\omega_0 = \pm i\frac{1}{\alpha_0^2}$, the coefficient of the space-dependent contribution $\pm i\omega_0 x$, with a complex function of time, $C = \frac{\dot{\alpha}}{\alpha} + i\frac{1}{\alpha^2}$, and its complex conjugate, where both complex quantities fulfil the Riccati equation (2.4). A different generalization of the creation/annihilation operators can be found that is also connected with Riccati equations (this time usually with real ones) in the supersymmetric formulation [1–3] of time-independent (TI) quantum mechanics. There, essentially the term linear in the coordinate x is replaced by a function of x, the so-called "superpotential"[1] $W(x)$, leading to the operators[2]

$$B^{\pm} = \mp i\frac{1}{\sqrt{2}}\left[\frac{p_{\mathrm{op}}}{\sqrt{m}} \pm i\, W\right] = \frac{1}{\sqrt{2}}\left[W(x) \mp i\frac{p_{\mathrm{op}}}{\sqrt{m}}\right]. \tag{3.1}$$

In this case, the term $\omega_0 x$ with constant ω_0 is replaced by a real, position-dependent function $W(x)$. The operators $B^{\pm}$ fulfil the commmutator and anti-commutator relations

$$[B^-, B^+]_- = \frac{\hbar}{\sqrt{m}}\frac{dW}{dx}, \quad [B^-, B^+]_+ = W^2 + \frac{p_{\mathrm{op}}^2}{m}. \tag{3.2}$$

---

[1] This nomenclature is somewhat misleading, as W has the dimension of the square-root of energy, not energy like a usual potential.

[2] In comparison with a and a^+ as defined in (2.200) and (2.201), a factor $\sqrt{\frac{1}{\hbar\omega_0}}$ is missing because the definitions in Sect. 2.11 refer to $\tilde{H}_{\mathrm{op}} = \frac{H_{\mathrm{op}}}{\hbar\omega_0}$. Further, B^- corresponds to a.

D. Schuch, *Quantum Theory from a Nonlinear Perspective*,
Fundamental Theories of Physics 191, https://doi.org/10.1007/978-3-319-65594-9_3

The supersymmetric Hamiltonian

$$H_{\text{SUSY}} = \begin{pmatrix} H_1 & 0 \\ 0 & H_2 \end{pmatrix} \tag{3.3}$$

can be expressed with the help of $B^{\pm}$ in the form

$$H_1 = B^+ B^- = -\frac{\hbar^2}{2m}\frac{d^2}{dx^2} + V_1(x), \tag{3.4}$$

$$H_2 = B^- B^+ = -\frac{\hbar^2}{2m}\frac{d^2}{dx^2} + V_2(x), \tag{3.5}$$

where H_1 represents the bosonic and H_2 the fermionic part.

A detailed discussion of the formalism of supersymmetry (SUSY) can be found, e.g. in [1, 3]. For our discussion, only the aspects mentioned in the following will be necessary. Important in this context is that the supersymmetric partner potentials $V_1(x)$ and $V_2(x)$ fulfil real Riccati equations, what follows directly from the definition of $B^{\pm}$, i.e.,

$$V_1 = \frac{1}{2}\left[W^2 - \frac{\hbar}{\sqrt{m}}\frac{dW}{dx}\right], \tag{3.6}$$

$$V_2 = \frac{1}{2}\left[W^2 + \frac{\hbar}{\sqrt{m}}\frac{dW}{dx}\right]. \tag{3.7}$$

The energy spectra of H_1 and H_2 are identical apart from the ground state. H_1 has the ground state energy $E_0^{(1)} = 0$ whereas the ground state energy $E_0^{(2)}$ of H_2 is identical to the energy of the first excited state, $E_1^{(1)}$, of H_1. The ground state wave function of H_1, $\Psi_0^{(1)}$, has no node and determines the superpotential via

$$W = -\frac{\hbar}{\sqrt{m}}\frac{\frac{d}{dx}\Psi_0^{(1)}}{\Psi_0^{(1)}}. \tag{3.8}$$

From Eqs. (3.6) and (3.7), the partner potentials V_1 and V_2 then follow. On the other hand, $\Psi_0^{(1)}$ is connected to V_1 via the solution of the equation $H_1\Psi_0^{(1)} = 0$, i.e.,

$$H_1\Psi_0^{(1)} = -\frac{\hbar^2}{2m}\frac{d^2}{dx^2}\Psi_0^{(1)} + \frac{1}{2}\left[W^2 - \frac{\hbar}{\sqrt{m}}\frac{dW}{dx}\right]\Psi_0^{(1)} = E_0^{(1)}\Psi_0^{(1)} = 0. \tag{3.9}$$

The connection between the spectra of H_1 and H_2, i.e., $E_n^{(1)}$ and $E_n^{(2)}$, and the corresponding wave functions $\Psi_n^{(1)}$ and $\Psi_n^{(2)}$, is determined via the generalized creation/annihilation operators $B^{\pm}$ according to

$$\Psi_{n+1}^{(1)} = \frac{1}{\sqrt{E_n^{(2)}}} B^+ \Psi_n^{(2)} \tag{3.10}$$

and

$$\Psi_n^{(2)} = \frac{1}{\sqrt{E_{n+1}^{(1)}}} B^- \Psi_{n+1}^{(1)} \tag{3.11}$$

where B^+ creates a node and B^- annihilates a node in the wave function. So, e.g., the first excited state $\Psi_1^{(1)}$ of H_1 (which has one node) can be obtained from the ground state $\Psi_0^{(2)}$ of H_2 (which has no node) by applying B^+ onto it as described in (3.10).

In order to obtain higher eigenvalues and eigenfunctions, $\Psi_0^{(1)}$ in definition (3.8) of $W = W_1$ must be replaced by $\Psi_0^{(2)}$, leading to $W_2 = -\frac{\hbar}{\sqrt{m}} \frac{\frac{d}{dx}\Psi_0^{(2)}}{\Psi_0^{(2)}}$ etc., i.e.,

$$W_s = -\frac{\hbar}{\sqrt{m}} \frac{\frac{d}{dx}\Psi_0^{(s)}}{\Psi_0^{(s)}} \tag{3.12}$$

with the corresponding operators

$$B_s^{\pm} = \frac{1}{\sqrt{2}}\left[W_s \mp \frac{\hbar}{\sqrt{m}}\frac{d}{dx}\right], \tag{3.13}$$

thus creating a hierarchy that provides all eigenvalues and eigenfunctions of the Hamiltonians H_1 and H_2.

In the context of this and the next section, only two systems with analytic solutions are considered explicitly, namely the one-dimensional HO (with constant frequency $\omega = \omega_0$) and the Coulomb problem. The latter case, a three-dimensional system with spherical symmetry ($V(\mathbf{r}) = V(r) = -\frac{e^2}{r}$), can be essentially reduced to a one-dimensional problem via separation of the radial and angular parts. Using the ansatz $\Phi_{nlm}(\mathbf{r}) = \frac{1}{r}\Psi_{nl}(r)Y_{nlm}(\vartheta, \varphi) = R(r)Y_{nlm}(\vartheta, \varphi)$ for the wave function (with n = total quantum number, l = azimuthal quantum number, m = magnetic quantum number, r, ϑ, φ = polar coordinates), the energy eigenvalues E_n of the system can be obtained from the radial Schrödinger equation (SE)

$$\left\{-\frac{\hbar^2}{2m}\frac{d^2}{dr^2} + V_{\text{eff}}\right\}\Psi_{nl}(r) = E_n \Psi_{nl}(r) \tag{3.14}$$

with the effective potential

$$V_{\text{eff}} = V(r) + \frac{l(l+1)\hbar^2}{2mr^2} = -\frac{e^2}{r} + \frac{l(l+1)\hbar^2}{2mr^2}. \tag{3.15}$$

The superpotential W, the energy eigenvalues E_n and the supersymmetric potential V_1 for the systems under consideration are given by:[3]

(a) HO: $V(x) = \frac{m}{2}\omega^2 x^2$ (eigenfunctions $\Psi_n(x)$: Hermite polynomials multiplied by a Gaussian function)

$$W = \sqrt{m}\omega_0 x \tag{3.16}$$

$$E_n = \hbar\omega_0 \left(n + \frac{1}{2}\right) \tag{3.17}$$

$$V_1 = \frac{m}{2}\omega_0^2 x^2 - \frac{\hbar}{2}\omega = V(x) - E_0. \tag{3.18}$$

(b) Coulomb potential: $V(r) = -\frac{e^2}{r}$ (eigenfunctions $\Psi_{nl}(r)$: Laguerre polynomials)

$$W = \frac{\sqrt{m}e^2}{(l+1)\hbar} - \frac{(l+1)\hbar}{\sqrt{m}r} \tag{3.19}$$

$$E_{n'} = -\frac{mc^2}{2}\left(\frac{e^2}{\hbar c}\right)^2 \frac{1}{(n'+l+1)^2} \tag{3.20}$$

$$V_1 = -\frac{e^2}{r} + \frac{l(l+1)\hbar^2}{2mr^2} + \frac{mc^2}{2}\left(\frac{e^2}{\hbar c}\right)^2 \frac{1}{(l+1)^2} = V_{\text{eff}} - E_0. \tag{3.21}$$

In the second case, the radial quantum number n' occurs which indicates the number of nodes in the wave function and is connected with the total quantum number n that actually characterizes the energy eigenvalue, via $n = n' + l + 1$.

Particularly the quantities V_1, given in Eqs. (3.18) and (3.21), shall be compared with similar expressions obtained in the next section where a nonlinear (NL) formulation of TI quantum mechanics is presented. In analogy with the time-dependent (TD) case, one can assume that the superpotential obtained in the way described in Eq. (3.8) is only a particular solution $\tilde{W}$ of the Riccati equation(s) (3.6, 3.7) and, again, the general solution can be written as $W(x) = \tilde{W}(x) + \Phi(x)$ where $\Phi(x)$ must now fulfil the Bernoulli equation

$$\frac{\hbar}{\sqrt{m}}\frac{d}{dx}\Phi + 2\tilde{W}\Phi + \Phi^2 = 0 \tag{3.22}$$

(written here for the plus-sign of the derivative). This can be solved in the same way, by linearization, as shown for the TD problem in Sect. 2.2 to finally yield the general solution

$$W(x) = \tilde{W}(x) + \frac{\hbar}{\sqrt{m}}\frac{d}{dx}\ln[\mathcal{I}(x) + \varepsilon]. \tag{3.23}$$

[3]Here E_0 is the ground state energy of the conventional solution of the problems.

The integral $\mathcal{I}(x)$ is formally identical to the one in the TD case, only t must be replaced by x and $\tilde{\mathcal{C}}$ by $\tilde{W}$. Also this solution depends on a (this time real) parameter ϵ (corresponding to $\Phi^{-1}(0)$). This generalized $W(x)$ gives rise to a one-parameter family of isospectral potentials (e.g., for $i = 1$)

$$V_1(x;\varepsilon) = \tilde{V}_1(x) - \frac{\hbar^2}{m}\frac{d^2}{dx^2}\ln[\mathcal{I}(x)+\varepsilon], \tag{3.24}$$

i.e., all members of this family possess the same energy spectrum although the shapes of these potentials may be significantly different.[4] Therefore, from an experiment measuring only the energy levels, it would not be possible to determine to which potential of the family the measurements belong; a problem also relevant to inverse scattering theory (for further details, see also [1, 3, 4]).[5]

As an example, for the HO the particular solution $\tilde{W}(x) = \sqrt{m}\omega_0 x$ leads to

$$\tilde{V}_{1/2} = \frac{m}{2}\omega_0^2 x_0^2 \mp \frac{\hbar}{2}\omega_0, \tag{3.25}$$

i.e., essentially the parabolic harmonic potential, only shifted by minus/plus the (conventional) ground state energy $\frac{\hbar}{2}\omega_0$. To find the general solution $W(x)$, the solution $\Phi(x)$ of the Bernoulli equation (3.22) is needed, requiring the determination of the integral $\mathcal{I}(x)$ which, in this case, has the form

$$\mathcal{I}(x) = 1 - \frac{1}{2}\,\mathrm{erfc}(\sqrt{\omega_0}x) \quad \text{with} \quad \mathrm{erfc}(x) := \frac{2}{\sqrt{\pi}}\int_x^\infty \mathrm{e}^{-t^2}\,dt. \tag{3.26}$$

Following (3.24), the one-parameter family $V_1(x;\epsilon)$ is then given by

$$V_1(x;\epsilon) = \frac{m}{2}\omega_0^2 x^2 - \frac{\hbar^2}{m}\frac{d^2}{dx^2}\ln\left[\epsilon + 1 - \frac{1}{2}\,\mathrm{erfc}\left(\sqrt{\omega_0}x\right)\right] - \frac{\hbar}{2}\omega_0. \tag{3.27}$$

The shape of the potentials is now, unlike the parabolic harmonic potential, no longer symmetric under the exchange $x \to -x$. For ϵ decreasing from ∞ to 0, a second minimum shows up which shifts towards $x = -\infty$. For $\epsilon = 0$, this attractive potential vanishes and the bound state corresponding to the ground state energy equal to zero gets lost. The potential for this limiting case $\epsilon = 0$ is called Pursey potential (for further details, see [1, 3]). Only for $\epsilon \to \infty$, the ln-term vanishes and the asymmetric double potential turns into the parabolic potential $\tilde{V}_1(x)$. So, obviously the parameter ϵ can have drastic qualitative consequences for the solution

[4]This "construction of families of potentials *strictly* isospectral to the initial (bosonic) one" can also be interpreted as a "double Darboux general Riccati" transformation of the inverse Darboux type, going in two steps from an initial bosonic to a deformed bosonic system; for details, see [4].

[5]Note that the notation in [3] and the one used here differs as the quantities with and without tilde are interchanged.

of the NL Riccati equation. This information is certainly hidden when only the linear TISE is considered, a situation similar to the TD case discussed in 2.2 where the (complex) parameter κ_0 plays a role similar to ϵ.

3.2 Nonlinear Version of Time-Independent Quantum Mechanics

In the last section, it has been shown that the superpotential $W(x)$ can be written according to (3.8) as a logarithmic derivative of the *real* ground state function $\Psi_0(x)$ thus linearizing the (real) NL Riccati equation(s) (3.6, 3.7) to (real) TISE(s).

This is like in the TD case where the variable $C(t)$ can be written according to (2.48) as a logarithmic derivative of the *complex* function $\lambda(t)$, this time linearizing the (complex) NL Riccati equation (2.4) to the (complex) Newtonian equation (2.49). In this TD problem it has been shown that amplitude $\alpha(t)$ and phase $\varphi(t)$ of the complex function $\lambda = \alpha e^{\mathrm{i}\varphi}$ (or real and imaginary parts of $\lambda(t)$ when written as $\lambda = u + \mathrm{i}z$) are not independent of each other but *coupled* by a kind of *conservation law*, Eq. (2.52).

The question arises if something similar is also possible in the TI case, i.e., can the real Riccati equation(s) (3.6, 3.7) be replaced by a complex version and the real ground state function Ψ_0 in the definition of the logarithmic derivative by some complex wave function? And what would be the relation between phase and amplitude of this complex wave function fulfilling a linear TISE? An answer to this question has been given by Reinisch [5, 6] in his NL formulation of (TI) quantum mechanics.

His starting point is Madelung's hydrodynamic formulation of quantum mechanics [7] that uses the polar ansatz

$$\Psi(\mathbf{r}, t) = a(\mathbf{r}, t) \exp\left\{-\frac{\mathrm{i}}{\hbar} S(\mathbf{r}, t)\right\} \tag{3.28}$$

for the (complex) wave function $\Psi(\mathbf{r}, t)$ (where $a^2(\mathbf{r}, t) = \varrho(\mathbf{r}, t) = \Psi^*\Psi$) turning the linear SE

$$\mathrm{i}\hbar \frac{\partial}{\partial t} \Psi(\mathbf{r}, t) = \left\{-\frac{\hbar^2}{2m} \Delta + V(\mathbf{r})\right\} \Psi(\mathbf{r}, t) \tag{3.29}$$

(with $\Delta = \nabla^2 =$ Laplace operator, $\nabla =$ Nabla operator and for *any* potential $V(\mathbf{r})$) into two coupled equations for the amplitude $a(\mathbf{r}, t)$ and the phase $S(\mathbf{r}, t)$, i.e., the continuity equation

$$\frac{\partial}{\partial t} a^2 + \frac{1}{m} \nabla (a^2 \nabla S) = 0 \tag{3.30}$$

and the Hamilton–Jacobi-type equation

$$\frac{\partial}{\partial t} S + \frac{1}{2m} (\nabla S)^2 + V - \frac{\hbar^2}{2m} \frac{\Delta a}{a} = 0. \tag{3.31}$$

Already here, the coupling of phase and amplitude can be seen clearly because the Hamilton–Jacobi equation for the phase S contains a term (misleadingly called "quantum potential", $V_{qu} = -\frac{\hbar^2}{2m}\frac{\Delta a}{a}$) depending on the amplitude $a(\mathbf{r}, t)$ and the continuity equation for the probability density $\varrho = a^2$ contains ∇S. It is shown in the following that, also in the TI case, this coupling is not arbitrary but related to a conservation law.

For stationary states, the energy of the system is related to the action S via $\frac{\partial}{\partial t} S = -E = \text{const.}$ and the density is TI, i.e., $\frac{\partial}{\partial t} a^2 = 0$. The continuity equation (3.30) then turns into

$$\nabla (a^2 \nabla S) = 0 \tag{3.32}$$

and the modified Hamilton–Jacobi equation (3.31) into

$$-\frac{\hbar^2}{2m} \Delta a + (V - E) a = -\frac{1}{2m} (\nabla S)^2 \, a. \tag{3.33}$$

Equation (3.32) is definitely fulfilled for $\nabla S = 0$, turning (3.33) into the usual TISE for the real wave function $a = |\Psi|$ with position-independent phase S,[6]

$$-\frac{\hbar^2}{2m} \Delta a + V a = E a. \tag{3.34}$$

However, Eq. (3.32) can also be fulfilled for $\nabla S \neq 0$ if only the conservation law

$$\nabla S = \frac{C}{a^2} \tag{3.35}$$

is fulfilled with constant (or, at least, position-independent) C.

This expression now shows explicitly the coupling between phase and amplitude of the wave function and is equivalent to Eq. (2.51) in the TD case. Inserting (3.35) into the rhs of Eq. (3.33) changes this into the *Ermakov* equation

$$\Delta a + \frac{2m}{\hbar^2} (E - V) a = \left(\frac{1}{\hbar} \nabla S \right)^2 a = \left(\frac{C}{\hbar} \right)^2 \frac{1}{a^3}, \tag{3.36}$$

equivalent to Eq. (2.16) in the TD case. The corresponding *complex Riccati* equation, equivalent to Eq. (2.4) in the TD case, is given here by [8]

$$\nabla \left(\frac{\nabla \Psi}{\Psi} \right) + \left(\frac{\nabla \Psi}{\Psi} \right)^2 + \frac{2m}{\hbar^2} (E - V) = 0, \tag{3.37}$$

---

[6] N.B.: The *kinetic energy* term divided by a is just identical to V_{qu}!

where the following substitutions must be made

$$\frac{\partial}{\partial t} \leftrightarrow \nabla,\ \left(\frac{2\hbar}{m}y\right) = C = \frac{\dot{\lambda}}{\lambda} \leftrightarrow \frac{\nabla\Psi}{\Psi},\ \lambda = \alpha e^{i\varphi} \leftrightarrow \Psi = a e^{i\frac{S}{\hbar}},\ \omega^2(t) \leftrightarrow \frac{2m}{\hbar^2}(E - V(\mathbf{r})). \tag{3.38}$$

Considering first the one-dimensional HO, and introducing the dimensionless variable ζ via $\zeta = |k_0|x$ with $\hbar k_0 = p_0 = \pm\sqrt{2mE}$, $\hat{V}(\zeta) = V[x(\zeta)]$ and $\ddot{a} = \frac{d^2}{d\zeta^2}a$, Eq. (3.36) acquires the familiar form[7]

$$\ddot{a} + \left(1 - \frac{\hat{V}}{E}\right)a = \frac{1}{a^3}. \tag{3.39}$$

Following the method described in [5], from the *real* solution $a_{\mathrm{NL}}(\zeta)$ of this NL Ermakov equation (3.39) the *complex* solution $a_{\mathrm{L}}(\zeta)$ of the *linear* SE (3.34) can be obtained via

$$a_{\mathrm{L}}(\zeta) = a_{\mathrm{NL}}(\zeta)\ \exp\left\{-\frac{i}{\hbar}S\right\} = a_{\mathrm{NL}}(\zeta)\exp\left\{-i\int_{\zeta_0}^{\zeta} d\zeta' \frac{1}{a_{\mathrm{NL}}^2(\zeta')}\right\}, \tag{3.40}$$

from which a *real* (not normalized) solution of the TISE can be constructed according to

$$\tilde{a}_{\mathrm{L}}(\zeta) = \mathcal{R}e[a_{\mathrm{L}}(\zeta)] = \frac{1}{2}\left[a_{\mathrm{NL}}e^{\frac{i}{\hbar}S} + a_{\mathrm{NL}}e^{-\frac{i}{\hbar}S}\right] = a_{\mathrm{NL}}\cos\left(\int_{\zeta_0}^{\zeta} d\zeta'\right). \tag{3.41}$$

So far, the energy E occurring in Eq. (3.39) is still a free parameter that can take any value. However, solving Eq. (3.39) numerically for arbitrary values of E leads, in general, to solutions a_{NL} that diverge for increasing ζ. Only if the energy E is appropriately tuned to any eigenvalue E_n of Eq. (3.34) does this divergence disappear and the integral in the cosine of Eq. (3.41) takes for $\zeta \to \infty$ the exact value $\frac{\pi}{2}$, i.e. the cosine vanishes at infinity. So, the quantization condition that is usually obtained from the requirement of the truncation of an infinite series in order to avoid divergence of the wave function is, in this case, obtained from the requirement of nondiverging solutions of the NL Ermakov equation (3.39) by variation of the parameter E. This has been verified numerically in the case of the one-dimensional HO and the Coulomb

[7] A similar formulation of the TISE in terms of this equation, but within a different context and with different applications, has also been given in [9]. In another paper [10, 11] the relation between the Ermakov equation (3.39) and the TISE has been extended to also include magnetic field effects and in [11] possible connections between SUSY and Ermakov theories are considered. The NL differential equation (3.39) has also been used to obtain numerical solutions of the TISE for single and double-minimum potentials as well as for complex energy resonance states; for details see [12, 13]. Here, however, we want to concentrate on the similarities between the TD and TI situation, in particular with respect to SUSY.

problem and there is the conjecture that this property is "*universal*" in the sense that it does not depend on the potential V (see [5]).

For comparison with the situation in SUSY quantum mechanics, the HO and the Coulomb problem can be written in the form

(a) HO: with $\mu = \left(\frac{\hbar\omega_0}{2E}\right)^2$, $E = E_n = \left(n + \frac{1}{2}\right)\hbar\omega_0 \rightarrow \mu_n = \frac{1}{(2n+1)^2}$ and $\mu\zeta^2 = \frac{\frac{m}{2}\omega_0^2 x^2}{E} = \frac{V}{E}$ follows:

$$\ddot{a} + \left(1 - \mu\zeta^2\right) a = \ddot{a} + \left(1 - \frac{\hat{V}}{E}\right) a = \ddot{a} - \frac{\tilde{U}_n}{E} a = \frac{1}{a^3}, \tag{3.42}$$

where

$$U_n = \frac{m}{2}\omega_0^2 x^2 - \hbar\omega_0 \left(n + \frac{1}{2}\right) = V(x) - E_n. \tag{3.43}$$

(b) Coulomb problem: with $a(r, \vartheta, \varphi) = R(r) Y_{lm}(\vartheta, \varphi)$ the radial part can be separated and, with the dimensionless variable $\zeta = |k_0| r$ with now $\hbar k_0 = p_0 = \pm\sqrt{2m(-E)}$, $(E < 0)$, the radial wave function can be written as $R(\zeta) = r(\zeta) X[r(\zeta)]$ which corresponds to $\Psi_{nl}(r)$ in SUSY quantum mechanics. This function again obeys an Ermakov equation, namely

$$\ddot{X} + \left(\frac{\hat{W}}{E} - 1\right) X = \ddot{X} + \frac{U_{n'}}{E} X = \frac{1}{X^3}, \tag{3.44}$$

where

$$\hat{W}(\zeta) = \hat{V}[r(\zeta)] + \frac{l(l+1)\hbar^2}{2mr^2(\zeta)} \mathrel{\hat{=}} V_{\text{eff}} \tag{3.45}$$

is simply the effective potential from SUSY (see Eqs. (3.15) and (3.21)) and $E = E_{n'} = -\frac{mc^2}{2}\left(\frac{e^2}{\hbar c}\right)^2 \frac{1}{n^2}$ with $n = n' + l + 1$ (where here e is the elementary charge). The coefficient of the term linear in X can again be expressed with the help of the potential-like expression $U_{n'}$,

$$U_{n'} = -\frac{e^2}{r} + \frac{l(l+1)\hbar^2}{2mr^2} + \frac{mc^2}{2}\left(\frac{e^2}{\hbar c}\right)^2 \frac{1}{(n'+l+1)^2} = V_{\text{eff}} - E_{n'}. \tag{3.46}$$

In both cases, the *ground state* ($n = 0$) wave functions are real, nodeless ($n' = 0$) and the phase does not depend on the spatial variables (i.e., $\nabla S = 0$). Therefore, the rhs of Eqs. (3.42) and (3.44) vanishes as $\frac{1}{a^3} \propto (\nabla S)^2 a = 0$, i.e., the NL Ermakov equations turn into the usual TISEs. In this case, comparison shows that, for the HO and the Coulomb problem, the potential-like terms U_0 are identical to the corresponding V_1 of SUSY. For $n > 0$ and $n' > 0$, however, U_n and $U_{n'}$ are different from V_1 and describe higher excited states. In SUSY, these states can only be obtained from the hierarchy described in the previous section. Here, the price that must be paid to include these excited states is the nonlinearity on the rhs of Eqs. (3.42) and (3.44).

Comparing the situation in this NL formulation of TI quantum mechanics with SUSY, one can see the following similarities:

The *real* superpotential $W = -\frac{\hbar}{\sqrt{m}} \left(\frac{\nabla \Psi_0}{\Psi_0} \right)$ is replaced by the *complex* "superpotential" $\mathcal{C}(\mathbf{r}) = -\frac{\hbar}{\sqrt{m}} \left[\frac{\nabla |\Psi|}{|\Psi|} + \mathrm{i} \frac{\nabla S}{\hbar} \right]$, i.e., the ground state Ψ_0 is replaced by the absolute value $|\Psi|$ of any eigenstate and an additional imaginary part depending on the phase $\frac{1}{\hbar} S$ of the wave function occurs, being responsible for the non-vanishing rhs of the Ermakov equations (3.42) and (3.44).

A comparison with the TD systems discussed in Chap. 2 will be given at the end of this section.

3.3 Complex Hamiltonians with Real Spectra

In this section, the ideas of the preceding two are combined to obtain a new class of complex potentials with real spectrum, a concept with growing interest for numerous reasons over the last years [14–23]. The factorization of the Hamiltonians $H_{1/2}$ in terms of the linear operators $B^{\pm}$, as shown in Eqs. (3.4) and (3.5) for SUSY quantum mechanics, is an example of this (factorization) method, the main idea of which is to reduce the second-order differential form of the one-dimensional Hamilton operator $H = -\frac{\hbar^2}{2m} \frac{d^2}{dx^2} + V(x)$ into the product of two first-order differential operators A, B, up to an additive constant E (proper units are assumed); i.e., $H = AB + E$. As mentioned previously, this method was already applied by Schrödinger [24] himself but can be traced back to Dirac's book in 1936 [25] and Fock's paper [26] on second quantization. The entire development of this method "shows chronological gaps and inconsistencies; the ideas emerge, disappear and re-emerge again" [27]. Renewed interest in the factorization method was definitely spurred by the development of SUSY quantum mechanics.

Replacing the ground state wave function Ψ in definition (3.8) of the superpotential W with any solution Ψ of the TISE

$$H\Psi(x) = \left\{ -\frac{\hbar^2}{2m} \frac{d^2}{dx^2} + V(x) \right\} \Psi(x) = E\Psi(x) \tag{3.47}$$

where, in this case, E can be any (real) eigenvalue, not necessarily $E_0 = 0$ as in Eq. (3.9), this generalized W can be written as

$$W = -\frac{\hbar}{\sqrt{m}} \frac{\frac{d}{dx}\Psi}{\Psi} = -\frac{\hbar}{\sqrt{m}} \frac{d}{dx} \ln \Psi \tag{3.48}$$

and used to express the potential $V(x)$ in (3.47) as

$$V = \frac{1}{2}\left[-\frac{\hbar}{\sqrt{m}}\frac{dW}{dx} + W^2 + 2E\right]. \tag{3.49}$$

In analogy with the operators $B^{\pm}$ defined in (3.1), one can now define the linear operators

$$A = \frac{1}{\sqrt{2}}\left[W - \frac{\hbar}{\sqrt{m}}\frac{d}{dx}\right], \quad B = \frac{1}{\sqrt{2}}\left[W + \frac{\hbar}{\sqrt{m}}\frac{d}{dx}\right]. \tag{3.50}$$

With these, the Hamiltonian operators

$$H = AB + E = -\frac{\hbar^2}{2m}\frac{d^2}{dx^2} + V(x) \tag{3.51}$$

and

$$\hat{H} = BA + E = -\frac{\hbar^2}{2m}\frac{d^2}{dx^2} + \hat{V}(x) \tag{3.52}$$

can be constructed where the first is identical to the one in (3.47), the second differs by the definition of its potential $\hat{V}$ as

$$\hat{V}(x) = V(x) + 2\frac{\hbar}{2\sqrt{m}}\frac{d}{dx}W = V(x) - 2\frac{\hbar^2}{2m}\frac{d^2}{dx^2}\ln\Psi(x) \tag{3.53}$$

or

$$V(x) - \hat{V}(x) = [A, B] = -2\frac{\hbar}{2\sqrt{m}}\frac{d}{dx}W = 2\frac{\hbar^2}{2m}\frac{d^2}{dx^2}\ln\Psi(x) \tag{3.54}$$

which gives the difference of the two Darboux-related[8] potentials.

Like in SUSY quantum mechanics, where V_2 (Eq. 3.7) is partner potential to V_1 (Eq. 3.6), here $\hat{V}$ can be considered as partner potential to V. In the SUSY case, both Hamiltonians H_1 and H_2 have the same eigenvalue spectrum (apart from the ground state of H_1); here the same applies to H and $\hat{H}$. The difference between the two Hamiltonians is essentially the difference between the two potentials $V - \hat{V}$ which, according to (3.54) solely depends on the derivatives of $\Psi(x)$. Usually, in the case of the TISE, wave functions are considered that are real (or at least their second-derivative is); so V and $\hat{V}$ are real as long as V is.

In Sect. 3.2 however, we have seen that also complex wave functions $\Psi(x)$ can actually be constructed as a solution of the TISE (with a non-vanishing second-derivative of the imaginary part or phase of the wave function). The logarithmic derivative of these functions fulfils the complex Riccati equation (3.37) that is equivalent to the real Ermakov equation (3.36) for the amplitude of the wave function

[8]For a survey of Darboux transformations, also in relation to the factorization method and SUSY quantum mechanics, see [4].

when written in polar form (3.28). Once the solution of this Ermakov equation is known, via (3.35), also the gradient of the phase can be determined that enters the expression for the partner potential $\hat{V}$ according to (3.53) via the logarithm.

That means, starting from a real potential $V(x)$ and constructing a complex wave function $\Psi(x)$ according to (3.40), the Darboux-transformed partner potential $\hat{V}(x)$ will be complex (and thus the Hamiltonian *non-Hermitian*) though having real eigenvalues as $V(x)$.

From the form of the *complex* solution of the *linear* TISE as given in Eq. (3.40) (where $a = a_{\rm NL}, a_{\rm L}(\xi)$ is equivalent to $\Psi(x)$ and the constant C from Eq. (3.35) has been kept for more generality) and written as $\Psi(x) = a(x)\exp\left\{-\mathrm{i}C\int^x dx'\frac{1}{a^2(x)}\right\}$, it follows for the now *complex* superpotential $W(x)$ that

$$W(x) = -\frac{\hbar}{\sqrt{m}}\left(\frac{\frac{d}{dx}a(x)}{a(x)} - \mathrm{i}\frac{C}{a^2(x)}\right), \tag{3.55}$$

leading to the potential $\hat{V}(x)$ in the form

$$\hat{V}(x) = V(x) - 2\frac{\hbar^2}{2m}\left[\frac{d^2}{dx^2}\ln a(x) - \mathrm{i}2C\frac{\frac{d}{dx}a(x)}{a^3(x)}\right] \tag{3.56}$$

with real constant parameter C. A more detailed discussion is given in [24]. In the following only an example for the most simple case, i.e. V = 0, will be given, showing the corresponding complex potential $\hat{V}(x)$.

For $V = 0$, the Ermakov equation (3.36) (in one dimension with $' = \frac{d}{dx}$ and $E = \frac{\hbar^2}{2m}k^2 = -\frac{\hbar^2}{2m}\frac{\kappa^2}{4}$, i.e., $k = \mathrm{i}\frac{\kappa}{2}$) takes the form

$$\frac{a''}{a} = \frac{\kappa^2}{4} + \frac{C^2}{a^4} \tag{3.57}$$

with the solution

$$a(x) = \sqrt{\cosh(\kappa x)} \tag{3.58}$$

and $C^2 = \frac{\kappa^2}{4}$, i.e. $C = \pm\frac{\kappa}{2}$. Choosing (without restriction of generality) the positive sign for C, the "superpotential" $W(x)$, according to (3.55), can be written as

$$W(x) = -\frac{\hbar}{\sqrt{m}}\left(\frac{\kappa}{2}\tanh(\kappa x) - \mathrm{i}\frac{\kappa}{2}\frac{1}{\cosh(\kappa x)}\right). \tag{3.59}$$

Inserting this $W(x)$ (and E as defined above) into Eq. (3.49) confirms $V(x) = 0$ whereas, for $\hat{V}(x)$, it follows from Eqs. (3.53) and (3.59) that

$$\hat{V}(x) = -\frac{\hbar^2}{2m}\frac{\kappa^2}{\cosh^2(\kappa x)} - \mathrm{i}\frac{\hbar^2}{2m}\kappa^2\frac{\sinh(\kappa x)}{\cosh^2(\kappa x)}$$
$$= [1 + \mathrm{i}\sinh(\kappa x)]\, V_{\mathrm{PT}} \qquad (3.60)$$

where the real part $V_{\mathrm{PT}} = -\frac{\hbar^2}{2m}\frac{\kappa^2}{\cosh^2(\kappa x)}$ represents the modified Pöschl–Teller potential, known as the SUSY partner potential to the free motion $V = 0$, but now also an additional imaginary part appears in $\hat{V}(x)$, nevertheless associated with the real energy $E = -\frac{\hbar^2}{2m}\frac{\kappa^2}{4}$.

A more detailed discussion is also given in [28].

3.4 Comparison of Time-Dependent and Time-Independent Systems

The generalization of the creation and annihilation operators provides a link to the TISE in the form of SUSY quantum mechanics. In the TD case the constant ω_0 of the HO is replaced by a complex TD function $C(t)$, in the SUSY TI case the term $\omega_0 x$ of the HO is replaced by a more general function of x, the so-called (real) "superpotential" $W(x)$. This superpotential now fulfils a *real* Riccati equation. Also in the TI case, the usual term proportional to $\omega_0 x$ in the creation/annihilation operators can be considered only a particular solution of the corresponding Riccati equation. Looking for the general solution leads, via the Bernoulli equation mentioned in the TD case, to the same solution (only for a real function and replacing t with x), depending on a parameter, ϵ, that can have drastic influence on the shape of the potential (as mentioned above) but not on the energy spectrum.

A complex generalization of this approach (Sect. 3.2) shows that, in this case, exactly the same formal structures can be found as in the TD case. Both systems, the TD and the TI ones, are compared in Fig. 3.1.

The TD complex function $\lambda(t)$ fulfilling a (linear) Newtonian equation (2.49) corresponds to the TI complex wave function $\Psi(x)$ fulfilling the (linear) TISE. The logarithmic derivatives of both functions fulfil the corresponding complex Riccati equations. The square of the TD frequency $\omega(t)$ corresponds to $\frac{2m}{\hbar^2}(E - V(\mathbf{r}))$, depending via $V(\mathbf{r})$, on the position $\mathbf{r}$. While the TD case was essentially restricted to quadratic Hamiltonians (with possible explicit time-dependence via $\omega(t)$), the TI case is valid for *any potential* $V(\mathbf{r})$.

In the same way as phase $\varphi(t)$ and amplitude $\alpha(t)$ (or real and imaginary parts) of λ are not independent of each other, also phase $\frac{1}{\hbar}S(\mathbf{r})$ and amplitude $a(\mathbf{r})$ of $\Psi(\mathbf{r})$ are not independent of each other but connected via a corresponding conservation law. Furthermore, in both cases the complex Riccati equations can be rewritten as real NL Ermakov equations. Using this information it is possible to start from a system with real potential and energy to proceed, via a Darboux transformation, to a system with complex partner potential, but still real energy.

	Time	Space
NL complex	$\frac{d}{dt}\left(\frac{\dot{\lambda}}{\lambda}\right) + \left(\frac{\dot{\lambda}}{\lambda}\right)^2 + \omega^2(t) = 0$	$\nabla\left(\frac{\nabla\Psi}{\Psi}\right) + \left(\frac{\nabla\Psi}{\Psi}\right)^2 + \frac{2m}{\hbar^2}\left(E - V(\underline{r})\right) = 0$
linearized	$\ddot{\lambda} + \omega^2\lambda = 0$	$\Delta\Psi + \frac{2m}{\hbar^2}(E - V)\Psi = 0$
complex variable	$\lambda = \alpha e^{i\varphi}$	$\Psi = a e^{iS/\hbar}$
conservation law	$\dot{\varphi} = \frac{1}{\alpha^2}$	$\nabla S = \frac{k}{a^2}$
NL real	$\ddot{\alpha} + \omega^2\alpha = \dot{\varphi}^2\alpha = \frac{1}{\alpha^3}$	$\Delta a + \frac{2m}{\hbar^2}(E - V)a = \frac{1}{\hbar^2}(\nabla S)^2 \cdot a = \frac{(k/\hbar)^2}{a^3}$

contains info about phase on rhs

Fig. 3.1 Comparison of time-dependent and space-dependent cases

In the next chapters it is shown how this whole formalism can be extended to also incorporate open quantum systems with irreversible time-evolution and dissipation of energy via (linear) velocity dependent friction forces. After the discussion of the TD case, also the TI situation will be considered and a comparison with the TD one will be made, similarly to the conservative case.

References

1. F. Cooper, A. Khare, U. Sukhatme, *Supersymmetry in Quantum Mechanics* (World Scientific, Singapore, 2001)
2. A. Khare, U. Sukhatme, Phase-equivalent potentials obtained from supersymmetry. J. Phys. A: Math. Gen. **22**, 2847 (1989); W.-Y. Keung, U.P. Sukhatme, Q. Wang, T.D. Imbo, Families of strictly isospectral potentials. J. Phys. A: Math. Gen. **22**, L987 (1989)
3. H. Kalka, G. Soff, *Supersymmetrie* (Teubner, Stuttgart, 1997)
4. H.C. Rosu, Short survey of Darboux transformations (1999). arXiv:quant-ph/9809056v3
5. G. Reinisch, Nonlinear quantum mechanics. Physica A **206**, 229–252 (1994)
6. G. Reinisch, Classical position probability distribution in stationary and separable quantum systems. Phys. Rev. A **56**, 3409 (1997)
7. E. Madelung, Quantentheorie in hydrodynamischer Form. Z. Phys. **40**, 322 (1927)
8. D. Schuch, Riccati and Ermakov equations in time-dependent and time-independent quantum systems. SIGMA **4**, 043, 16 pp (2008). arXiv:0805.1687
9. R.A. Lee, Quantum ray equations. J. Phys. A: Math. Gen. **15**, 2761–2774 (1982)
10. R.S. Kaushal, Quantum analogue of Ermakov systems and the phase of the quantum wave function. Int. J. Theor. Phys. **40**, 835–847 (2001)
11. R.S. Kaushal, D. Parashar, Can quantum mechanics and supersymmetric quantum mechanics be the multidimensional Ermakov theories? J. Phys. A: Math. Gen. **29**, 889–893 (1996)

12. H.J. Korsch, H. Laurent, Milne's differential equation and numerical solutions of the Schrödinger equation. I. Bound-state energies for single- and double-minimum potentials. J. Phys. B: At. Mol. Phys. **14**, 4213 (1981)
13. H.J. Korsch, H. Laurent, R. Möhlenkamp, Milne's differential equation and numerical solutions of the Schrödinger equation. II. Complex energy resonance states. J. Phys. B: At. Mol. Phys. **15**, 1 (1982)
14. D. Baye, G. Lévai, J.M. Sparenberg, Phase-equivalent complex potentials. Nucl. Phys. A **599**, 435–456 (1996)
15. A.A. Andrianov, M.V. Ioffe, F. Cannata, J.P. Dedonder, Quantum mechanics with complex superpotentials and real energy spectra. Int. J. Mod. Phys. A **14**, 2675–2688 (1999). arXiv:quant-ph/9806019
16. B. Bagchi, S. Mallik, C. Quesne, Generating complex potentials with real eigenvalues in supersymmetric quantum mechanics. Int. J. Mod. Phys. A **16**, 2859 (2001). arXiv:quant-ph/0102093
17. F. Cannata, G. Junker, J. Trost, Schrödinger operators with complex potential but real spectrum. Phys. Lett. A **246**, 219–226 (1998). arXiv:quant-ph/9805085.
18. D.J. Fernández, R. Muñoz, A. Ramos, Second order SUSY transformations with 'complex energies'. Phys. Lett. A **308**, 11–16 (2003). arXiv:quant-ph/0212026
19. O. Rosas-Ortiz, R. Muñoz, Non-Hermitian SUSY hydrogen-like Hamiltonians with real spectra. J. Phys. A: Math. Gen. **36**, 8497 (2003). arXiv:quant-ph/0302190
20. O. Rosas-Ortiz, Gamow vectors and supersymmetric quantum mechanics. Rev. Mex. Fis. **53**(S2), 103–109 (2007). arXiv:0810.2283
21. N. Fernández-García, O. Rosas-Ortiz, Optical potentials using resonance states in supersymmetric quantum mechanics. J. Phys. Conf. Ser. **128**, 012044 (2008)
22. N. Fernández-García, O. Rosas-Ortiz, Gamow–Siegert functions and Darboux-deformed short range potentials. Ann. Phys. **323**, 1397–1414 (2008). arXiv:0810.5597
23. C.M. Bender, S. Boettcher, Real spectra in non-Hermitian Hamiltonians having PT symmetry. Phys. Rev. Lett. **80**, 5243 (1998)
24. E. Schrödinger, A method of determining quantum-mechanical eigenvalues and eigenfunctions. Proc. R. Irish Acad. A **46**, 9–16 (1940)
25. P.A.M. Dirac, *The Principles of Quantum Mechanics*, 2nd edn. (Clarendon, Oxford, 1935)
26. V. Fock, Konfigurationsraum und zweite Quantelung. Z. Phys. **75**, 622–647 (1932)
27. B. Mielnik, O. Rosas-Ortiz, Factorization: little or great algorithm? J. Phys. A: Math. Gen. **37**, 10007–10035 (2004)
28. O. Rosas-Ortiz, O. Castaños, D. Schuch, New supersymmetry-generated complex potentials with real spectra. J. Phys. A: Math. Theor. **48**, 445302 (2015)

Chapter 4
Dissipative Systems with Irreversible Dynamics

Classical Hamiltonian mechanics and quantum mechanics describe isolated systems with reversible dynamics and (for not explicitly time-dependent (TD) potentials) conservation of energy. However, realistic physical systems are always in contact with some kind of environment and this coupling usually introduces the phenomena irreversibility and dissipation (not necessarily always both simultaneously). How then can this be taken into account in the formalism of classical (Hamiltonian or Lagrangian) mechanics and, especially, in a quantum mechanical context?

The fundamental equations of classical mechanics (Lagrange, Hamilton) as well as of quantum mechanics (Schrödinger, Heisenberg) are invariant under time-reversal, i.e. the form does not change under the simultaneous replacements $t \to -t$, $v \to -v$, $p \to -p$ (and complex conjugation in quantum mechanics). Furthermore, the forces are assumed to be derived from a potential, thus guaranteeing for time-independent (TI) potentials $V(\mathbf{r})$ conservation of energy.

In classical mechanics, the time-evolution can be traced back to canonical transformations; in quantum mechanics to unitary transformations, essentially representing pure rotations, i.e., angular changes of the wave function or state vector in an abstract Hilbert space with conserved norm, i.e., length of the vector.

Considering natural evolution processes as observed in the macroscopic world around us, growth processes (in two or more dimensions) are usually connected with a combination of angular and radial changes. The shell of a nautilus or the horn of a ram, e.g., undergo these combined changes when growing, leading to forms like (often logarithmic) spirals.

So, not only rotation, but also radial expansion (or contraction) are essential elements of these kinds of irreversible evolutions thus offering also the possibility of defining a direction of time (e.g., larger radius = later in time, or vice versa). Another characteristic feature of macroscopic evolution processes is the loss of energy by transfer from the dynamical system to the environment due to dissipation, caused for example by friction forces, transforming (mechanical) energy into heat.

D. Schuch, *Quantum Theory from a Nonlinear Perspective*,
Fundamental Theories of Physics 191, https://doi.org/10.1007/978-3-319-65594-9_4

Irreversibility and dissipation are usually not elements contained in the description of a single system in terms of the conventional Hamiltonian formalism.

The aspect of irreversibility also cannot be introduced by changing to a statistical description in terms of distribution functions (or density operators) as the corresponding equations of motion are also invariant under time-reversal, i.e., in the classical case the Liouville equation[1]

$$\frac{\partial}{\partial t}\mathsf{W}_{\text{cl}}(x,p,t) = -\dot{x}\frac{\partial}{\partial x}\mathsf{W}_{\text{cl}} - \dot{p}\frac{\partial}{\partial p}\mathsf{W}_{\text{cl}} = -\{H,\mathsf{W}_{\text{cl}}\} \tag{4.1}$$

for Hamiltonian systems with

$$\dot{x} = \frac{\partial H}{\partial p},\ \dot{p} = -\frac{\partial H}{\partial x}, \tag{4.2}$$

or its quantum mechanical version, the von Neumann equation, where the classical distribution function W_{cl} is replaced by the density operator and the Poisson brackets $\{\ ,\ \}$ by $\frac{1}{\mathrm{i}\hbar}[\ ,\ ]_-$, with $[\ ,\ ]_-$ being the commutator.

Explaining the origin of irreversibility by introducing our subjective (human) ignorance of the microscopic details via a kind of coarse-graining argument (Gibbs, ink-drop example, see e.g. [1]), degrading irreversibility to a mere illusion, is also unsatisfactory.

4.1 Different Approaches for Treating Open Dissipative Systems

Alternative approaches that include irreversibility and dissipation to be discussed subsequently are:

1. phenomenological equations (e.g. Langevin, Fokker–Planck) (in this subsection);
2. system-plus-reservoir approaches (Sect. 4.2);
3. modifications of classical and quantum mechanical equations of motion (leading to non-canonical/non-unitary transformations, NL evolution equations, etc.) (Sects. 4.3–4.5).

In a phenomenological description, a Brownian motion-type situation, i.e. a macroscopic body that is moving in a viscous liquid (many-body problem) is considered. However, the degrees of freedom of the bath/environment are not taken into account explicitly, only their effect on the observable macroscopic system. This effect is

[1] This is the form for a one-particle distribution function. For a many-body system W_{cl} actually depends on $3N$ spatial variables $x_1, \ldots x_{3N}$ and $3N$ momentum variables $p_1, \ldots p_{3N}$ and the corresponding derivatives must be considered where for a macroscopic system N is of the order 10^{24}.

twofold: (1) the environment causes a slowing down of the motion that can be described by a friction force proportional to velocity $\mathbf{v}(t)$ (or momentum); (2) even if the average velocity is zero, the system still does not come to a complete rest but still fulfils a rapidly-fluctuating (Brownian) motion around its final position which can be described by a purely random stochastic force $\mathcal{F}(t)$; i.e., an artificial decomposition of the environmental effect into two contributions is assumed.

In the trajectory picture, the time-evolution of the system can be described by a modified Newtonian equation of motion, the *Langevin equation*

$$m\dot{\mathbf{v}} + m\gamma\mathbf{v} + \mathcal{F}(t) + \nabla V = 0 \tag{4.3}$$

with the friction force $-m\gamma\mathbf{v}$ where γ is the friction coefficient and $\mathcal{F}$ the stochastic force (which vanishes on average,[2] $\langle\mathcal{F}(t)\rangle_{\text{cl}} = 0$) and V is an external potential.

Due to the friction force, the velocity decreases on average according to

$$\langle\mathbf{v}\rangle_{\text{cl}}(t) = \mathbf{v}_0 \,\mathrm{e}^{-\gamma t}. \tag{4.4}$$

Consequently, the average kinetic energy decreases like

$$\langle T\rangle_{\text{cl}} = \frac{m}{2}\langle v^2\rangle_{\text{cl}} = \frac{m}{2} v_0^2 \mathrm{e}^{-2\gamma t} \tag{4.5}$$

if the fluctuating force is not being taken into account. This force adds a contribution to the remaining energy even if the average velocity $\langle\mathbf{v}\rangle_{\text{cl}}$ has become zero. At this point an additional assumption is introduced, explicitly, that the final state of the systems evolution is thermal equilibrium. Using the equipartition theorem, the contribution of the fluctuations to the system's (kinetic) energy is assumed to be $\frac{1}{2}\mathsf{k}_{\mathrm{B}}\mathsf{T}$ (per degree of freedom), with $\mathsf{k}_{\mathrm{B}} =$ Boltzmann's constant and $\mathsf{T} =$ temperature.

If the Langevin equation is a description in the trajectory picture, with a modified Newtonian equation of motion, there is also an equivalent description of the same scenario in terms of a single-particle distribution function with a modified continuity or Liouville equation. For a one-dimensional system, this discribution function $\mathsf{W}_{\text{cl}}(x, v, t)$ in phase space fulfils the *Fokker–Planck equation*

$$\frac{\partial}{\partial t}\mathsf{W}_{\text{cl}}(x, v, t) + v\frac{\partial}{\partial x}\mathsf{W}_{\text{cl}} + \frac{\partial}{\partial v}\left[\left(-\gamma v + \frac{F(x)}{m}\right)\mathsf{W}_{\text{cl}}\right] - \frac{\gamma \mathsf{k}_{\mathrm{B}}\mathsf{T}}{m}\frac{\partial^2}{\partial v^2}\mathsf{W}_{\text{cl}} = 0, \tag{4.6}$$

where the second and third terms on the lhs correspond to the convection terms in the Liouville equation with $v = \dot{x}$ and $-\gamma v + \frac{F(x)}{m} = \dot{v}$ and the last term on the rhs is an irreversible diffusion term with diffusion coefficient $\frac{\gamma \mathsf{k}_{\mathrm{B}}\mathsf{T}}{m}$.

Note: The time-reversal symmetry of this equation is already broken by the convection term due to the friction contribution $-\gamma v$!

[2]The bracket $\langle\cdots\rangle_{\text{cl}}$ denotes classical (ensemble) averages; for details, see also [2].

In his famous paper from 1940 [3] Kramers used a trick [4] to derive a non-trivial position-space version of this equation, the so-called *Smoluchowski equation*

$$\frac{\partial}{\partial t}\varrho(x,t) + \frac{\partial}{\partial x}\left(\frac{F(x)}{m\gamma}\varrho\right) - \frac{\mathrm{k_B T}}{m\gamma}\frac{\partial^2}{\partial x^2}\varrho = 0 \tag{4.7}$$

for the distribution function $\varrho(x,t)$, again with a convection and a diffusion term where now the diffusion coefficient obeys the Einstein relation

$$D = \frac{\mathrm{k_B T}}{m\gamma}. \tag{4.8}$$

A further step towards a description of the non-equilibrium many-body problem defined above is the *Boltzmann equation*. This is essentially a modification of the Liouville equation (4.1) in phase space where, on the rhs, gain and loss terms are added thus changing the equation into an irreversible rate equation or so-called *master equation*, in this case, again for a one-particle distribution function. The major advantage of the Boltzmann equation versus the afore-mentioned approaches is that the final state of thermal equilibrium for the system is not just put in by hand but is a direct result of this approach. Attempts to obtain a quantum mechanical version of the Boltzmann equation finally led to the *generalized master equation* [5]. In this case, the system-plus-reservoir approach [6] is applied that will be considered in the next subsection.

4.2 System-Plus-Reservoir Approaches

In the system-plus-reservoir approaches, the system of interest is coupled to an environment where the system and the environment together are considered to be a closed Hamiltonian system.

The conventional way is to couple the system of interest to an environment with many (in the limit infinitely many) degrees of freedom (e.g., coupled linearly to a bath of harmonic oscillators (HOs) [7]). Via averaging over the environmental degrees of freedom and other procedures (for details see e.g., [6]), an equation of motion for the system of interest including a friction force can finally be obtained. In the following, the frequently-applied approach of Caldeira and Leggett [7] and modifications thereof will be mentioned.

Afterwards, a different approach will be discussed where the environment consists of only *one* additional degree of freedom. In this approach by Bateman [8] the equation of motion including the damping friction force (and a complementary one with an equally-strong accelerating force for a second degree of freedom) can be obtained from a Hamiltonian that is also a constant of motion.

4.2.1 *Caldeira–Leggett Model and Kossakowski–Lindblad Generators*

As one of the simplest examples of a dissipative system, a HO (linearly) coupled to an environment represented by a bath of HOs has been subject of numerous studies (e.g., [7, 9–15]). In the literature it is usually called the Caldeira–Leggett model. In this model, a HO with coordinate x, momentum p and mass m is interacting with a bath of N HOs with coordinates q_i, momenta p_i, masses m_i and frequencies ω_i. The Hamiltonian for the composite system can then be split into three contributions,

$$H = H_S + H_R + H_I, \tag{4.9}$$

where the Hamiltonian of the system of interest is[3]

$$H_S = \frac{p^2}{2m} + \frac{m}{2}\omega_0^2 x^2, \tag{4.10}$$

and the reservoir of N HOs is described by

$$H_R = \sum_{i=1}^{N} \left(\frac{p_i^2}{2m_i} + \frac{1}{2} m_i \omega_i^2 q_i^2 \right). \tag{4.11}$$

The interaction Hamiltonian H_I could be written as

$$H_I = -x \sum_i c_i q_i + \Delta V(x). \tag{4.12}$$

The last term on the rhs, $\Delta V(x)$, depends only on the coordinate of the system but not on q_i and can be interpreted as a shift of the systems potential by an amount $\Delta V(x)$. In the literature[4] [6] it is argued that "$\Delta V(x)$ may serve to compensate frequency-renormalization effects induced by the first term in the expression for H_I. If we choose $\Delta V(x) = 0$, the minimum of the potential surface of the system plus environment for given x is when $q_i = \frac{c_i x}{m_i \omega_i^2}$ for all i and the "effective" potential is then given by"

$$V_{\text{eff}}(x) = V(x) - \sum_{i=1}^{N} \frac{c_i^2}{2m_i \omega_i^2} x^2. \tag{4.13}$$

In our case, "the second term in (4.13) causes a negative shift $\Delta\omega^2 = -\sum_i \frac{c_i^2}{m m_i \omega_i^2}$ in the squared frequency ω_0^2 of small oscillations about the minimum. Such coupling-induced renormalization effects can be very large, and if $\omega_{\text{eff}}^2 = \omega_0^2 + \Delta\omega^2 < 0$ they

---

[3]Here the HO-potential is chosen, but in general any potential $V(x)$ is possible.

[4]Compared with Ref. [6], here x and q are interchanged.

even change the potential qualitatively" from an attractive to a repulsive one. In order to eliminate such effects and only study the effect of dissipation, one must add

$$\Delta V(x) = \sum_{i=1}^{N} \frac{c_i^2}{2m_i\omega_i^2}x^2 \tag{4.14}$$

to the potential $V(x)$ which means for a HO as system to change ω_0^2 into $(\omega_0^2+|\Delta\omega^2|)$.

From this Hamiltonian (4.9) one obtains a coupled set of equations of motion for the system,

$$m\ddot{x}(t) + m\omega_0^2 x(t) - \sum_{i=1}^{N} c_i \left(q_i(t) - \frac{c_i}{m_i\omega_i^2}x(t)\right) = 0, \tag{4.15}$$

and the environmental degrees of freedom,

$$m_i\ddot{q}_i(t) + m_i\omega_i^2 q_i(t) - c_i x(t) = 0. \tag{4.16}$$

After Fourier transformation, elimination of the environmental degrees of freedom, assumption of an Ohmic spectral density etc., one finally arrives at an equation of motion for the system including a linear velocity dependent friction force as in the Langevin equation.

In a quantum mechanisch context, the same model is used, only now the system is usually described by the reduced density matrix (or operator) ϱ_{op}. After the afore-mentioned elimination of the environmental degrees of freedom etc., the von Neumann equation, with the Hamiltonian (4.9) in the commutator, finally leads to the corresponding (high-temperature) master equation

$$\frac{\partial}{\partial t}\varrho_{\mathrm{op}} = \frac{1}{\mathrm{i}}\mathcal{L}_{\mathrm{op}}\varrho_{\mathrm{op}} = \frac{1}{\mathrm{i}\hbar}[H_S', \varrho_{\mathrm{op}}]_- - \frac{\gamma}{\hbar}\left\{\frac{mk_B\mathrm{T}}{\hbar}[x,[x,\varrho_{\mathrm{op}}]_-]_- + \frac{\mathrm{i}}{2}[x,[p,\varrho_{\mathrm{op}}]_+]_-\right\} \tag{4.17}$$

where $\mathcal{L}_{\mathrm{op}} = \frac{1}{\hbar}[H,\ ]_-$ is the generator of the dynamics and H_S' is H_S plus correction terms, as mentioned above. However, this equation is "known to violate the positivity requirement for the density operator" [15] and therefore, in certain cases, leads to unphysical results.

An alternative approach that uses quantum dynamical semi-groups, proposed by Kossakowski et al. [16] and Lindblad [17] with generators of the form

$$\mathcal{L}_{\mathrm{KL}}[\varrho_{\mathrm{op}}] = \frac{1}{\mathrm{i}\hbar}[H, \varrho_{\mathrm{op}}]_- + \frac{1}{2\hbar}\sum_{i=1}^{\infty}\left([V_i\varrho_{\mathrm{op}}, V_i^+]_- + [V_i, \varrho_{\mathrm{op}}V_i^+]_-\right) \tag{4.18}$$

(where the V_i's are bounded operators on the Hilbert space of the Hamiltonian, usually linear combinations of position and momentum operators) does not have this problem. The specific mathematical form guarantees positivity but the particular choice of the operators occurring therein is usually "guided by intuition" [18] as no obvious physical guideline exists.

Gao [15] tried linking these two approaches in order to eliminate the shortcomings and to obtain a master equation that can also be applied in the low-temperature regime. But also this approach has some shortcomings like a wrong friction force in the Ehrenfest equation of motion, as pointed out in [19]. This problem does not occur in a similar approach by Diosi [20], but this approximation diverges at low temperatures.

One drawback in employing the system-plus-reservoir method is the large number of environmental degrees of freedom that must be considered for a realistic description (without revivals and other artefacts) in the beginning (though they are eliminated in the end). This leads to large, unwieldy and costly calculations.

As pointed out above, in its quantized version this approach is usually applied to the density operator (or matrix) causing the computational effort to scale at least quadratically with the number of degrees of freedom, thus limiting drastically the systems that can be treated. In contrast, methods based on pure states (like WPs) have the chance, in the most favourable case, of scaling linearly with the number of degrees of freedom. Therefore, subsequent to the discussion of the Bateman model, in the quantum mechanical context only Schrödinger equations (SEs) for "one-particle" wave functions/wave packets (WPs) shall be considered for the description of dissipative systems.

This choice is also supported by a statement made by Davidson [21] that classical mechanics has entered the domain of NL and chaotic dynamics, the study of which usually requires intense numerical simulation. A rigorous treatment of dissipation in this context defies simulation on a quantum mechanical level when considering a large number of degrees of freedom, as the largest and fastest computer(s) would be required to calculate even the simplest NL problems. "Hence there is a need to find some practical way to simulate a lossy quantum system" [21].

Can our many-body problem then be reduced to an effective few or even one-body problem? This is somehow similar to the philosophy of density functional theory where the solution of a many-body problem can also be reduced to the solution of a representative one-body problem. Although this procedure might contain aspects that could be criticised from an *ab initio* point of view, the success of the theory shows, at least from a pragmatic viewpoint, that this kind of approach can provide a useful alternative.

In the next subsection, the environmental degrees of freedom that are explicitly considered are not (yet) totally eliminated but at least reduced to the minimum of one in the Bateman model.

4.2.2 Bateman Hamiltonian

The number of environmental degrees of freedom is drastically reduced to one in an approach by Bateman [8] describing the damped HO. In order to be able to apply the canonical formalism, the phase-space dimension must be doubled to obtain a kind of effective description. The new degree of freedom can be considered a collective one for the bath that absorbs the energy dissipated by the damped oscillator. The variable of the dual system that fulfils a time-reversed equation with an acceleration force of the same magnitude as the friction force of the Langevin equation, but with the opposite sign, looks like a position variable and its relation to, and interpretation in terms of, physical position and momentum (or velocity) is discussed in the following.

After the rediscovery of the Bateman dual Hamiltonian by Morse and Feshbach [22] and Bopp [23] various different features of it were studied, also in recent years. So squeezed states for the Bateman Hamiltonian were considered in [24] and [25] and a quantum field theoretical approach was used by Vitiello et al. [26]. The same author also applied the dual approach as a dissipative quantum model of the brain [27]. Quantization using Feynman's path integral method was discussed by Blasone and Jizba [28, 29] and the Bateman system was also studied by the same authors and Vitiello [29, 30] as a (toy) model for 't Hooft's proposal of a deterministic version of quantum mechanics [31]. More recently, together with Scardigli, these authors considered a composite system of two classical Bateman oscillators as a particle in an effective magnetic field [32]. Complex eigenvalues of the quantized version of Bateman's Hamiltonian in connection with resonances and two-dimensional parabolic potential barriers are discussed in [33, 34]. Also the Wigner function for the Bateman system on non-commutative phase space [35] and the inclusion of a TD external force [36] have been studied. The Bateman approach (as well as that of Caldirola [37] and Kanai [38] that is considered subsequently) is also discussed in an attempt to reformulate a dissipative system in terms of an infinite number of non-dissipative ones [39]. A different method for the description of dissipative systems that seems to have some advantages in the high energy regime has been compared with the Bateman approach [40] and shown to be equivalent to it locally. Finally, in a more recent paper [41] by Bender et al., the Bateman Hamiltonian enlarged by a quadratic term in the two dual coordinates is studied as a model for two coupled optical resonators. Despite the age of Bateman's approach, this shows that there is still considerable interest in, and potential applicability of, this model.

Since the position and momentum variables (and likewise quantities depending on them or their time- derivatives such as Lagrangians and Hamiltonians) of the Bateman approach obey the rules of conventional classical canonical Lagrangian or Hamiltonian mechanics but can have a meaning quite different from their physical counterparts, a distinction will be made between canonical and physical quantities in the following by supplying canonical quantities with a hat "$.\hat{.}.$".

The Bateman Hamiltonian $\hat{H}_B$, expressed in terms of the canonical position variables $\hat{x}$ and $\hat{y}$ and the canonical momenta $\hat{p}_x$ and $\hat{p}_y$, has the form

$$\hat{H}_B = \frac{1}{m}\hat{p}_x\hat{p}_y + \frac{\gamma}{2}(\hat{y}\hat{p}_y - \hat{x}\hat{p}_x) + m\left(\omega^2 - \frac{\gamma^2}{4}\right)\hat{x}\hat{y} = \hat{H}_\Omega + \hat{D} \tag{4.19}$$

with $\hat{D} = \frac{\gamma}{2}(\hat{y}\hat{p}_y - \hat{x}\hat{p}_x)$. The Poisson brackets of $\hat{H}_B$ with $\hat{D}$ as well as with $\hat{H}_\Omega$ vanish, $\{\hat{H}_\Omega, \hat{H}_B\} = \{\hat{D}, \hat{H}_B\} = 0$, so both are constants of motion (in the quantized version, the three corresponding operators commute).

The Hamiltonian equations of motion are

$$\begin{aligned}
\frac{\partial \hat{H}_B}{\partial \hat{p}_x} &= \frac{1}{m}\hat{p}_y - \frac{\gamma}{2}\hat{x} = \dot{\hat{x}}, \quad \frac{\partial \hat{H}_B}{\partial \hat{p}_y} = \frac{1}{m}\hat{p}_x + \frac{\gamma}{2}\hat{y} = \dot{\hat{y}} \\
\frac{\partial \hat{H}_B}{\partial \hat{x}} &= -\frac{\gamma}{2}\hat{p}_x + m\left(\omega^2 - \frac{\gamma^2}{4}\right)\hat{y} = -\dot{\hat{p}}_x\,, \\
\frac{\partial \hat{H}_B}{\partial \hat{y}} &= \frac{\gamma}{2}\hat{p}_y + m\left(\omega^2 - \frac{\gamma^2}{4}\right)\hat{x} = -\dot{\hat{p}}_y,
\end{aligned} \tag{4.20}$$

where, from (4.20), $\hat{p}_x$ and $\hat{p}_y$ can be expressed as

$$\hat{p}_y = m\left(\dot{\hat{x}} + \frac{\gamma}{2}\hat{x}\right), \tag{4.21}$$

$$\hat{p}_x = m\left(\dot{\hat{y}} - \frac{\gamma}{2}\hat{y}\right). \tag{4.22}$$

From there, and with the help of Eqs. (4.20), the equations of motion for $\hat{x}$ and $\hat{y}$ can be obtained as

$$\ddot{\hat{x}} + \gamma\dot{\hat{x}} + \omega^2\hat{x} = 0, \tag{4.23}$$

$$\ddot{\hat{y}} - \gamma\dot{\hat{y}} + \omega^2\hat{y} = 0. \tag{4.24}$$

Equation (4.23) is just the equation for the damped HO with friction force $-m\gamma\dot{x}$ whereas, in the time-reversed equation for $\hat{y}$, the accelerating force $+m\gamma\dot{\hat{y}}$ occurs. From Eqs. (4.23), (4.24) and (4.21), (4.22) it is clear that the $(\hat{x}, \hat{p}_x, \hat{y}, \hat{p}_y)$-space splits into two invariant subspaces: the one of variables $(\hat{x}, \hat{p}_y)$ undergoing a damped oscillator motion and the one of variables $(\hat{y}, \hat{p}_x)$ with time-reversed (accelerated) behaviour.

Using these equations of motion, it can also be confirmed that

$$\frac{d}{dt}\hat{H}_B = 0, \tag{4.25}$$

i.e. $\hat{H}_B$ is a dynamical invariant which, in a first naive attempt, could be interpreted as such that the energy dissipated by the damped system is gained by the accelerated one. Rewritten in terms of $\hat{x}$, $\hat{y}$ and the corresponding velocities $\dot{\hat{x}}$ and $\dot{\hat{y}}$, the terms depending on the friction (or acceleration) coefficient γ cancel out (although the Lagrangian does contain terms in γ) and it remains that

$$\hat{H}_B \mathrel{\hat{=}} m(\dot{\hat{x}}\dot{\hat{y}} + \omega^2\hat{x}\hat{y}). \tag{4.26}$$

In fact, the individual energies and their changes in time for both systems, written in terms of the velocities, take the form

$$\hat{E}_x = \frac{m}{2}\dot{\hat{x}}^2 + \frac{m}{2}\omega^2\hat{x}^2, \tag{4.27}$$

with

$$\frac{d}{dt}\hat{E}_x = -\gamma m \dot{\hat{x}}^2, \tag{4.28}$$

(in agreement with the energy dissipated by a physical system with the same friction force, i.e. $\frac{d}{dt}E = -2m\gamma T$ where $E = T + V$) and

$$\hat{E}_y = \frac{m}{2}\dot{\hat{y}}^2 + \frac{m}{2}\omega^2\hat{y}^2, \tag{4.29}$$

with

$$\frac{d}{dt}\hat{E}_y = +\gamma m \dot{\hat{y}}^2. \tag{4.30}$$

So, the sum of $\hat{E}_x$ and $\hat{E}_y$ would be constant and (apart from another constant term) could be equal to H_{B} if

$$\frac{d}{dt}(\hat{E}_x + \hat{E}_y) = \gamma\, m(\dot{\hat{y}}^2 - \dot{\hat{x}}^2) = 0, \tag{4.31}$$

which is fulfilled only for $\dot{\hat{y}} = \pm\dot{\hat{x}}$; so $\hat{y}$ and $\hat{x}$ could differ, at most, by a constant and $\hat{H}_{\text{B}}$, as given in (4.26), (again apart from a constant term) would turn into $\hat{H}_{\text{B}} \rightarrow m(\dot{\hat{x}}^2 + \omega^2\, \hat{x}^2)$, i.e., the energy of two undamped HO.

However, $\dot{\hat{y}}$, derived from the solution of Eq. (4.24), differs by more than just its sign from $\dot{\hat{x}} = \dot{x}$ derived from the solution of Eq. (4.23). The energy of the damped system is decreasing to zero for time going to infinity whereas the energy of the accelerated system would grow to infinity in the same limit, yielding that the sum of both cannot achieve the constant value of the Bateman Hamiltonian. So one must be careful with the simple picture of energy exchange between the $\hat{x}$- and $\hat{y}$-systems, when considering $\hat{y}$ as a physical position variable like $\hat{x} = x$.

Clarification can be found by eliminating the second (dual) degree of freedom by imposing some constraints (which is not possible in a unique way) leading to a description in terms of canonical variables for the system of interest alone without additional degrees of freedom. This kind of effective canonical descriptions for the dissipative system is discussed in the next subsection. Further details on the connection between these approaches and the Bateman model are given at the end of Sect. 4.5 and in Appendix D.

4.3 Effective Models Within the Canonical Formalism

In this subsection two approaches are presented that are able to describe the damped HO in the framework of Lagrangian and Hamiltonian mechanics. They use only one canonical position and momentum variable (for a one-dimensional system) without *any* degree of freedom of the environment, but the canonical quantities are chosen in a way that they include the effect of the environment, like damping by friction. These *canonical* variables, however, are connected with the *physical* position and momentum via *non − canonical* transformations. In the following, *canonical* variables and corresponding Lagrangians/Hamiltonians will again be characterized by a hat $(.\hat{.}.)$. This also applies to the canonical operators and wave functions in the quantized version.

4.3.1 Caldirola–Kanai Hamiltonian

Historically, the first and most frequently-used approach of that kind was proposed by Caldirola [37] and Kanai [38] and uses an explicitly TD Lagrangian/Hamiltonian. The application of a TD Hamiltonian leading to a modified (explicitly TD) *linear* SE should then permit the direct use of standard schemes of quantization. The classical version starts with the explicitly TD Lagrangian

$$\hat{L}_{\text{CK}} = \left[\frac{m}{2}\dot{\hat{x}}^2 - V\left(\hat{x}\right)\right] \mathrm{e}^{\gamma t}, \tag{4.32}$$

leading, via the Euler–Lagrange equation

$$\frac{d}{dt}\frac{\partial \hat{L}_{\text{CK}}}{\partial \dot{\hat{x}}} - \frac{\partial \hat{L}_{\text{CK}}}{\partial \hat{x}} = 0, \tag{4.33}$$

to the equation of motion

$$m\ddot{\hat{x}} + m\gamma\dot{\hat{x}} + \frac{\partial}{\partial \hat{x}} V\left(\hat{x}\right) = 0 \tag{4.34}$$

which is identical to the Langevin equation for the HO (4.3) (in one dimension) without stochastic force $\mathcal{F}(t)$ but including the linear velocity-dependent friction force, providing $\hat{x}$ is identified with the *physical* position variable x, i.e., $\hat{x} = x$. Therefore, if not otherwise specified, the hat will be omitted for this variable in the following.

From $\hat{L}_{\text{CK}}$, the *canonical* momentum $\hat{p}$ can be obtained in the usual way as

$$\frac{\partial}{\partial \dot{x}} \hat{L}_{\text{CK}} = m\dot{x}\,\mathrm{e}^{\gamma t} = p\,\mathrm{e}^{\gamma t} = \hat{p}. \tag{4.35}$$

Note that this *canonical* momentum $\hat{p}$ is different from the *physical* kinetic momentum $p = m\dot{x}$ and the transition from the physical variables (x, p) to the canonical variables $(\hat{x} = x, \hat{p} = p e^{\gamma t})$ is given by a *non−canonical* transformation (the Jacobian determinant is different from one).

The corresponding Hamiltonian $\hat{H}_{CK}(x, \hat{p})$ can be obtained straightforwardly in the form

$$\hat{H}_{CK} = e^{-\gamma t} \frac{\hat{p}^2}{2m} + e^{\gamma t} V(x) \tag{4.36}$$

which also supplies the correct equations of motion including the friction force,

$$\frac{\partial \hat{H}_{CK}}{\partial \hat{p}} = \frac{1}{m} \hat{p}\, e^{-\gamma t} = \dot{x}, \tag{4.37}$$

$$\frac{\partial \hat{H}_{CK}}{\partial \hat{x}} = m\omega^2 x\, e^{\gamma t} = -\dot{\hat{p}} = -(m\ddot{x} + m\gamma\dot{x})\, e^{\gamma t}. \tag{4.38}$$

$\hat{H}_{CK}$ has been criticised as not being a constant of motion. However, the same could be said about the usual Hamiltonian for the parametric oscillator with $\omega = \omega(t)$.

In conventional classical mechanics, the Hamiltonian not only determines the dynamics of the system via the Hamiltonian equations of motion but also represents the energy of the system, $H = T + V$. The Caldirola–Kanai (CK) Hamiltonian is not identical to the energy of the system but is at least uniquely related to it via $\hat{H}_{CK} \hat{=} \left(\frac{p^2}{2m} + V(x)\right) e^{\gamma t}$. So, from a pragmatic point of view, this approach is at least useful on the classical level. Let us therefore proceed to the quantum mechanical one.

Canonical quantization, i.e. replacing the *canonical* momentum $\hat{p}$ with the operator $\hat{p}_{op} = \frac{\hbar}{i} \frac{\partial}{\partial x}$ and applying the resulting (linear but explicitly TD) Hamiltonian operator $\hat{H}_{CK,op}$ to the *canonical* wave function $\hat{\Psi}_{CK}(x, t)$, leads to the modified SE

$$\begin{aligned} i\hbar \frac{\partial}{\partial t} \hat{\Psi}_{CK}(x, t) &= \hat{H}_{CK,op} \hat{\Psi}_{CK}(x, t) \\ &= \left\{ e^{-\gamma t} \left(-\frac{\hbar^2}{2m} \frac{\partial^2}{\partial x^2} \right) + e^{\gamma t} V(x) \right\} \hat{\Psi}_{CK}(x, t) . \end{aligned} \tag{4.39}$$

In the cases discussed for the TDSE in the conservative case in Chap. 2, i.e. for oscillator potentials $V(x) = \frac{m}{2}\omega^2 x^2$, this equation also possesses exact Gaussian WP solutions in the form

$$\hat{\Psi}_{CK}(x, t) = \hat{N}_{CK}(t) \exp\left\{ i \left[\hat{y}_{CK}(t)\tilde{x}^2 + \frac{1}{\hbar} \langle \hat{p} \rangle_{CK} \tilde{x} + \hat{K}_{CK}(t) \right] \right\} \tag{4.40}$$

with complex TD parameter $\hat{y}_{CK}(t)$. Here, $\langle \cdots \rangle_{CK}$ indicates the mean value is calculated using $\hat{\Psi}_{CK}(x, t)$.

The equation of motion for the WP maximum is simply the one for the classical trajectory including the friction force, in the notation of Sect. 2.1,

$$\ddot{\eta} + \gamma\,\dot{\eta} + \omega^2\eta = 0. \tag{4.41}$$

The modified complex Riccati equation for the complex coefficient $\hat{y}_{CK}(t)$ of the quadratic term in the exponent of the WP, or $\hat{C}_{CK}(t) = \frac{2\hbar}{m}\hat{y}_{CK}(t)$,

$$\dot{\hat{C}}_{CK} + \hat{C}^2_{CK}e^{-\gamma t} + \omega^2 e^{\gamma t} = 0, \tag{4.42}$$

can, with the help of the definition

$$\hat{C}_{CK,I} = \frac{\hbar}{2m\langle\tilde{x}^2\rangle_{CK}} = \frac{1}{\alpha^2_{CK}(t)}, \tag{4.43}$$

also be transformed, as in the conservative case, into a real Ermakov-type equation,

$$\ddot{\alpha}_{CK} + \gamma\dot{\alpha}_{CK} + \omega^2\alpha_{CK} = \frac{e^{-2\gamma t}}{\alpha^3_{CK}}. \tag{4.44}$$

This equation, together with Eq. (4.41), forms the system of equations of motion, coupled via ω, that possesses an exact Ermakov-type invariant, now given in the form

$$\hat{I}_{CK} = \frac{1}{2}\left[e^{2\gamma t}\left(\dot{\eta}\,\alpha_{CK} - \eta\,\dot{\alpha}_{CK}\right)^2 + \left(\frac{\eta}{\alpha_{CK}}\right)^2\right] = \text{const.} \tag{4.45}$$

However, there are some serious points of criticism against this quantized version of the approach.

If one tries to express the operator of the *physical* momentum (which corresponds to the quantity that can actually be measured) simply by $p_{op} = e^{-\gamma t}\hat{p}_{op}$, the commutation relation and the uncertainty product of position and *physical* momentum both decay exponentially,

$$[x, p_{op}]_- = [x, \hat{p}_{op}]_- e^{-\gamma t} = i\hbar\, e^{-\gamma t} \to 0, \tag{4.46}$$

$$U_{CK,phys} = \langle\tilde{x}^2\rangle_{CK}\langle\tilde{p}^2\rangle_{CK} = \langle\tilde{x}^2\rangle_{CK}\langle\tilde{\hat{p}}^2\rangle_{CK}\, e^{-2\gamma t} \to 0. \tag{4.47}$$

This apparent violation of the uncertainty principle has been criticised by several authors [42–44] and attempts to justify or modify the CK approach [45–47] in order to avoid this unphysical behaviour are not convincing. Nevertheless, this approach is still used frequently and physical conclusions are drawn from it that sometimes totally contradict the results of other approaches.

The situation becomes even more puzzling after Yu and Sun [13, 14] showed that the CK-Hamiltonian *operator* can be derived from the conventional system-plus-reservoir approach of Caldeira–Leggett, discussed in Sect. 4.2.1. This puzzle can be solved if the relation between this approach and another effective model, using a (logarithmic) NLSE (see Sect. 4.4.3) on the physical level is clarified (see Sect. 4.5). Consequently, further effective approaches for the description of dissipative quantum systems are discussed in the following.

4.3.2 Expanding Coordinate System

A point of criticism raised against the CK-approach is that it is not a constant of motion. However, it can be turned into such an invariant by adding a term $\frac{\gamma}{2}xp\ \mathrm{e}^{\gamma t}$ to $\hat{H}_{\mathrm{CK}}$, leading to (in the following for a potential quadratic in x and constant ω)

$$\left(\frac{1}{2m}p^2 + \frac{\gamma}{2}xp + \frac{m}{2}\omega^2 x^2\right)\mathrm{e}^{\gamma t} = \text{const.} \tag{4.48}$$

if $x(t)$ obeys the equation of motion including the friction force. This invariant can be rewritten in a form like a conventional Hamiltonian if a new expanding (canonical) coordinate and the corresponding (canonical) momentum are introduced via [48–51]

$$\hat{Q} = x\ \mathrm{e}^{\frac{\gamma}{2}t}, \tag{4.49}$$

$$\hat{P} = m\dot{\hat{Q}} = m\left(\dot{x} + \frac{\gamma}{2}x\right)\mathrm{e}^{\frac{\gamma}{2}t}. \tag{4.50}$$

The Hamiltonian then takes the form [48, 50, 51]

$$\hat{H}_{exp} = \frac{1}{2m}\hat{P}^2 + \frac{m}{2}\left(\omega^2 - \frac{\gamma^2}{4}\right)\hat{Q}^2 = \text{const.} \mathrel{\hat{=}} \frac{m}{2}\left[\dot{x}^2 + \gamma\dot{x}x + \omega^2x^2\right]\mathrm{e}^{\gamma t} \tag{4.51}$$

and is not only an invariant, but for $x_0 = 0$ or $p_0 = m\dot{x}(0)$, is even identical to the initial energy $E_0 = \frac{1}{2m}p_0^2 + \frac{m}{2}\omega^2 x_0^2$ of the system.

Hamiltonian (4.51) looks like that of an undamped HO with shifted frequency $\Omega = (\omega^2 - \frac{\gamma^2}{4})^{\frac{1}{2}}$ and the corresponding equation of motion for $\hat{Q}$ is consequently

$$\ddot{\hat{Q}} + \Omega^2\hat{Q} = 0 \tag{4.52}$$

which, expressed in physical coordinates, provides again the averaged Langevin equation (here for the HO)

$$\ddot{x} + \gamma\dot{x} + \omega^2 x = 0. \tag{4.53}$$

The same result is also obtained consistently from the corresponding Hamiltonian equations of motion

$$\frac{\partial \hat{H}_{exp}}{\partial \hat{P}} = \frac{1}{m}\hat{P} = \dot{\hat{Q}} = \left(\dot{x} + \frac{\gamma}{2}x\right) e^{\frac{\gamma}{2}t}, \tag{4.54}$$

$$\frac{\partial \hat{H}_{exp}}{\partial \hat{Q}} = m\Omega^2 \hat{Q} = m\left(\omega^2 - \frac{\gamma^2}{4}\right) x\, e^{\frac{\gamma}{2}t} = -\dot{\hat{P}} = -m\left(\ddot{x} + \gamma\dot{x} + \frac{\gamma^2}{4}x\right) e^{\frac{\gamma}{2}t}. \tag{4.55}$$

Although the dissipative system in the expanding *canonical variables* $(\hat{Q}, \hat{P})$ can be described via the usual *canonical formalism*, this is obviously *not* possible in terms of the *physical variables* (x, p).

However, as the relations between the physical and the canonical variables are known, it is possible to transform the canonical results into the ones on the physical level.

In particular, if the *time* – *evolution* of any phase space function $\hat{F}(\hat{Q}, \hat{P}, t)$ on the *canonical* level is given by

$$\frac{d}{dt}\hat{F} = \{\hat{F}, \hat{H}_{exp}\}_{(\mathrm{Q,P})} + \frac{\partial}{\partial t}\hat{F}, \tag{4.56}$$

(where the subscript $(\hat{Q}, \hat{P})$ indicates the variables to which the derivatives are taken), on the *physical* level additional Poisson brackets $\{\ ,\ \}_-$ and anti-Poisson brackets $\{\ ,\ \}_+$ (with derivatives with respect to the physical variables (x, p)) occur,

$$\frac{d}{dt}F(x,p,t) = \{F, H\}_- + \{F, \frac{\gamma}{2}xp\}_- - \{F, \frac{\gamma}{2}xp\}_+ + \frac{\partial}{\partial t}F, \tag{4.57}$$

where H is given by $H = \frac{p^2}{2m} + \frac{m}{2}\omega^2 x^2$ (for further details see [51] and Appendix C).

This modification of the conventional formalism is due to the fact that also the transformation between the *canonical* variables $(\hat{Q}, \hat{P})$ and the *physical* variables (x, p) is a *non-canonical* transformation.

So what is the connection between the CK-variables and the expanding ones, i.e., between $(\hat{x}, \hat{p})$ and $(\hat{Q} = \hat{x}\, e^{\frac{\gamma}{2}t}, \hat{P} = \hat{p}\, e^{-\frac{\gamma}{2}t} + m\frac{\gamma}{2}\hat{x}\, e^{\frac{\gamma}{2}t})$?

The determinant in Eq. (4.58), the Jacobian

$$D = \begin{vmatrix} \frac{\partial \hat{Q}}{\partial \hat{x}} & \frac{\partial \hat{Q}}{\partial \hat{p}} \\ \frac{\partial \hat{P}}{\partial \hat{x}} & \frac{\partial \hat{P}}{\partial \hat{p}} \end{vmatrix} = \begin{vmatrix} e^{\frac{\gamma}{2}t} & 0 \\ m\frac{\gamma}{2}\, e^{\frac{\gamma}{2}t} & e^{-\frac{\gamma}{2}t} \end{vmatrix} = 1 \tag{4.58}$$

shows that these two approaches are connected via a *canonical* transformation (on the canonical level).

Why is $\hat{H}_{exp}$ a constant of motion whereas $\hat{H}_{\mathrm{CK}}$ is not? Because the generating function $\hat{F}_2(\hat{x}, \hat{P}, t)$ leading to $\hat{H}_{exp}$ via

$$\hat{H}_{exp} = \hat{H}_{\mathrm{CK}} + \frac{\partial}{\partial t}\hat{F}_2 \tag{4.59}$$

is explicitly TD,

$$\hat{F}_2(\hat{x}, \hat{P}, t) = \hat{x}\hat{P}\, \mathrm{e}^{\frac{\gamma}{2}t} - m\frac{\gamma}{4}\hat{x}^2\, \mathrm{e}^{\gamma t}. \tag{4.60}$$

The explicit time-derivative $\frac{\partial}{\partial t}\hat{F}_2$ yields the missing contribution that turns $\hat{H}_{\mathrm{CK}}$ into the invariant $\hat{H}_{exp}$.

Consequently, also the action function $\hat{S}_{exp}$ and $\hat{S}_{\mathrm{CK}}$ are different which turns out to be important in the quantum mechanical case due to the relation between the action and the wave function (further details are given subsequently), and are related via

$$\hat{S}_{\mathrm{CK}} = \hat{S}_{exp} - m\frac{\gamma}{4}\hat{x}^2\mathrm{e}^{\gamma t}. \tag{4.61}$$

As we are working on the canonical level, quantization can be achieved in the usual way, e.g., in position space, by keeping the position operator $\hat{Q}_{\mathrm{op}} = \hat{Q}$ as a c-number and replacing the momentum operator with $\hat{P}_{\mathrm{op}} = \frac{\hbar}{\mathrm{i}}\frac{\partial}{\partial \hat{Q}}$. The resulting canonical SE then has the form

$$\mathrm{i}\hbar\frac{\partial}{\partial t}\hat{\Psi}_{exp}\left(\hat{Q}, t\right) = \left\{-\frac{\hbar^2}{2m}\frac{\partial^2}{\partial \hat{Q}^2} + \frac{m}{2}\left(\omega^2 - \frac{\gamma^2}{4}\right)\hat{Q}^2\right\}\hat{\Psi}_{exp}\left(\hat{Q}, t\right). \tag{4.62}$$

The analytic solution in form of a Gaussian WP can be written as

$$\hat{\Psi}_{exp}(\hat{Q}, t) = \hat{N}_{exp}(t)\exp\left\{\mathrm{i}\left[\hat{y}_{exp}(t)\tilde{Q}^2 + \frac{1}{\hbar}\langle\hat{P}\rangle_{exp}\tilde{Q} + \hat{K}_{exp}(t)\right]\right\} \tag{4.63}$$

with $\tilde{Q} = \hat{Q} - \langle\hat{Q}\rangle_{exp}$, $\langle\hat{P}\rangle_{exp} = m\langle\dot{\hat{Q}}\rangle_{exp}$ where $\langle\cdots\rangle_{exp}$ now indicates that the mean values are calculated with $\hat{\Psi}_{exp}(\hat{Q}, t)$ and $\hat{y}_{exp}(t)$ is again a complex function of time. The normalization function $\hat{N}_{exp}(t)$ and the phase factor $\hat{K}_{exp}(t)$ are purely TD and not relevant for the equations of motion determining the evolution of the maximum and width of the WP which are the two parameters completely describing the Gaussian function .

Inserting WP (4.63) into Eq. (4.62) provides the equation of motion for the maxium as

$$\frac{d^2}{dt^2}\langle\hat{Q}\rangle_{exp} + \left(\omega^2 - \frac{\gamma^2}{4}\right)\langle\hat{Q}\rangle_{exp} = 0, \tag{4.64}$$

or, bearing in mind the fact that the mean values follow the classical equations of motion on the physical as well as on the canonical level, this can also be expressed in terms of the physical position variable $\langle x\rangle = \eta(t)$ according to (4.53) like in Sect. 4.3.1 as

$$\ddot{\eta} + \gamma\,\dot{\eta} + \omega^2\eta = 0,$$

i.e., a damped HO.

The equation of motion for the WP width depends on the complex variable $\hat{y}_{exp}(t)$ and can be expressed in terms of the slightly modified variable $\hat{C}_{exp}(t) = \frac{2\hbar}{m}\hat{y}_{exp}(t)$ in the form of the (complex) Riccati equation

$$\dot{\hat{C}}_{exp} + \hat{C}^2_{exp} + \left(\omega^2 - \frac{\gamma^2}{4}\right) = 0, \tag{4.65}$$

where the imaginary part is connected to the mean square deviation of position (or position uncertainty) via $\hat{C}_{exp,\mathrm{I}} = \frac{\hbar}{2m\langle\tilde{Q}^2\rangle_{exp}}$ with $\langle\tilde{Q}^2\rangle_{exp} = \langle\hat{Q}^2\rangle_{exp} - \langle\hat{Q}\rangle^2_{exp}$. Equation (4.65) is identical to Eq. (2.4) in Sect. 2.1 only that C is replaced by $\hat{C}_{exp}$ and ω^2 by $\left(\omega^2 - \frac{\gamma^2}{4}\right) = \Omega^2$. Therefore, by introducing a new variable $\hat{\alpha}_{exp}(t)$ via $\hat{C}_{exp,\mathrm{I}} = \frac{1}{\hat{\alpha}^2_{exp}}$, in the same way as described in Sect. 2.1, the complex Riccati Eq. (4.65) can be transformed into the real NL Ermakov equation

$$\ddot{\hat{\alpha}}_{exp} + \left(\omega^2 - \frac{\gamma^2}{4}\right)\hat{\alpha}_{exp} = \frac{1}{\hat{\alpha}^3_{exp}}. \tag{4.66}$$

Again, following the procedure outlined in Sect. 2.1 and via elimination of $\left(\omega^2 - \frac{\gamma^2}{4}\right)$ between this equation and Eq. (4.64), one obtains a dynamical Ermakov invariant of the form

$$\hat{I}_{exp} = \frac{1}{2}\left[\left(\langle\dot{\hat{Q}}\rangle_{exp}\,\hat{\alpha}_{exp} - \langle\hat{Q}\rangle_{exp}\,\dot{\hat{\alpha}}_{exp}\right)^2 + \left(\frac{\langle\hat{Q}\rangle_{exp}}{\hat{\alpha}_{exp}}\right)^2\right] = \text{const.} \tag{4.67}$$

Expressing $\langle\hat{Q}\rangle_{exp}$ and $\langle\dot{\hat{Q}}\rangle_{exp}$ in terms of the physical variables $\eta = \langle x\rangle$ and $\dot{\eta} = \langle\dot{x}\rangle$, this can be rewritten as

$$\hat{I}_{exp} = \frac{1}{2}\,\mathrm{e}^{\gamma t}\left[\left(\dot{\eta}\,\hat{\alpha}_{exp} - \left(\dot{\hat{\alpha}}_{exp} - \frac{\gamma}{2}\hat{\alpha}_{exp}\right)\eta\right)^2 + \left(\frac{\eta}{\hat{\alpha}_{exp}}\right)^2\right] = \text{const.} \tag{4.68}$$

In the form (4.67), it becomes obvious that this invariant is not only independent of ω, i.e., also existing for $\omega = \omega(t)$, but also *independent of* γ (apart from the definition of $\hat{Q}$), i.e., also existing for $\gamma = \gamma(t)$! This form is identical to the one in the conservative case, only $\eta = \langle x\rangle$ is replaced by $\hat{Q} = \langle\hat{Q}\rangle_{exp}$ and α by $\hat{\alpha}_{exp}$. The meaning of the replacement $\eta \to \hat{Q}$ has been explained above; the meaning of the replacement $\alpha \rightarrow \hat{\alpha}_{exp}$ will become clear from the effective quantum mechanical description on the physical level, presented in Sect. 4.4 and its connection with the canonical level, as shown in Sect. 4.5.

Here, the discussion shall be restricted to connections within the canonical level. The connection between the classical CK approach and the one in expanding coordinates via the generating function $\hat{F}_2(\hat{x}, \hat{P}, t)$ has already been given in Eqs. (4.59) and (4.60), and also the corresponding connection between the action functions $\hat{S}_{\text{CK}}$ and $\hat{S}_{exp}$ has been given in Eq. (4.61). This, however, can be used to define the unitary transformation, corresponding to the canonical transformation of the classical variables that transforms the wave functions of the CK and the expanding approaches into each other. For this purpose, Schrödinger's original definition [52] of the wave function Ψ in terms of the action function S,

$$S = \frac{\hbar}{\mathrm{i}} \ln \Psi, \tag{4.69}$$

can be used (further details will be given in Sect. 4.5).

According to this definition, relation (4.61) translates into

$$\hat{\Psi}_{\text{CK}} = \exp\left\{-\frac{im}{2\hbar}\frac{\gamma}{2}\hat{x}^2 \mathrm{e}^{\gamma t}\right\}\hat{\Psi}_{exp}, \tag{4.70}$$

thus defining the unitary transformation between the two approaches.

One might ask, what happens to the problems with the commutator and the uncertainty product in the approach in expanding coordinates? Again, like in the CK-approach, the canonical variables pose no problem, $[\hat{P}_{\text{op}}, \hat{Q}_{\text{op}}]_- = \left[\frac{\hbar}{\mathrm{i}}\frac{\partial}{\partial \hat{Q}}, \hat{Q}\right]_- = \frac{\hbar}{\mathrm{i}}$. But what happens if this is expressed in terms of physical position and momentum? This means, particularly $\frac{\hbar}{\mathrm{i}}\frac{\partial}{\partial \hat{Q}}$ must be expressed in terms of $\frac{\hbar}{\mathrm{i}}\frac{\partial}{\partial x} = p_{\text{op}}$. Taking into account that, in position space, derivatives $\frac{\partial}{\partial p}$ do not contribute to the commutator, the commutator of physical position and momentum can be written as

$$[p_{\text{op}}, x_{\text{op}}]_- = \mathrm{e}^{\frac{\gamma}{2}t}\left[\frac{\hbar}{\mathrm{i}}\frac{\partial}{\partial \hat{Q}}, \hat{Q}\right]_- \mathrm{e}^{-\frac{\gamma}{2}t} = \left[\frac{\hbar}{\mathrm{i}}\frac{\partial}{\partial \hat{Q}}, \hat{Q}\right]_- = \frac{\hbar}{\mathrm{i}}, \tag{4.71}$$

thus no longer violating the commutator relation for the physical quantities. A similar situation is found for the uncertainty principle. But warning! As it has just been shown, the CK approach and the one in expanding coordinates are connected, classically, via a canonical and, quantum mechanically, via a unitary transformation. Assuming that on the canonical level, like on the physical level, these transformations do not change the physics of the system would solely create another puzzle were it to be considered a solution of the problem associated with the CK approach. The real solution will be discussed in Sect. 4.5; the absence of the CK-problem in the expanding approach might be more by chance.

Finally, it should be mentioned that, on the canonical level, there are further approaches like the one by Lemos [48] with the variables

$$\hat{R} = x\,\mathrm{e}^{\frac{\gamma}{2}t}\,, \quad \hat{\Pi} = p\,\mathrm{e}^{\frac{\gamma}{2}t}, \tag{4.72}$$

the Lagrangian

$$\hat{L}_{\mathrm{L}} = \frac{m}{2}\dot{\hat{R}}^2 - \frac{m}{2}\omega^2\hat{R}^2 - \frac{m}{2}\gamma\dot{\hat{R}}\hat{R} \tag{4.73}$$

and the Hamiltonian

$$\hat{H}_{\mathrm{L}}(\hat{R}, \hat{\Pi}, t) = \frac{1}{2m}\hat{\Pi}^2 + \frac{m}{2}\omega^2\hat{R}^2 + \frac{\gamma}{2}\hat{\Pi}\hat{R} \tag{4.74}$$

which, expressed in terms of the physical variables, is identical to the constant given in Eq. (4.48). As this approach (and others as well) is connected with the ones of CK and the expanding coordinates via canonical transformations, it does not provide any new aspects concerning the physics of the system and so will not be discussed further in the forthcoming (sub-)section. (For more details, see [53].)

4.4 Effective Models Using Nonlinear Modifications of the Schrödinger Equation

The effective models discussed in the last subsection were based on the classical canonical Lagrange and Hamilton formalism. Particular modifications of the conventional form were considered that provided a linear velocity or momentum dependent friction force, $-m\gamma\dot{x}$ or $-\gamma p$, respectively. This cannot be achieved by simply adding to the Lagrangian or Hamiltonian function a kind of "friction potential", $W(x, \dot{x}) = m\gamma\dot{x}x$ or $W(x, p) = \gamma p x$, whose negative derivative $-\frac{\partial}{\partial x}W$ would yield the desired friction force. In the Lagrangian formalism, adding such a term provides, via the Euler–Lagrange equation, an equation of motion for $x(t)$ without any friction force. In the Hamiltonian formalism, one of the Hamiltonian equations of motion, $\frac{\partial H}{\partial p} = \frac{1}{m}p + \gamma x \neq \dot{x}$, would no longer be correct. Therefore, the treatments discussed before were necessary.

However, there are attempts to break the time-reversal symmetry and/or to include such friction forces, at least on the *quantum mechanical* level, by adding such a "potential". Different criteria are used to determine the specific form of this term of which three distinct ones are discussed in the following. These selected approaches possess a mutual characteristic in that they all lead to a NL modification of the SE.

4.4.1 Models Based on Ehrenfest's Theorem and the Langevin Equation

The basic requirement of these approaches is that, according to Ehrenfest's theorem, the classical equation of motion, including the damping force, should be valid on an average,

$$\frac{d}{dt}\langle p\rangle + \gamma\langle p\rangle + \left\langle \frac{\partial}{\partial x}V\right\rangle = 0, \tag{4.75}$$

i.e., a kind of averaged[5] Langevin Eq. (4.3) where the contribution from the stochastic force $\mathcal{F}$ vanishes.

The correspondingly modified Hamiltonian operator would then have the form[6]

$$H_{\mathrm{NL}} = -\frac{\hbar^2}{2m}\frac{\partial^2}{\partial x^2} + V(x) + W = H_{\mathrm{L}} + W. \tag{4.76}$$

The dissipative term[7] W should be compatible with the equation of motion according to

$$\frac{d}{dt}\langle p\rangle = \left\langle \frac{\mathrm{i}}{\hbar}[H_{\mathrm{NL}}, p]_-\right\rangle = \left\langle -\frac{\mathrm{i}}{\hbar}\frac{\hbar}{\mathrm{i}}\frac{\partial}{\partial x}(V+W)\right\rangle = -\left\langle \frac{\partial}{\partial x}V\right\rangle - \left\langle \frac{\partial}{\partial x}W\right\rangle. \tag{4.77}$$

Comparison with Eq. (4.75) gives as criterion for the determination of W

$$\left\langle \frac{\partial}{\partial x}W\right\rangle = \gamma\langle p\rangle. \tag{4.78}$$

This definition of W is obviously not unique as in (4.78):

1. only the derivative of W occurs;
2. actually, not even this, but only the mean value of this quantity occurs;
3. this is compared only with the mean value of the momentum (or velocity) operator.

Therefore, various (different) "potentials" W exist in the literature that fulfil condition (4.78) and are thus in agreement with equation of motion (4.75). The best-studied and most frequently-used ones are discussed subsequently.

Before this, a problem arising from point (3), i.e. the comparison with $\langle p\rangle$ or $\langle v\rangle$, is considered.

There are obviously two different definitions of velocity occurring in quantum mechanics. One is the momentum operator, divided by mass m, i.e. $v_{op} = \frac{\hbar}{\mathrm{i}m}\frac{\partial}{\partial x}$. The

---

[5] In the context of the Langevin equation, classical mean values were considered, where now $\langle\cdots\rangle = \int dx\Psi^* \ldots \Psi$ denotes quantum mechanical ones.

[6] The subscript "NL" already indicates that this will lead to nonlinear modifications of the TDSE.

[7] This W should not be confused with the Wigner function $W(x, p, t)$ discussed in Sects. 2.8 and 2.9.

other is related to the velocity field in the continuity Eq. (2.43), given as [see Eq. (2.44)]

$$v_- = \frac{\hbar}{2mi}\left(\frac{\frac{\partial}{\partial x}\Psi}{\Psi} - \frac{\frac{\partial}{\partial x}\Psi^*}{\Psi^*}\right) = \frac{\hbar}{2mi}\frac{\partial}{\partial x}\ln\frac{\Psi}{\Psi^*}.$$

The difference between these two is connected with the "imaginary velocity"

$$v_+ = \frac{\hbar}{2mi}\left(\frac{\frac{\partial}{\partial x}\Psi}{\Psi} + \frac{\frac{\partial}{\partial x}\Psi^*}{\Psi^*}\right) = \frac{\hbar}{2mi}\frac{\partial}{\partial x}\ln\varrho = \frac{\hbar}{2mi}\frac{\frac{\partial}{\partial x}\varrho}{\varrho} \tag{4.79}$$

with $\varrho = \Psi^*\Psi$.

As $(v_- + v_+) = \frac{\hbar}{mi}\frac{\frac{\partial}{\partial x}\Psi}{\Psi}$, it follows that

$$(v_- + v_+)\Psi = \frac{\hbar}{mi}\frac{\partial}{\partial x}\Psi = v_{op}\Psi. \tag{4.80}$$

The difference between the two definitions of velocity in quantum mechanics is not immediately obvious as the mean value of v_+ according to

$$\langle v_+\rangle = \int_{-\infty}^{+\infty} dx\ \Psi^* v_+ \Psi = \frac{\hbar}{2mi}\int_{-\infty}^{+\infty} dx\left(\frac{\partial}{\partial x}\varrho\right) = \varrho(x)|_{-\infty}^{+\infty} = 0 \tag{4.81}$$

always vanishes, similar to the contribution of the fluctuating force to the averaged velocity in the Langevin equation.

However, this does not imply that v_+ has no influence on the physics of the system. There is a similarity with the contribution of the fluctuating force to the energy of the Langevin system via its square, leading to the energy that remains after the averaged motion has come to an end, i.e., the energy $\frac{1}{2}k_B T$ of the final thermal equilibrium. In the case of v_+, the so-called "quantum potential" mentioned in Sect. 4.2, $V_{qu} = -\frac{\hbar}{2m}\frac{\frac{\partial^2}{\partial x^2}\sqrt{\varrho}}{\sqrt{\varrho}}$, can be rewritten as

$$V_{qu} = \frac{m}{2}v_+^2 - i\frac{\hbar}{2}\frac{\partial}{\partial x}v_+ \tag{4.82}$$

with non-vanishing mean value[8] for the Gaussian WP, e.g., it is

$$\langle V_{qu}\rangle = \frac{\hbar^2}{8m\langle\tilde{x}^2\rangle} \neq 0. \tag{4.83}$$

[8]However, the mean value of its derivative $\frac{\partial}{\partial x}V_{qu} = -\frac{\hbar^2}{4m\langle\tilde{x}^2\rangle}(x - \langle x\rangle)$ vanishes, $\langle\frac{\partial}{\partial x}V_{qu}\rangle = 0$, i.e., it does not contribute to the averaged equation of motion.

From Eq. (4.81) follows that considering just the mean value of velocity $\langle v \rangle$, it cannot be distinguished whether v corresponds only to v_- or to $v_+ + v_-$ (which is equivalent to the definition of the velocity or momentum operator).

In the most frequently-cited, and to my knowledge first approach in this direction, Kostin [54] had chosen the velocity v_- as given in (2.44) for his definition of the "friction potential". From

$$\frac{\partial}{\partial x} W_{\mathrm{K}} = \gamma \frac{\hbar}{2\mathrm{i}} \ln \frac{\Psi}{\Psi^*} = \gamma m v_- \tag{4.84}$$

follows by integration and choosing the integration constant so that $\langle W_{\mathrm{K}} \rangle = 0$ is valid, Kostin's damping term in the form

$$W_{\mathrm{K}} = \gamma \frac{\hbar}{2\mathrm{i}} \left(\ln \frac{\Psi}{\Psi^*} - \left\langle \ln \frac{\Psi}{\Psi^*} \right\rangle \right). \tag{4.85}$$

Due to $\langle W_{\mathrm{K}} \rangle = 0$, the mean value of energy is still given by

$$\langle H \rangle \; = \; \langle T \rangle + \langle V \rangle, \tag{4.86}$$

but now with mean values calculated with the solutions $\Psi_{\mathrm{K}}(x, t)$ of the SE with the NL Hamiltonian (4.76) where W has the form (4.85) of W_{K}.

This NL Hamiltonian has subsequently been rederived by several authors [55–60] with different methods like stochastic quantization etc.. For potentials at most quadratic in x, analytic Gaussian WP solutions can also be obtained for this NLSE. Although this approach (by definition) provides the correct averaged equation of motion for the position, there are still several points of criticism:

1. the solutions of the damped HO contain the unshifted frequency ω_0 instead of the reduced frequency $\Omega = \left(\omega_0^2 - \frac{\gamma^2}{4}\right)^{\frac{1}{2}}$, known from the classical case;

2. all solutions of the undamped HO, which are real functions, i.e. $\Psi = \Psi^*$ and hence $\ln \frac{\Psi}{\Psi^*} = \ln 1 = 0$, are also solutions of the problem including friction! This is difficult to interprete;

3. as W_{K} is *real*, it can be considered an addition to the potential $V(x)$, i.e. it does not occur in the continuity Eq. (2.43) for $\varrho(x, t)$. Therefore, the equation of motion for $\varrho(x, t)$ still remains reversible although the averaged classical equation of motion describes an irreversible process, which appears to be inconsistent!

There have been several attempts [61] to overcome these shortcomings but none is convincing. So, one can presume that this approach is a step in the right direction, but does not yet give a complete description of the system.

Cho [60] made an attempt to eliminate the above-mentioned inconsistencies by adding a kind of contribution from a "diffusion velocity" according to

$$W_{\mathrm{Cho}} = \gamma \frac{\hbar}{2} \left\{ \frac{1}{\mathrm{i}} \ln \frac{\Psi}{\Psi^*} + \ln(\Psi \Psi^*) \right\} - \gamma f(t) \tag{4.87}$$

where "$f(t)$ is taken so that at stationary states the friction operator does not depend on time" [62]. The additional term[9] $\gamma\frac{\hbar}{2}\ln(\Psi\Psi^*)$ displays similarity with a term whose derivative would provide v_+, but is real, whereas v_+ is purely imaginary.

Due to the additional term in Cho's approach, this operator certainly does no longer permit the undamped solutions. However, as the additional term is real, the problem with the reversible continuity equation for $\varrho(x,t)$ still remains. Furthermore, in this case also the frequency of the damped oscillator, namely $\omega = \left(\omega_0^2 + \frac{\gamma^2}{4}\right)^{\frac{1}{2}} - \frac{\gamma}{2}$, is different from the classical damped frequency $\Omega = \left(\omega_0^2 - \frac{\gamma^2}{4}\right)^{\frac{1}{2}}$.

Around the same period of time, approaches in this direction were also made in order to find an effective description for deep inelastic scattering in nuclear physics. The underlying idea was to construct the "friction potential" W fulfilling (4.78) as a product of the operators of position and momentum and their mean values, taking into account that these operators do not commute. (Remember that in the classical case such a term did not work!)

Approaches in this direction were made by Süssmann [65], Albrecht [66] and Hasse [67]. The general "potential" of that kind can be written as

$$W_G = \gamma\langle p\rangle(x - \langle x\rangle) + \frac{\gamma}{2}\, C\,[(x - \langle x\rangle), (p - \langle p\rangle)]_+ \tag{4.88}$$

with C being a real constant (or purely TD function). It can be shown straightforwardly that W_G fulfils condition (4.78) for *any* value of C.

The choice $C = 1$ has been considered by Süssmann [65], leading to

$$W_{Sü} = \frac{\gamma}{2}\,[(x - \langle x\rangle), p]_+ \,, \tag{4.89}$$

the choice $C = 0$ by Albrecht leading to

$$W_{Al} = \gamma\langle p\rangle(x - \langle x\rangle). \tag{4.90}$$

Still, these two approaches, in certain aspects, display unphysical behaviour, e.g. they provide the wrong frequency for the damped HO:

$$\text{Süssmann}: \quad \Omega_{Sü} = \left(\omega_0^2 - \gamma^2\right)^{\frac{1}{2}} \tag{4.91}$$

$$\text{Albrecht}: \quad \Omega_{Al} = \omega_0 \tag{4.92}$$

$$\text{General}: \quad \Omega_G = \left(\omega_0^2 - C^2\gamma^2\right)^{\frac{1}{2}}. \tag{4.93}$$

[9]Under the different aspect, if linear quantum mechanics should be replaced by a NL modification when approaching mesoscopic dimensions, Bialynicki-Birula and Mycielski [63] investigated a NLSE with logarithmic nonlinearity of the form $-b\ln(\Psi\Psi^*)$ with real $b > 0$. However, a connection between this nonlinearity and a physically-justified dissipative term in the SE is not given (see, e.g., Enz [64]).

The W_G-term would provide the correct frequency $\Omega = \left(\omega_0^2 - \frac{\gamma^2}{4}\right)^{\frac{1}{2}}$ for $C = \pm\frac{1}{2}$! These cases have been considered by Hasse [67]. The choice $C = -\frac{1}{2}$ would lead to unphysical results (like WPs with positive exponent, i.e., diverging for increasing x). For $C = +\frac{1}{2}$, one obtains well-behaved results and Hasse's NL friction term takes the form

$$W_{\text{Has}} = \frac{\gamma}{2} \left[(x - \langle x \rangle), (p + \langle p \rangle) \right]_+ . \tag{4.94}$$

Comparison with the approaches by Süssmann (4.89) and Albrecht (4.90) shows that

$$W_{\text{Has}} = \frac{1}{2} \left(W_{\text{Sü}} + W_{\text{Al}} \right) . \tag{4.95}$$

Due to the "semi-empirical" fitting of the constant C, the solutions of Hasse's NLSE (also Gaussians like above) for the damped HO obviously have the correct frequency Ω. Also the undamped solutions are no longer admitted. Consequences for the continuity equation for the density $\varrho(x,t)$ are not immediately obvious but will become clearer when this approach is compared with the complex logarithmic one in Sect. 4.4.3. This will also clarify, in Hasse's approach, why the mean value of W_{Has} does not disappear,[10]

$$\langle W_{\text{Has}} \rangle \neq 0, \tag{4.96}$$

i.e., $\langle H \rangle \neq \langle T \rangle + \langle V \rangle$.

4.4.2 *Models Based on Non-unitary Time-Evolution*

A different approach by Gisin [68, 69] is initially not focussing on the dissipative Langevin equation. The major assumption here is that the quantum system is always in a *pure state* no matter if it exchanges energy with its environment or not. Therefore, the system can always be described by a normalizable complex state vector $\Psi(t)$.

Following the procedure for obtaining the generalized master equation for the density operator, Gisin arrived at a corresponding evolution equation for $\Psi(t)$,

$$\mathrm{i}\hbar \frac{\partial}{\partial t} \Psi(t) = \left\{ H_{\text{L}} + \frac{\hbar}{\mathrm{i}} B \right\} \Psi(t). \tag{4.97}$$

Due to the *imaginary* term, the time-reversal symmetry is broken but this equation does not conserve the norm of the state vector!

[10] This non-vanishing energy contribution is actually given by $\langle W_{\text{Has}} \rangle = \frac{\gamma}{4} \langle [\tilde{x}, \tilde{p}]_+ \rangle = \gamma \frac{\hbar}{4} \left(\dot{\alpha}\alpha - \frac{\gamma}{2} \alpha^2 \right)$ for a Gaussian WP and is different from zero even for $\dot{\alpha} = 0$.

However, normalization can be achieved if the mean value of B is subtracted, i.e., $B \to B - \langle B \rangle$, leading to the modified SE

$$i\hbar \frac{\partial}{\partial t} \Psi(t) = \left\{ H_{\rm L} + \frac{\hbar}{i} \left(B - \langle B \rangle \right) \right\} \Psi(t) \tag{4.98}$$

which is, due to $\langle B \rangle = \int_{-\infty}^{+\infty} dx\ \Psi^* B \Psi$, a NLSE.

The specific choice of B does not follow from this ansatz (it is only assumed that B is Hermitian). Gisin's original choice, $B = kH$, with real constant k, leads to a wrong energy dissipation, i.e. a decrease of energy not proportional to the kinetic energy, $\dot{E} \neq 2\gamma T$, in contrast to the classical case.

As mentioned before, an imaginary term in the Hamiltonian influences also the continuity equation for $\varrho(x,t)$ and can introduce irreversibility. However it has *no* influence on the (real) Ehrenfest equation of motion and therefore cannot induce dissipation of energy. In Gisin's case with $B = kH$, the application of H on Ψ produces real and imaginary terms proportion to Ψ. Due to the factor $\frac{\hbar}{i}$ in front of B, the imaginary terms turn into real ones and can contribute to dissipation. Therefore this approach with $B = kH$ is not only irreversible but also dissipative. However, since the choice of B seems to be quite arbitrary, this approach essentially indicates how to circumvent the problem with a time-reversible continuity equation, i.e., including an imaginary term in the modified Hamiltonian.

In a similar approach, Beretta [70, 71] tried to describe non-equilibrium systems (without dissipation, in agreement with the afore-mentioned) where the imaginary term is supposed to be related to the entropy of the system that should increase during time-evolution. The particular form of his NLSE is

$$i\hbar \frac{\partial}{\partial t} \Psi = \left\{ H_{\rm L} + \frac{\gamma}{2} \frac{\hbar}{i} \left(\ln \varrho - \langle \ln \varrho \rangle \right) \right\} \Psi, \tag{4.99}$$

a form that will appear again in another approach in Sect. 4.4.3. However, Beretta's approach was mainly discussed in the context of density operators. Applications can be found in connection with "quantum thermodynamics" [72].

The same logarithmic term has also been discussed recently by Nassar [73] in the context of WP dynamics under continuous measurement. A clearer picture of the physical meaning of this term and its origin is presented in the following.

4.4.3 *Models Based on a Smoluchowski Equation for the Probability Density*

Kostin's approach involving only the phase of the wave function via $\frac{\hbar}{2i} \ln \frac{\Psi}{\Psi^*} = S$ for $\Psi = \varrho^{\frac{1}{2}} e^{\frac{i}{\hbar} S}$ provided the correct (averaged) equation of motion including dissipative friction but no irreversibility for the equation of motion for $\varrho(x,t)$. Beretta's purely imaginary term proportional to $\frac{\hbar}{2i}\ln(\Psi\Psi^*)$ only breaks the time-reversal symmetry of the reversible continuity equation for $\varrho(x,t)$, but has no dissipative effect in the form of a friction force. Due to terms like $p_{\text{op}}\Psi$ or $H_{\text{op}}\Psi$ with complex Ψ, approaches such as those by Gisin or Hasse have real and imaginary contributions in their NL terms but it is not quite clear how the terms and their influence on the dynamics of the system can be distinguished.

The models discussed in Sect. 4.4.1 were based on a kind of Langevin-type description of the open (quantum) system. As mentioned before, an equivalent description can also be given in terms of a Fokker–Planck or, in position space, Smoluchowski equation [see Eq. (4.7)]. Further, we have seen in Sect. 3.2 that (at least in the TISE) phase and amplitude of the wave function are not independent of each other. This shall be used in the following to start on the level of the probability density $\varrho(x,t) = \Psi^*\Psi$ to introduce irreversibility by adding a diffusion term as in the classical Eq. (4.7). Then, a method will be applied that was introduced by Madelung [74] and Mrowka [75] to "rederive" the SE without making use of any Lagrangian or Hamiltonian form of classical mechanics. Using only (an averaged) Newton's equation of motion, it separates the (real) equation for $\varrho(x,t)$ into two complex conjugate equations for $\Psi(x,t)$ and $\Psi^*(x,t)$. These equations for the complex functions should then also contain a contribution corresponding to the (separated) diffusion term.

In a first step, the original method of Madelung and Mrowka for conservative systems is outlined in the following. Afterwards, the necessary modifications due to the additional diffusion term are discussed.[11]

Around the middle of the last century Madelung [74] and Mrowka [75] established a formalism to "rederive" the conventional SE more or less for didactical reasons. The axiomatic basis for this method is given by three principles taken from experimental experience:

(1) uncertainty principle or complementarity, respectively;
(2) the occurrence of interference phenomena in experiments with material systems, and
(3) the correspondence principle, specified in the form of Ehrenfest's theorem.

The mathematical form of a theory taking into account this empirical knowledge follows from the structure of these principles. (In the following, the method is outlined only to an extent necessary for showing how irreversibility can be introduced. For further details see Refs. [74–77].)

Owing to the uncertainty principle, exact initial conditions in the sense of classical particle mechanics cannot be given. Therefore, it is only possible to develop a theory where mean values $< Q >= \int d\mathbf{r}\; Q\; \varrho(\mathbf{r},t)$ of quantities Q are determined with the aid of a distribution function $\varrho(\mathbf{r},t)$.

---

[11] The general discussions are given for the three-dimensional case; the more detailed ones concerning the resulting NLSE are then again restricted to the case in one-dimension though a generalization to higher dimensions is straightforward in most cases.

In order to guarantee the conservation of ϱ (in a global sense), $\int d\mathbf{r}\ \varrho(\mathbf{r}, t) =$ const., Madelung and Mrowka assumed the function $\varrho(\mathbf{r}, t)$ to fulfil the continuity equation

$$\frac{\partial}{\partial t}\varrho + \nabla \mathbf{j} = \frac{\partial}{\partial t}\varrho + \nabla\left(\varrho \mathbf{v}_-\right) = 0. \tag{4.100}$$

Bearing in mind the second axiom and analogous to optics (where intensity is a quadratic function of the amplitudes), the bilinear ansatz

$$\varrho = a \cdot b \geq 0 \tag{4.101}$$

with the complex field functions $a(\mathbf{r}, t)$ and $b(\mathbf{r}, t)$ is used. With these functions also the current density $\mathbf{j} = \varrho \mathbf{v}_-$ can be defined as a bilinear form,

$$\mathbf{j} = \varrho \mathbf{v}_- = C(b\nabla a - a\nabla b), \tag{4.102}$$

where

$$\mathbf{v}_- = C\left(\frac{\nabla a}{a} - \frac{\nabla b}{b}\right) = C\ \nabla \ln \frac{a}{b} \tag{4.103}$$

with $C =$ constant.

For the evaluation of mean values, the distribution function $\varrho(\mathbf{r}, t)$ must be known, i.e., the continuity equation (4.100) must be solved. However, the time-derivative of ϱ depends on $\mathbf{j}$. So, according to Eqs. (4.102) and (4.103), not only the absolute value must be known but also the phase of the field function, i.e., the complex functions $a(\mathbf{r}, t)$ and $b(\mathbf{r}, t)$ have to be determined.

For this purpose, the definitions of ϱ and $\mathbf{j}$ are inserted into continuity Eq. (4.100). Introducing a function $f(\mathbf{r}, t)$ that is independent of a and b, it is possible to separate the continuity equation into two equations containing only a or b, respectively,

$$\frac{\partial}{\partial t}a + C\Delta a + fa = 0, \tag{4.104}$$

$$\frac{\partial}{\partial t}b - C\Delta b - fb = 0. \tag{4.105}$$

The physical meaning of the separation function $f(\mathbf{r}, t)$ becomes obvious when the third axiom is applied regarding the mean value of Newton's equation of motion with force F,

$$\int d\mathbf{r}\ \varrho\ \mathbf{F} = m\frac{d}{dt}\int d\mathbf{r}\ \varrho\ \mathbf{v}_- = m\int d\mathbf{r}\ 2C\left(\frac{\partial}{\partial t}b\ \nabla a - \frac{\partial}{\partial t}a\ \nabla b\right), \tag{4.106}$$

and $\frac{\partial}{\partial t}a$ and $\frac{\partial}{\partial t}b$ are replaced by the quantities defined in Eqs. (4.104) and (4.105). After some straightforward calculations one finally obtains

$$\int d\mathbf{r}\, \varrho\, \mathbf{F} = \int d\mathbf{r}\, \varrho\, \{-\nabla(2mCf)\} = \int d\mathbf{r}\, \varrho\, \{-\nabla V\}, \tag{4.107}$$

which is the mean value of a conservative force derived from a potential $V = 2mCf$. So, with the help of the correspondence principle one finds that the separation function $f(\mathbf{r}, t)$ is proportional to the potential V.

Replacing a with Ψ, where the choice $b = a^* = \Psi^*$ guarantees positive real ϱ, Eq. (4.104) multiplied by $\mathrm{i}\hbar$ attains the well-known form of the TDSE

$$\mathrm{i}\hbar \frac{\partial}{\partial t} \Psi = -\frac{\hbar^2}{2m} \Delta\Psi + V\Psi, \tag{4.108}$$

describing the quantum mechanical dynamics of a system exerted to the same force as given in the corresponding Newton-type equation of motion.

In this case, the constant C was chosen to be $C = \frac{\hbar}{2m\mathrm{i}}$ based on spectroscopic results.[12]

In describing irreversible dynamics, the conservation law expressed by the continuity equation must be replaced by another that is no longer symmetric in time. As mentioned above, a diffusion process is a well-suited candidate for this purpose. Therefore, adding a diffusion current density $\mathbf{j}_{\mathrm{D}} = -D\nabla\varrho = -D(b\nabla a + a\nabla b)$ (with diffusion coefficient D) to the convection current density (4.102), the time-reversible continuity Eq. (4.100) turns into the irreversible Smoluchowski equation

$$\frac{\partial}{\partial t}\varrho + \nabla(\mathbf{j} + \mathbf{j}_{\mathrm{D}}) = \frac{\partial}{\partial t}\varrho + \nabla\,(\varrho\, \mathbf{v}_-) - D\Delta\varrho = 0. \tag{4.109}$$

Following the scheme outlined above for conservative systems, this equation must be separated into two equations for the two conjugate complex functions a and b (or Ψ and Ψ^*, respectively). However, this is not generally possible due to the structure of the additional diffusion term

$$-D\Delta\varrho = -D\{a\,\Delta b + b\,\Delta a + 2(\nabla a)(\nabla b)\}, \tag{4.110}$$

where the product term $(\nabla a)(\nabla b)$ destroys the separability in a general sense. However one may nevertheless look for special cases where separation is still possible.

Separation of the Smoluchowski Eq. (4.109) can only be achieved if the additional condition

$$-D\frac{\Delta\varrho}{\varrho} = F_1(a) + F_2(b) \tag{4.111}$$

is fulfilled.

---

[12]This choice is not the only possible one as the constant C is not determined unequivocally by the three basic assumptions but must be taken from experimental experience. Therefore, from the viewpoint of classical field theory, a choice $C = \frac{1}{2\mu\mathrm{i}}$, i.e. a constant without $\hbar$ and m (m = mass of *one particle*) is also possible (see also Refs. [76, 77]).

An ansatz satisfying this requirement is given by

$$-D\frac{\Delta\varrho}{\varrho} = \gamma\,(\ln\varrho + Z) \qquad (4.112)$$

with Z = function independent of a and b and γ = constant (or at least not position dependent; for further details see Refs. [76, 77]). It can be shown that the choice $Z = 0$ would not allow the normalization of the function ϱ (similar to Gisin's Eq. (4.97)), whereas the choice $Z = -\langle\ln\varrho\rangle$ is in agreement with the normalizabililty condition (as in Gisin's case). Therefore, the separability condition, at least as a mathematical possibility, can be written as

$$-D\frac{\Delta\varrho}{\varrho} = \gamma\,(\ln\varrho - \langle\ln\varrho\rangle)\,. \qquad (4.113)$$

After separation, an equation to determine $a(\mathbf{r}, t)$ is again obtained which now has the form

$$\frac{\partial}{\partial t}a + C\Delta a + \gamma(\ln a - \langle\ln a\rangle)a + fa = 0, \qquad (4.114)$$

or, written in terms of $\Psi(x,t) = \Psi_{\mathrm{NL}}(x,t)$, as the NLSE

$$\mathrm{i}\hbar\frac{\partial}{\partial t}\Psi_{\mathrm{NL}} = \left\{-\frac{\hbar^2}{2m}\,\Delta + V + \gamma\frac{\hbar}{\mathrm{i}}\,(\ln\Psi_{\mathrm{NL}} - \langle\ln\Psi_{\mathrm{NL}}\rangle)\right\}\Psi_{\mathrm{NL}} \qquad (4.115)$$

with a complex logarithmic logarithmic nonlinearity that can be written as

$$\tilde{W}_{\mathrm{SCH}} = \gamma\frac{\hbar}{\mathrm{i}}\,(\ln\Psi_{\mathrm{NL}} - \langle\ln\Psi_{\mathrm{NL}}\rangle)\,. \qquad (4.116)$$

So far, this additional term is only a consequence of the additional diffusion term in Eq. (4.109) and the *mathematical* separability condition (4.113).

To elucidate the physical meaning of this mathematical condition and the resulting nonlinearity, the third axiom, Ehrenfests theorem, must again be applied. Now, however, (4.114) instead of (4.104) is substituted for $\frac{\partial}{\partial t}a$ in the mean value of Newton's equation. With the help of the definition of $\mathbf{v}_-$ [see (4.103)], this leads to

$$\int d\mathbf{r}\,\varrho\,\mathbf{F} = \int d\mathbf{r}\,\varrho\,\{-\nabla V - m\gamma\mathbf{v}\}$$

$$\text{or}\quad \langle\mathbf{F}\rangle = m\frac{d}{dt}\langle\mathbf{v}\rangle = \langle-\nabla V\rangle - m\gamma\langle\mathbf{v}\rangle \qquad (4.117)$$

(where $\mathbf{v} = \mathbf{v}_- + \mathbf{v}_+$ with $\langle\mathbf{v}_+\rangle = 0$; see above).

So, in addition to Eq. (4.107), the mean value of the linear velocity dependent friction force, $-m\gamma\mathbf{v}$ with friction coefficient γ, occurs and is uniquely connected to the diffusion term of the Smoluchowski equation via the additional separability condition and the resulting logarithmic nonlinearity.

The additional NL term $\tilde{W}_{\rm SCH}$ can be written as real and imaginary contributions in the form

$$\tilde{W}_{\rm SCH} = \tilde{W}_{\rm R} + \tilde{W}_{\rm I} = \frac{\gamma}{2}\frac{\hbar}{\rm i}\left(\ln\frac{\Psi_{\rm NL}}{\Psi^*_{\rm NL}} - \left\langle \ln\frac{\Psi_{\rm NL}}{\Psi^*_{\rm NL}}\right\rangle\right)$$
$$+\frac{\gamma}{2}\frac{\hbar}{\rm i}\left(\ln\varrho_{\rm NL} - \langle\ln\varrho_{\rm NL}\rangle\right), \qquad (4.118)$$

where the real part only depends on the phase of the wave function and provides the friction force in the averaged equation of motion. Comparison with the aforementioned approaches shows that this real part is just identical to Kostin's term (4.85).

The imaginary part, however, resolves all the problems mentioned in connection with Kostin's approach. By definition it breaks the time-symmetry of the equation for $\varrho_{\rm NL}(\mathbf{r}, t)$ via the corresponding diffusion term. In addition, it changes the frequency of the HO into the correct one known from the classical case, $\Omega = \left(\omega_0^2 - \frac{\gamma^2}{4}\right)^{\frac{1}{2}}$, without any fitting. Also the undamped solutions of the HO problem no longer solve the corresponding NLSE. Formally, it is equivalent to Beretta's term (for pure states) and the one discussed by Nassar in [73].

Furthermore, this imaginary term introduces a non-unitary time-evolution and turns the Hamiltonian into a non-Hermitian one while still guaranteeing normalizable wave functions and real energy mean values as its own mean value vanishes.

Also from the real part $\tilde{W}_{\rm R}$ no additional contribution to the energy mean value occurs, so this is still represented by the mean values of the operators of kinetic and potential energies. Yet this real part is not arbitrary but uniquely determined by the separation condition and provides the correct dissipative friction force in the equation of motion for the mean values. Besides, the ratio of energy dissipation (for the classical contribution) is in agreement with the classical counterpart and arises because the mean values are calculated with $\Psi_{\rm NL}$ instead of $\Psi_{\rm L}$ (the solution of the linear SE).

The real part alone would provide dissipation but retain a unitary time-evolution of the wave function, whereas the imaginary part, on its own, would provide irreversibility via a non-unitary time-evolution but no dissipation. Consequently, only the combination of real and imaginary parts provides all the desired properties of the quantum system under consideration. The reason for this is the coupling of phase and amplitude of the wave function (as $\tilde{W}_{\rm R}$ depends on the phase and $\tilde{W}_{\rm I}$ on the amplitude), as mentioned in Sect. 3.2.

With respect to formal similarity, an interesting interpretation of $\tilde{W}_{\rm I}$ can be found if, according to Grössing et al. [78], one identifies the Einstein diffusion coefficient with the quantum mechanical one (providing the SE is considered a diffusion equation with imaginary diffusion coefficient), i.e., $D = \frac{k_{\rm B}T}{m\gamma} = \frac{\hbar}{2m}$. Then $\tilde{W}_{\rm I}$ turns into

$$\tilde{W}_{\mathrm{I}} = -\mathrm{i}\mathsf{k}_{\mathrm{B}}\mathsf{T}\left(\ln \varrho_{\mathrm{NL}} - \langle \ln \varrho_{\mathrm{NL}} \rangle\right), \tag{4.119}$$

where $-\mathsf{k}_{\mathrm{B}}\langle \ln \varrho_{\mathrm{NL}} \rangle = -\mathsf{k}_{\mathrm{B}} \int_{-\infty}^{+\infty} d\mathbf{r}\ \varrho_{\mathrm{NL}} \ln \varrho_{\mathrm{NL}}$ has a form like the definition of entropy, $\mathcal{S}$. So the mean value of the linear Hamiltonian that still represents the energy of the system, $\langle H_{\mathrm{L}} \rangle = E$, together with the second term of (4.119) would resemble $E - \mathrm{i}\mathsf{T}\mathcal{S}$ which is similar to the expression for free energy. Only, here again, the imaginary unit i appears in the quantum mechanical context. This point (still) requires further investigation.

The separation condition (4.113) is certainly compatible with wave functions like plane waves but, more interestingly, also with Gaussian functions. Choosing therefore $\varrho(x,t)$ (in agreement with the notation used for the WP (2.1) in one dimension) as

$$\varrho(x,t) = NN^* \exp\left\{-2y_{\mathrm{I}}\tilde{x}^2\right\} = \left(\frac{1}{2\pi\langle \tilde{x}^2 \rangle}\right)^{\frac{1}{2}} \exp\left\{-\frac{\tilde{x}^2}{2\langle \tilde{x}^2 \rangle}\right\} \tag{4.120}$$

enables one to determine the diffusion coefficient D in the Smoluchowski Eq. (4.109) via the separation condition (4.113) in the form

$$D = \frac{\gamma}{2}\langle \tilde{x}^2 \rangle, \tag{4.121}$$

which is compatible with similar expressions in the classical theory of Brownian motion (see, e.g., [79]). Note that for TD WP widths also D is TD.

In comparison with Grössing's choice $D = \frac{\hbar}{2m}$, this would lead to $\frac{\gamma}{2} = \frac{\hbar}{2m\langle \tilde{x}^2 \rangle} = \frac{1}{\alpha^2} = \dot{\varphi}$. Without dissipation, for the HO with constant width, this would imply a "resonance-type" condition, $\frac{\gamma}{2} = \omega_0$, however, in the damped case, $\dot{\varphi} \neq \omega_0$. Further work will be necessary for a better understanding of this connection.

Comparison with the Einstein relation $D = \frac{\mathsf{k}_{\mathrm{B}}\mathsf{T}}{m\gamma}$ finally provides

$$\mathsf{k}_{\mathrm{B}}\mathsf{T} = \frac{m}{2}\gamma^2\langle \tilde{x}^2 \rangle, \tag{4.122}$$

an expression like the quantum mechanical contribution to the potential energy of the HO [see (2.75)], only replacing ω_0 by γ. Also this needs further clarification.

As the separation condition allows for Gaussian functions as solutions, it is not surprising that the NLSE (4.115) also possesses analytic solutions in the form of Gaussian WPs for at most quadratic potentials. Details will be discussed in Chap. 5. Here (for comparison with the approaches considered before), it is only mentioned that the maximum of the Gaussian WP, as given in Eq. (2.1), follows the Newtonian equation of motion including the linear friction term (4.41),

$$\ddot{\eta} + \gamma\,\dot{\eta} + \omega^2\eta = 0,$$

in other words, as in the CK-approach.

Similarly, the coefficient $y_{\rm NL}(t)$ of the quadratic term in the exponent of the WP, or $\mathcal{C}_{\rm NL}(t) = \frac{2\hbar}{m} y_{\rm NL}(t)$, fulfils the modified complex Riccati equation

$$\dot{\mathcal{C}}_{\rm NL} + \gamma \mathcal{C}_{\rm NL} + \mathcal{C}_{\rm NL}^2 + \omega^2 = 0, \tag{4.123}$$

which, with the definition

$$\mathcal{C}_{\rm NL,I} = \frac{\hbar}{2m\langle \tilde{x}^2 \rangle_{\rm NL}} = \frac{1}{\alpha_{\rm NL}^2}, \tag{4.124}$$

can now be transformed into the real Ermakov-type equation

$$\ddot{\alpha}_{\rm NL} + \left(\omega^2 - \frac{\gamma^2}{4}\right) \alpha_{\rm NL} = \frac{1}{\alpha^3{}_{\rm NL}}. \tag{4.125}$$

Eliminating ω^2 between Eqs. (4.41) and (4.125) finally leads to the exact Ermakov-type invariant

$$I_{\rm NL} = \frac{1}{2} e^{\gamma t} \left[\left(\dot{\eta}\, \alpha_{\rm NL} - \left(\dot{\alpha}_{\rm NL} - \frac{\gamma}{2} \alpha_{\rm NL} \right) \eta \right)^2 + \left(\frac{\eta}{\alpha_{\rm NL}} \right)^2 \right] = \text{const.} \tag{4.126}$$

which is identical to (4.68), the invariant for the canonical description in expanding coordinates, expressed in physical variables. These connections will be clarified further in Sect. 4.5.

Before addressing this, another approach should be mentioned which is based on the same Smoluchowski equation for $\varrho(\mathbf{r}, t)$ that confirms the additional diffusion term on group-theoretical grounds [80, 81]. In this approach by Doebner and Goldin the authors do not try to separate this term but add half of it to the modified SE and the other half to its complex conjugate, leading to a NLSE of the form

$$i\hbar \frac{\partial}{\partial t} \Psi(\mathbf{r}, t) = \left\{ -\frac{\hbar^2}{2m} \Delta + V(\mathbf{r}) + i\, \mathcal{I}(F[\Psi]) + \mathcal{R}(F[\Psi]) \right\} \Psi(\mathbf{r}, t) \tag{4.127}$$

where

$$\mathcal{I}(F[\Psi]) = \hbar \frac{D}{2} \frac{\Delta \varrho}{\varrho} \tag{4.128}$$

originates from the diffusion term. It is assumed that a real part $\mathcal{R}(F[\Psi])$ can also be added which can also be a function of Ψ, thus leading to another nonlinearity; "but there is no information on the real part" in their approach [80] that could specify the form of this real part. With certain additional assumptions, a whole family of NLSEs was obtained with this approach.

For Gaussian WPs, the diffusion term is obviously identical to the $\ln \varrho$-terms of $\tilde{W}_{SCH}$. One price that must be paid for the "bisection" according to (4.128) is that, in the equation for Ψ, also Ψ^* occurs and vice versa.

The mathematical and physical properties of the NLSE (4.127) have been investigated in many papers (for references, see e.g. [82]). Attempts have also been made to linearize this NLSE via NL gauge transformations [83, 84] leading to a form similar to that of the CK approach. In the next subsection, it will be shown how linearization of the logarithmic NLSE (4.115) can be achieved via a non-unitary transformation. Similarities and differences between these two methods are discussed in [85]. For the purpose of the problems discussed in this work it will be sufficient and relevant to consider the non-unitary transformations.

4.5 Non-unitary Connections Between the Canonical and Nonlinear Approaches

In order to establish the connection between the explicitly TD (linear) canonical approach of Caldirola and Kanai (4.39) and the NLSE (4.115) with complex logarithmic nonlinearity, referral is made to Schrödinger's first communication on wave mechanics [52] where he starts with the Hamilton–Jacobi equation, here written in the form (in one dimension) as

$$\frac{\partial}{\partial t} S_c + H\left(x, \frac{\partial}{\partial x} S_c, t\right) = 0 \tag{4.129}$$

with action function S_c and momentum $p_c = \frac{\partial}{\partial x} S_c$. He introduces the wave function via the definition

$$S_c = \frac{\hbar}{\mathrm{i}} \ln \Psi \tag{4.130}$$

and arrives, via a variational ansatz, at the Hamiltonian operator

$$H_{\mathrm{L}} = -\frac{\hbar^2}{2m} \frac{\partial^2}{\partial x^2} + V(x) \,. \tag{4.131}$$

The subscript c (added here) indicates that this action (different from the classical one) is a *complex* function as Ψ, in general, is a complex function (a fact that Schrödinger did not like at all in the beginning [86] and he also omitted the i in definition (4.130) but changed this afterwards without any explanation (see also [87]).

Now, Schrödinger's procedure is reversed. Starting with Eq. (4.115) (divided by Ψ which causes no problems, at least for Gaussian WPs) and, using the definition of S_c, one arrives at[13]

$$\left(\frac{\partial}{\partial t}+\gamma\right) S_c + H = \gamma \langle S_c \rangle . \tag{4.132}$$

Of course this is as little rigorous as Schrödinger's first attempt was. However, it follows his idea of connecting the classical Hamilton–Jacobi theory with a wave (mechanical) equation. The purely TD term $\gamma\langle S_c\rangle$ is necessary mainly for normalization purposes (it can therefore be absorbed by the normalization coefficient) and will be neglected for the moment.

Multiplying the remaining Eq. (4.132) by $e^{\gamma t}$ and using the definitions[14]

$$\hat{S}_c = e^{\gamma t} S_c \quad \text{and} \quad \hat{H} = e^{\gamma t} H_L \tag{4.133}$$

it can be rewritten as (canonical) Hamilton–Jacobi equation

$$\frac{\partial}{\partial t}\hat{S}_c + \hat{H} = 0. \tag{4.134}$$

From the definition of the action function, it follows that the wave function $\hat{\Psi}(x,t)$ on the transformed (canonical) level is connected with the wave function $\Psi_{NL}(x,t)$ on the physical level via the *non-unitary* relation

$$\ln \hat{\Psi} = e^{\gamma t} \ln \Psi_{NL} . \tag{4.135}$$

Consequently, the (complex)[15] momenta in the two systems are connected via

$$\hat{p}_c = \frac{\hbar}{i}\frac{\partial}{\partial x} \ln \hat{\Psi} = e^{\gamma t}\, \frac{\hbar}{i}\frac{\partial}{\partial x} \ln \Psi_{NL} = e^{\gamma t} p_c , \tag{4.136}$$

which is equivalent to the connection between the canonical and the physical (kinetic) momentum in the approach by Caldirola and Kanai [see Eq. (4.35)]. The *non-canonical* connection between the classical variables (x, p) and $(\hat{x} = x, \hat{p} = e^{\gamma t} p)$ corresponds to the *non-unitary* transformation between Ψ_{NL} and $\hat{\Psi}$.

Note: Although Ψ_{NL} and $\hat{\Psi}$ depend explicitly on the same variables x and t, the two wave functions are analytically different functions of x and t and have different physical meanings due to the non-unitary transformation (4.135).

Expressing $\hat{H}$ in terms of the canonical momentum, in the classical case this would lead to the CK-Hamiltonian (4.36) and, following Schrödinger's quantization procedure, but now using the canonical momentum $\hat{p}_c$, finally leads to the modified SE

---

[13] In Appendix E it is shown how this can be used for a "hydrodynamic" derivation of the logarithmic NLSE (see also [88]).

[14] Quantities with a hat again denote canonical ones in contrast to their physical counterparts.

[15] Compare the complex velocity $v_- + v_+$ discussed in Sect. 4.4.1

$$
\begin{aligned}
\mathrm{i}\hbar\frac{\partial}{\partial t}\,\hat{\Psi}\,(x,t) &= \hat{H}_{\mathrm{op}}\,\hat{\Psi}\,(x,t) \\
&= \left\{\mathrm{e}^{-\gamma t}\left(-\frac{\hbar^2}{2m}\frac{\partial^2}{\partial x^2}\right)+\mathrm{e}^{\gamma t}V\right\}\hat{\Psi}\,(x,t)\,.
\end{aligned} \tag{4.137}
$$

The Hamiltonian operator $\hat{H}_{\mathrm{op}}$ in (4.137) is identical to the one of Caldirola and Kanai in (4.39), $\hat{H}_{\mathrm{CK,op}}$. However, the important difference is, that in the CK approach it is assumed that $\hat{H}_{\mathrm{CK,op}}$ acts on a wave function in physical space (like $\Psi(x,t)$) whereas it actually acts on a canonical wave function $\hat{\Psi}(x,t)$. This fact is hidden because, in this particular non-canonical transformation between the physical and canonical levels, $x = \hat{x}$. So it is misleadingly giving the impression that $\Psi(x,t)$ and $\hat{\Psi}(x,t)$ are identical. But, to be consistent it is necessary to transform operators *and* wave functions simultaneously to avoid unphysical results[16] like the violation of the uncertainty principle. This will be demonstrated explicitly following a short comment concerning the term $-\gamma\langle S_c\rangle$ on the rhs of Eq. (4.132) that has been ignored so far.

As the connection between Ψ_{NL} and $\hat{\Psi}$ is not unitary, it follows that if the solution $\hat{\Psi}$ of Eq. (4.137) [or (4.39)] is normalized, the solution Ψ after the transformation, i.e., the solution of the logarithmic NLSE without the term $\gamma\langle S_c\rangle$, is not normalized. So, for normalization purposes, $\gamma\langle S_c\rangle$ must be taken into account on the rhs of the equation for Ψ_{NL}, thus leading to the NLSE (4.115).

Now the uncertainty products in canonical and physical descriptions are considered. For the cases with analytic Gaussian WP solutions of the CK-SE, this WP $\hat{\Psi}_{\mathrm{CK}}(x,t)$ is proportional to

$$
\hat{\Psi}_{\mathrm{CK}}(x,t) \;\propto\; \exp\left\{\mathrm{i}\;\hat{y}_{\mathrm{CK}}\,(x-\langle x\rangle_{\mathrm{CK}})^2+\frac{\mathrm{i}}{\hbar}\langle\hat{p}\rangle_{\mathrm{CK}}x\right\}. \tag{4.138}
$$

For this purpose, the imaginary part of the complex quantity $\hat{y}_{\mathrm{CK}}(t) = \hat{y}_{\mathrm{CK,R}} + \mathrm{i}\hat{y}_{\mathrm{CK,I}}$ is relevant as it is connected with the position uncertainty via

$$
\left\langle\tilde{x}^2\right\rangle_{\mathrm{CK}} = \frac{1}{4\;\hat{y}_{\mathrm{CK,I}}}, \tag{4.139}
$$

where the subscript "CK" indicates the mean value is calculated using $\hat{\Psi}_{\mathrm{CK}}$. Similarly, the uncertainty of the *canonical* momentum is given by

$$
\left\langle\tilde{\hat{p}}^2\right\rangle_{\mathrm{CK}} = \hbar^2\left(\frac{\hat{y}_{\mathrm{CK,R}}^2+\hat{y}_{\mathrm{CK,I}}^2}{\hat{y}_{\mathrm{CK,I}}}\right), \tag{4.140}
$$

---

[16] Although this was published quite some time ago [89], still papers with the wrong interpretation and unphysical results like localization, i.e., shrinking of WPs to delta functions are considered and published.

hence the uncertainty product

$$\hat{U}_{\rm CK} = \langle \tilde{x}^2 \rangle_{\rm CK} \left\langle \tilde{\hat{p}}^2 \right\rangle_{\rm CK} = \frac{\hbar^2}{4} \left[1 + \left(\frac{\hat{y}_{\rm CK,R}}{\hat{y}_{\rm CK,I}} \right)^2 \right] \geq \frac{\hbar^2}{4} \tag{4.141}$$

obviously does not violate Heisenberg's principle.

The violation seems to occur if the canonical momentum $\hat{p}_{\rm op} = \frac{\hbar}{\rm i} \frac{\partial}{\partial x}$ is replaced by the physical (kinetic) momentum, according to $p_{\rm op} = {\rm e}^{-\gamma t} \hat{p}_{\rm op}$, and the momentum uncertainty is calculated with the *same* WP $\hat{\Psi}_{\rm CK}$ as used for $\hat{p}_{\rm op}$, i.e., $\langle \tilde{p}^2 \rangle_{\rm CK} = \langle \tilde{\hat{p}}^2 \rangle_{\rm CK} {\rm e}^{-2\gamma t}$. However, the operator $p_{\rm op}$ is not defined on the canonical space to which $\hat{\Psi}_{\rm CK}$ belongs but on the physical space. Therefore, to be consistent, if the operators are transformed, the WPs must be transformed accordingly.

The WPs that apply to the physical level are, as shown above, the solutions of the logarithmic NLSE (4.115). These WPs can be written in a form equivalent to (4.138),

$$\Psi_{\rm NL}(x,t) \propto \exp\left\{ {\rm i}\, y_{\rm NL} \left(x - \langle x \rangle_{\rm NL} \right)^2 + \frac{\rm i}{\hbar} \langle \hat{p} \rangle_{\rm NL} x \right\}, \tag{4.142}$$

but, according to (4.135), $\hat{y}_{\rm CK}$ must be replaced by $y_{\rm NL} = \hat{y}_{\rm CK}\, {\rm e}^{-\gamma t}$.

The uncertainties of position and physical momentum, calculated with $\Psi_{\rm NL}(x,t)$ can now be given in terms of $y_{\rm NL,R}$ and $y_{\rm NL,I}$ and expressed in terms of the canonical uncertainties as

$$\langle \tilde{x}^2 \rangle_{\rm NL} = \frac{1}{4\, y_{\rm NL,I}} = {\rm e}^{\gamma t} \frac{1}{4\, \hat{y}_{\rm CK,I}} = {\rm e}^{\gamma t} \langle \tilde{x}^2 \rangle_{\rm CK}, \tag{4.143}$$

$$\langle \tilde{p}^2 \rangle_{\rm NL} = \hbar^2 \left(\frac{y_{\rm NL,R}^2 + y_{\rm NL,I}^2}{y_{\rm NL,I}} \right) = {\rm e}^{-\gamma t}\, \hbar^2 \left(\frac{\hat{y}_{\rm CK,R}^2 + \hat{y}_{\rm CK,I}^2}{\hat{y}_{\rm CK,I}} \right) = {\rm e}^{-\gamma t} \left\langle \tilde{\hat{p}}^2 \right\rangle_{\rm CK}. \tag{4.144}$$

So the uncertainty product

$$U_{\rm NL} = \langle \tilde{x}^2 \rangle_{\rm NL} \langle \tilde{p}^2 \rangle_{\rm NL} = \frac{\hbar^2}{4} \left[1 + \left(\frac{y_{\rm NL,R}}{y_{\rm NL,I}} \right)^2 \right] = \frac{\hbar^2}{4} \left[1 + \left(\frac{\hat{y}_{\rm CK,R}}{\hat{y}_{\rm CK,I}} \right)^2 \right] \geq \frac{\hbar^2}{4} \tag{4.145}$$

is identical on the physical and the canonical levels and does not violate Heisenberg's principle, providing both operators *and* wave functions are transformed simultaneously.

As the link between the physical and canonical levels has now been established for the classical as well as for the quantum mechanical case, any approach on the canonical level that is related to that of Caldirola and Kanai by a (classical) canonical or (quantum mechanical) unitary transformation can also be uniquely linked to the physical level. This can be done in two steps via the CK-model or directly via one non-canonical or non-unitary transformation [89, 90].

In Figs. 4.1 and 4.2 this is demonstrated using the approach with expanding coordinates discussed in Sect. 4.3.2 as an example.

In the classical case (Fig. 4.1), the transition between the expanding system and the CK-approach is given by a canonical transformation, as shown in Sect. 4.3.2. A subsequent non-canonical transformation then leads to the physical level. These two transformations, however, can be combined to one non-canonical transformation that directly links the expanding system to the physical one.

Considering the quantum mechanical case, the key quantity is the action function and its relation to the wave function according to Schrödinger's definition $S = \frac{\hbar}{\mathrm{i}} \ln \Psi$. In position-space representation, in all approaches the momentum operator is defined as $\frac{\hbar}{\mathrm{i}}$ times the derivative with respect to the corresponding canonical position variable, thus providing the Hamiltonian operator when the Hamiltonian function is given. If the action functions of two approaches are different, the corresponding wave functions are also different, can have different meanings and can lead to different mean values calculated with them even if the functions are expressed in the same variables as in the other approach (as in the case of $\Psi_{\mathrm{NL}}(x, t)$ and $\hat{\Psi}_{\mathrm{CK}}(x, t)$). On the canonical level (and similarly on the physical level), a difference in the action functions can be compensated for by a unitary transformation (as shown in Sect. 4.3.2 between $\hat{\Psi}_{\mathrm{CK}}$ and $\hat{\Psi}_{exp}$). However, between the canonical and physical levels the non-unitary transformation discussed in the transition between the CK-approach and the logarithmic NLSE is necessary. Again, as in the classical case, different approaches with different action functions on the canonical level can be linked directly to the physical level via one non-unitary transformation.

So, depending on the problem under consideration, it might be favourable in some cases to exploit the mathematical advantage of linearity on the canonical level and, via a non-unitary transformation, subsequently translate the results into the corresponding physical quantities. In other cases, it might be better to use the immediate physical image, even for the price of putting up with nonlinearities of the equations.

In the case of the damped HO, all three approaches discussed in detail in this subsection possess Gaussian WP solutions (see (4.40) for $\hat{\Psi}_{\mathrm{CK}}$, (4.63) for $\hat{\Psi}_{exp}$ and (2.1) for Ψ_{NL}). The (Newton-type) equation of motion for the maximum of these WPs as well as the complex Riccati equations or equivalent real NL Ermakov equations and the corresponding Ermakov invariants (that exist in all these cases) might look different when expressed in the canonical variables of the respective approach (see Fig. 4.3). However, using the above-mentioned transformations they can all be transformed into the equation for the damped HO, (4.41), describing the motion of the maximum and thus the trajectory of the classical position and, for the dynamics of the width, the Riccati Eq. (4.123) or equivalent Ermakov Eq. (4.125) with the corresponding invariant 4.126) obtained on the physical level from the logarithmic NLSE (4.115). Therefore, this approach will be used in the next section to demonstrate the similarities and differences in comparison with the conservative case if the

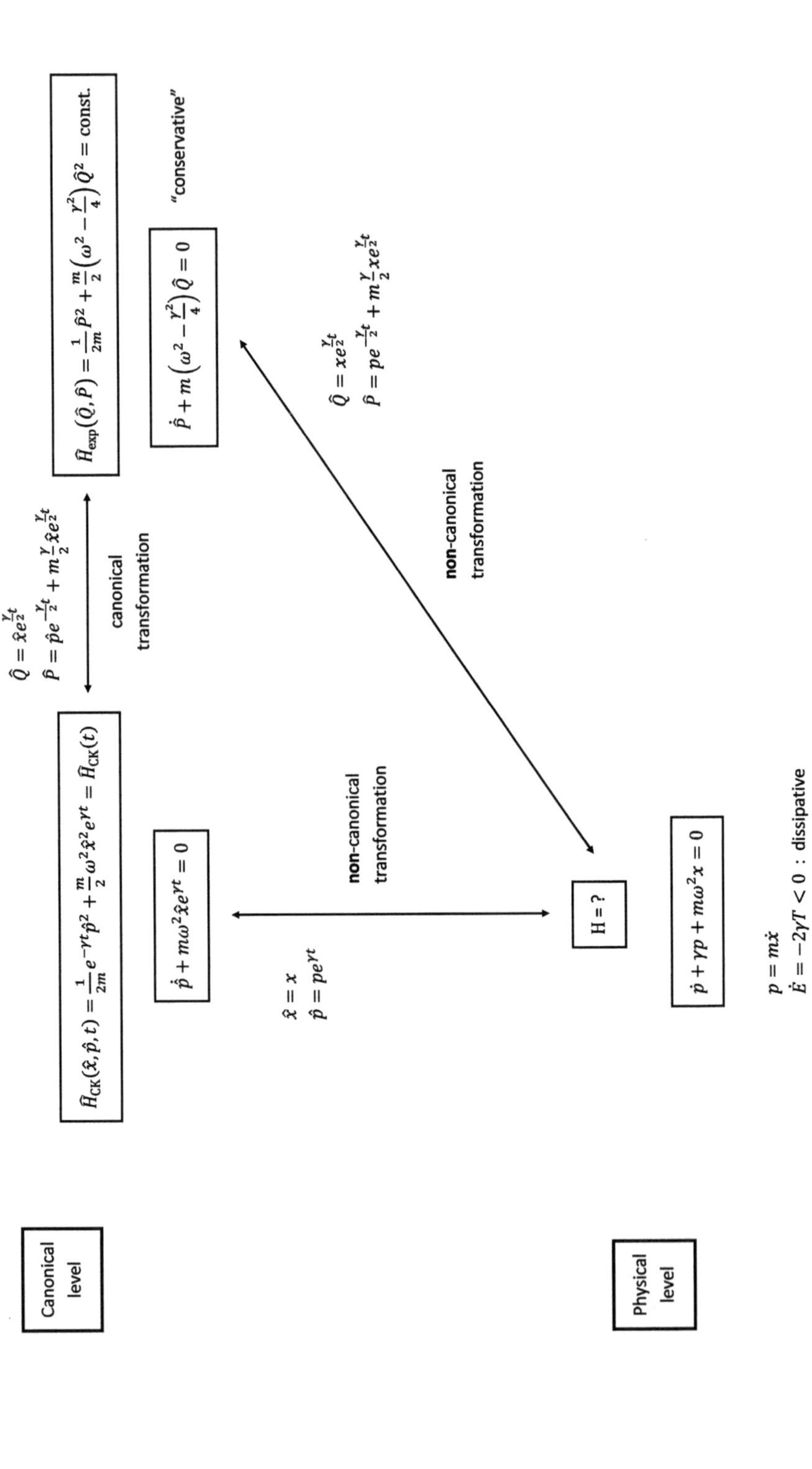

Fig. 4.1 Canonical transformation between the dissipative approaches on the canonical level and their non-canonical transformations to the physical level in the classical case

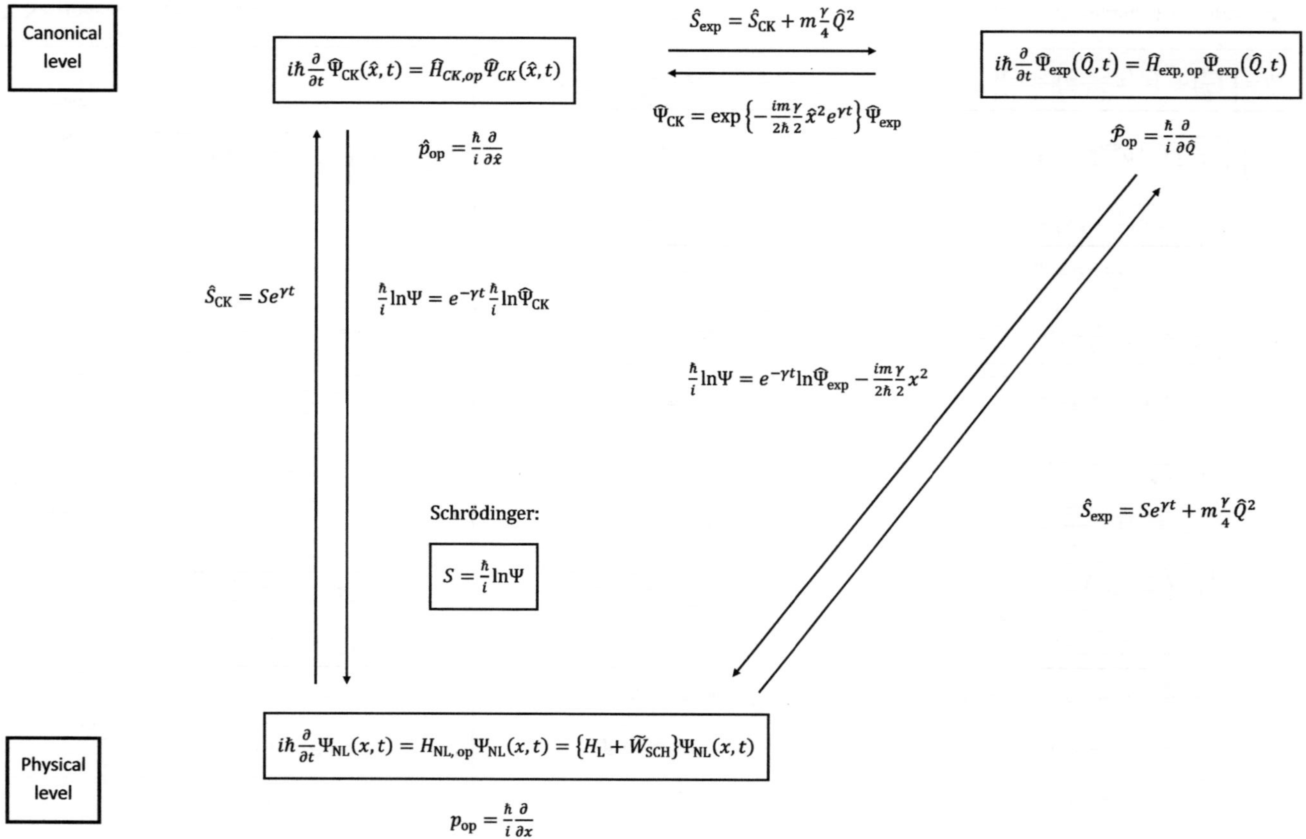

Fig. 4.2 Unitary transformation between the dissipative approaches on the canonical level and their non-unitary transformations to the physical level in the quantum mechanical case

	Physical	Canonical	
	(x,p)	CK: $(\hat{x},\hat{p})$	exp: $(\hat{Q},\hat{P})$
Newtonian equations of motion	$\dot{p}+\gamma p+m\omega^2 x=0$ $x=\eta\,,\ p=m\,\dot{\eta}$ $\boxed{\ddot{\eta}+\gamma\dot{\eta}+\omega^2\eta=0}$	$\dot{\hat{p}}+m\omega^2\hat{x}e^{\gamma t}=0$ $\hat{x}=\eta\,,\ \hat{p}=m\,\dot{\eta}e^{\gamma t}$ $\boxed{\ddot{\eta}+\gamma\dot{\eta}+\omega^2\eta=0}$	$\dot{\hat{P}}+m\left(\omega^2-\frac{\gamma^2}{4}\right)\hat{Q}=0$ $\hat{Q}=\eta e^{\frac{\gamma}{2}t}\,,\ \hat{P}=m\left(\dot{\eta}+\frac{\gamma}{2}\eta\right)e^{\frac{\gamma}{2}t}$ $\boxed{\ddot{\eta}+\gamma\dot{\eta}+\omega^2\eta=0}$
Riccati equations	$\boxed{\dot{C}+\gamma C+C^2+\omega^2=0}$ $C_I=\frac{1}{\alpha^2}$ $C_R=\frac{\dot{\alpha}}{\alpha}$ $C_I=\frac{\hbar}{2m\langle\tilde{x}^2\rangle}$	$\boxed{\dot{\hat{C}}_{CK}+\hat{C}_{CK}^2\,e^{-\gamma t}+\omega^2 e^{\gamma t}=0}$ $\hat{C}_{I,CK}=\frac{1}{\hat{\alpha}_{CK}^2}$ $\hat{C}_{R,CK}=\frac{\dot{\hat{\alpha}}_{CK}}{\hat{\alpha}_{CK}}e^{\gamma t}$ $\hat{C}_{CK}=Ce^{\gamma t}$	$\boxed{\dot{\hat{C}}_{exp}+\hat{C}_{exp}^2+\left(\omega^2-\frac{\gamma^2}{4}\right)=0}$ $\hat{C}_{I,exp}=\frac{1}{\hat{\alpha}_{exp}^2}=\hat{C}_{I,CK}e^{-\gamma t}$ $\hat{C}_{R,exp}=\hat{C}_{R,CK}e^{-\gamma t}+\frac{\dot{\hat{\alpha}}_{CK}}{\hat{\alpha}_{CK}}e^{\gamma t}$ $\hat{C}_{exp}=\hat{C}_{CK}e^{-\gamma t}+\frac{\gamma}{2}$
Ermakov equation	$\boxed{\ddot{\alpha}+\left(\omega^2-\frac{\gamma^2}{4}\right)\alpha=\frac{1}{\alpha^3}}$	$\ddot{\hat{\alpha}}_{CK}+\gamma\dot{\hat{\alpha}}_{CK}+\omega^2\hat{\alpha}_{CK}=\frac{e^{-2\gamma t}}{\hat{\alpha}_{CK}^3}$ $\hat{\alpha}_{CK}=\alpha e^{-\frac{\gamma}{2}t}$ $\rightarrow\boxed{\ddot{\alpha}+\left(\omega^2-\frac{\gamma^2}{4}\right)\alpha=\frac{1}{\alpha^3}}$	$\ddot{\hat{\alpha}}_{exp}+\left(\omega^2-\frac{\gamma^2}{4}\right)\hat{\alpha}_{exp}=\frac{1}{\hat{\alpha}_{CK}^3}$ $\hat{\alpha}_{exp}=\hat{\alpha}_{CK}\,e^{\frac{\gamma}{2}t}=\alpha$ $\rightarrow\boxed{\ddot{\alpha}+\left(\omega^2-\frac{\gamma^2}{4}\right)\alpha=\frac{1}{\alpha^3}}$
Ermakov invariant	$\boxed{I_{NL}=\frac{1}{2}e^{\gamma t}\left[\left(\dot{\eta}\alpha-\left(\dot{\alpha}-\frac{\gamma}{2}\alpha\right)\eta\right)^2+\left(\frac{\eta}{\alpha}\right)^2\right]}$	$\hat{I}_{CK}=\frac{1}{2}\left[e^{2\gamma t}(\dot{\eta}\hat{\alpha}_{CK}-\dot{\hat{\alpha}}_{CK}\eta)^2+\left(\frac{\eta}{\hat{\alpha}_{CK}}\right)^2\right]$ $\boxed{\hat{I}_{CK}\mathrel{\hat{=}}\frac{1}{2}e^{\gamma t}\left[\left(\dot{\eta}\alpha-\left(\dot{\alpha}-\frac{\gamma}{2}\alpha\right)\eta\right)^2+\left(\frac{\eta}{\alpha}\right)^2\right]=I_{NL}}$	$\hat{I}_{exp}=\frac{1}{2}\left[\left(\dot{\hat{Q}}\hat{\alpha}_{exp}-\dot{\hat{\alpha}}_{exp}\hat{Q}\right)^2+\left(\frac{\hat{Q}}{\hat{\alpha}_{exp}}\right)^2\right]$ $\boxed{\hat{I}_{exp}\mathrel{\hat{=}}\frac{1}{2}e^{\gamma t}\left[\left(\dot{\eta}\alpha-\left(\dot{\alpha}-\frac{\gamma}{2}\alpha\right)\eta\right)^2+\left(\frac{\eta}{\alpha}\right)^2\right]=I_{NL}}$

Fig. 4.3 Newtonian, complex Riccati and Ermakov equations as well as Ermakov invariant in terms of the physical and canonical variables of the three approaches discussed for the description of the dissipative system

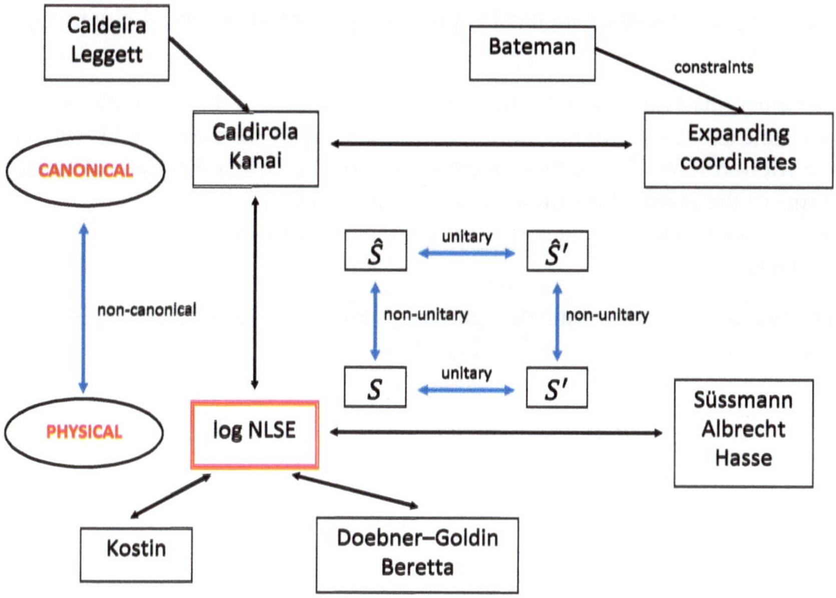

Fig. 4.4 Interrelations amongst canonical and physical approaches for the description of dissipative systems

dynamics of the dissipative system is expressed with the help of the complex Riccati equation, like shown in Chap. 2.

On the physical level, the logarithmic NLSE combines the dissipative aspect of Kostin's approach with the irreversible one of Doebner and Goldin and Beretta in a consistent manner and, apart from a purely TD contribution that does not affect the dynamics of the WP solutions, leads to identical results like the approach by Hasse that combines attempts by Süssmann and Albrecht (see Fig. 4.4). A closer comparison between Hasse's form of the friction term and the logarithmic nonlinearity will be given in Chap. 6 where the dissipative version of TI NL quantum mechanics is discussed.

The connection to the canonical level has been discussed above, as well as possible transitions between different approaches on this level via canonical or unitary transformations.

In the work of Sun and Yu [13, 14], also the transition between the system-plus-reservoir approach of Caldeira and Leggett and the Hamiltonian operator of Caldirola and Kanai has been supplied (though using the wrong interpretation of the corresponding wave function), providing the link between the system-plus-reservoir approaches and the effective models discussed above.

What is still missing in the context of the approaches mentioned in this work is the connection between the canonical Bateman model and, e.g., the effective canonical models. This gap will finally be closed by showing the connection between the system with expanding coordinates and the Bateman approach. In order to connect the

Bateman approach to the canonical approaches presented in Sect. 4.3, it is required that

1. the equation of motion (4.23) for the position variable of the dissipative system is the same as the equation of motion for the (physical) position variable (including the friction force) in the two canonical approaches when these are expressed in terms of the physical position variable x (or $\eta(t)$);
2. the Bateman Hamiltonian represents a constant of motion with the dimension of an energy.

The two descriptions of the dissipative system can be connected if it is assumed that the conserved quantity $\hat{H}_{\mathrm{B}}$ [see Eq. (4.19)]

$$\hat{H}_{\mathrm{B}} = \frac{1}{m}\hat{p}_x\hat{p}_y + \frac{\gamma}{2}(\hat{y}\hat{p}_y - \hat{x}\hat{p}_x) + m\left(\omega^2 - \frac{\gamma^2}{4}\right)\hat{x}\hat{y} = \text{const.}$$

is identical to the conserved quantity $\hat{H}_{exp}$ [see Eq. (4.51)],

$$\hat{H}_{exp} = \frac{1}{2m}\,\hat{P}^2 + \frac{m}{2}\left(\omega^2 - \frac{\gamma^2}{4}\right)\hat{Q}^2 = \text{const.} \mathrel{\hat{=}} \frac{m}{2}\left[\dot{x}^2 + \gamma\dot{x}x + \omega^2x^2\right]\mathrm{e}^{\gamma t} = \text{const.}$$

(with $x = \hat{x}$) and some constraints are imposed so that the dual variable $\hat{y}$ and the corresponding momentum are eliminated. From Eq. (4.21), $\hat{p}_y = m\left(\dot{\hat{x}} + \frac{\gamma}{2}\hat{x}\right)$, it follows that none of the product terms of $\hat{x}$ and $\hat{p}_y$ with one of the other variables $\hat{y}$ and $\hat{p}_x$ in (4.19) contains the exponential factor $\mathrm{e}^{\gamma t}$ that is common in Eq. (4.51).

Following the course outlined above, the Hamiltonian $\hat{H}_{\mathrm{B}}$ (in the form (4.26), i.e., $\hat{H}_{\mathrm{B}} \mathrel{\hat{=}} m(\dot{\hat{x}}\dot{\hat{y}} + \omega^2\hat{x}\hat{y})$) is equated with $\hat{H}_{exp}$ (and, as $\hat{x} = x =$ physical position variable, the hat is omitted for x):

$$\frac{m}{2}\,\mathrm{e}^{\gamma t}\left[\dot{x}^2 + \gamma\dot{x}x + \omega^2x^2\right] = \hat{p}_x\dot{x} + m\frac{\gamma}{2}\hat{y}\dot{x} + m\omega^2x\hat{y}, \tag{4.146}$$

which is only possible if $\hat{y}$ and $\hat{p}_x$ are expressed in terms of x and $\dot{x}$. For this purpose the ansatz

$$\hat{p}_x = \mathrm{e}^{\gamma t}\,(a\,\dot{x} + b\,x) \tag{4.147}$$

and

$$\hat{y} = \mathrm{e}^{\gamma t}\,(c\,\dot{x} + d\,x) \tag{4.148}$$

is inserted into (4.146) and the coefficients of $\dot{x}^2$, $\dot{x}x$ and x^2-terms are equated, leading to

$$d = \frac{1}{2} \tag{4.149}$$

$$a = \frac{m}{2}(1 - \gamma c) \tag{4.150}$$

$$b = m\left(\frac{\gamma}{4} - \omega^2 c\right), \tag{4.151}$$

where a, b and c must still be determined. As only two equations (4.150, 4.151) are given, one parameter is still free to be chosen. In Appendix D a more detailed discussion of the cases 1) $c = 0$, 2) $a = 0$ and 3) $b = 0$ are given.

Generally, one parameter can be eliminated leaving a condition for the relation of the other two, e.g., a and b that must be fulfilled for any choice of c, etc.,

$$a = \frac{1}{2\omega^2}\left(m\left(\omega^2 - \frac{\gamma^2}{4}\right) + \gamma\, b\right) \tag{4.152}$$

or

$$b = \frac{2}{\gamma}\left(\omega^2\, a - \frac{m}{2}\left(\omega^2 - \frac{\gamma^2}{4}\right)\right). \tag{4.153}$$

Note that, in expressing (4.147) and (4.148) in terms of canonical variables,

$$\hat{p}_x = e^{\gamma t}\,(a\,\dot{x} + b\,x) = e^{\gamma t}\left(\frac{a}{m}\,\hat{p}_y + \left(b - \frac{\gamma}{2}a\right)x\right) \tag{4.154}$$

$$\hat{y} = e^{\gamma t}\,(c\,\dot{x} + d\,x) = e^{\gamma t}\left(\frac{c}{m}\,\hat{p}_y + \left(d - \frac{\gamma}{2}c\right)x\right), \tag{4.155}$$

an explicit TD character of the constraints shows up although they are compatible with the equations of motion, that is, the total time-derivative of the constraints is zero. The explicit dependence of the constraints on time is traced back to the fact that they have non-vanishing Poisson brackets with the Hamiltonian. From the examples discussed in Appendix D, it becomes obvious that, in general, i.e. independent of the choice of the third parameter, and after imposing the constraints, the Hamiltonian $\hat{H}_B$ is no longer a Hamiltonian in the sense that it would provide the correct equations of motion as the constraints contain an explicit time-dependence (for further details, see [91]).

Once the relation between the variables of the Bateman system and the ones in expanding coordinates is given explicitly (in terms of the physical position x and velocity $\dot{x}$ of the damped system), the connection with the CK-model[17] can be achieved via the TD canonical transformation, or, with the physical level, via the non-canonical transformation specified above.

Now the connections between all the approaches specified in Fig. 4.4 are established. However, many other approaches exist for the description of dissipative systems (like the interesting one by Dekker using complex variables [93] that are not

[17] A different way of embedding the TD constraints in a TD canonical transformation for getting directly from the Bateman Hamiltonian to the CK-Hamiltonian has been shown in [92].

mentioned here. It is not because they are considered unimportant or (even) wrong, but because their connections with the above-mentioned approaches are not yet so clear to the author that they can be specified in detail. (However, this may be subject of further investigations.)

Similarly, the connection between the approaches for the pure states and those for the density matrix/operator, particularly in the case of the logarithmic NLSE, has not yet been specified. Particularly the question of compatibility of a term corresponding to the logarithmic nonlinearity with the requirements of Kossakowski/Lindblad will be considered in forthcoming works.

The following discussion, as mentioned above, focuses on the modifications of the description of quantum dynamics in terms of complex Riccati equations, or equations equivalent to it (like real NL Ermakov equations), for systems with exact WP solutions, providing a dissipative friction force $-m\gamma\dot{x}$ is involved. Particularly, the logarithmic NLSE (4.115) is discussed; the relations with other effective approaches can be found using Fig. 4.4.

References

1. R.J. Jellito, *Dynamik und Thermodynamik* (Franz Steiner Verlag, Wiesbaden, 1983); R. Balescu, *Equilibrium and Nonequilibrium Statistical Mechanics* (Wiley, New York, 1975), Appendix 6
2. F. Reif, *Fundamentals of Statistical and Thermal Physics* (McGraw-Hill, Tokyo, 1965) Chap. 15
3. H.A. Kramers, Brownian motion in a field of force and the diffusion model of chemical reactions. Physica **7**, 284–304 (1940)
4. R. Becker, *Theorie der Wärme* (Springer-Verlag, Heidelberg, 1966), p. 275
5. E. Fick and G. Sauermann, *Quantenstatistik Dynamischer Prozesse, Band 1: Generelle Aspekte* (Verlag Harri Deutsch, Thun, 1993), Chap. 7
6. U. Weiss, *Quantum Dissipative Systems* (World Scientific, Singapore, 1993)
7. A.O. Caldeira and A.J. Leggett, Influence of Dissipation on Quantum Tunneling in Macroscopic Systems. Phys. Rev. Lett. **46**, 211 (1981); A.O. Caldeira and A.J. Leggett, Quantum tunnelling in a dissipative system. Ann. Phys. (N.Y.) **149**, 374-456 (1983); ibid **153**, 445 (E) (1984); A.O. Caldeira and A.J. Leggett, Path integral approach to quantum Brownian motion. Physica A **121**, 587-616 (1983)
8. H. Bateman, On Dissipative Systems and Related Variational Principles. Phys. Rev. **38**, 815 (1931)
9. I.R. Senitzky, Dissipation in quantum mechanics. The Harmonic Oscillator. Phys. Rev. **119**, 670 (1960)
10. G.W. Ford, M. Kac, P. Mazur, Statistical mechanics of assemblies of coupled oscillators. J. Math. Phys. **6**, 504 (1965)
11. P. Ullersma, An exactly solvable model for Brownian motion: I. Derivation of the Langevin equation, 27–55 II. Derivation of the Fokker-Planck equation and the master equation, 56–73 III. Motion of a heavy mass in a linear chain, 74–89 IV. Susceptibility and Nyquist's theorem, 90–96. Physica **32**, 27–96 (1966)
12. R.J. Rubin, Statistical Dynamics of Simple Cubic Lattices. Model for the Study of Brownian Motion. J. Math. Phys. **1**, 309 (1960); Statistical Dynamics of Simple Cubic Lattices. Model for the Study of Brownian Motion. II. J. Math. Phys. **2**, 373 (1961)
13. L.H. Yu, C.P. Sun, Evolution of the wave function in a dissipative system. Phys. Rev. A **49**, 592 (1994)

14. C.P. Sun, L.H. Yu, Exact dynamics of a quantum dissipative system in a constant external field. Phys. Rev. A **51**, 1845 (1995)
15. S. Gao, Dissipative quantum dynamics with a Lindblad functional. Phys. Rev Lett. **79**, 3101 (1997)
16. V. Gorini, A. Kossakowski, E.C.G. Sudarshan, Completely positive dynamical semigroups of N-level systems. J. Math. Phys. **17**, 821 (1976)
17. G. Lindblad, On the generators of quantum dynamical semigroups. Comm. Math. Phys. **48**, 119 (1976)
18. V. May and O. Kühn, *Charge and Energy Transfer Dynamics in Molecular Systems* 2nd ed. (Wiley-VCh, Berlin, 2003), Sect. 3.12.3
19. H.M. Wiseman, W.J. Munro, Comment on "dissipative quantum dynamics with a Lindblad functional". Phys. Rev. Lett. **80**, 5702 (1998)
20. L. Diosi, On high-temperature markovian equation for quantum Brownian motion. Europhys. Lett. **22**, 1 (1993)
21. A. Davidson, Damping in Schrödinger's equation for macroscopic variables. Phys. Rev. A **41**, 3395 (1990)
22. P.M. Morse, H. Feshbach, *Methods of Theoretical Physics*, vol. 1 (McGraw-Hill, New York, 1953)
23. F. Bopp, Quantisierung des gedämpften harmonischen Oszillators. Sitz.-Ber. Bayer. Akad. Wiss., Math.-Natur. Kl., pp. 67–77 (1974)
24. A.L. De Brito, B. Baseia, Bateman Hamiltonian and squeezing. Phys. Rev. A **40**, 4097 (1989)
25. J. Aliaga, G. Crespo, A.N. Proto, Squeezed states for the Bateman Hamiltonian. Phys. Rev. A **43**, 595 (1991)
26. E. Celeghini, M. Rasetti, G. Vitiello, Quantum dissipation. Ann. Phys. (NY) **215**, 156–170 (1992)
27. G. Vitiello, *My Double Unveiled - The Dissipative Model of Brain* (John Benjamin, Amsterdam, 2001)
28. M. Blasone, P. Jizba, Quantum mechanics of the damped harmonic oscillator. Can. J. Phys. **80**, 645–660 (2002)
29. M. Blasone, P. Jizba, Bateman's dual system revisited: quantization, geometric phase and relation with the ground-state energy of the linear harmonic oscillator. Ann. Phys. (NY) **312**, 354–397 (2004)
30. M. Blasone, P. Jizba, G. Vitiello, Dissipation and quantization. Phys. Lett. A **287**, 205–210 (2001)
31. G. 't Hooft, in: *Basics and Highlights of Fundamental Physics* Erice, hep-th/0003005 (1999)
32. M. Blasone, P. Jizba, F. Scardigli, G. Vitiello, Dissipation and quantization for composite systems. Phys. Lett. A **373**, 4106–4112 (2006)
33. D. Chruściński, J. Jarkowski, Quantum damped oscillator I: dissipation and resonances. Ann. Phys. (NY) **321**, 854–874 (2006)
34. D. Chruściński, Quantum damped oscillator II: Batemans Hamiltonian vs. 2D parabolic potential barrier. Ann. Phys. (NY) **321**, 840–853 (2006)
35. T.-H. Heng, B.-S. Lin, S.-C. Jing, Wigner Functions for the Bateman System on Noncommutative Phase Space. Chin. Phys. Lett. **27**, 090302 (2010)
36. H. Majima, A. Suzuki, Quantization and instability of the damped harmonic oscillator subject to a time-dependent force. Ann. Phys. (NY) **326**, 3000–3012 (2011)
37. P. Caldirola, Forze non conservative nella meccanica quantistica. Nuovo Cimento **18**, 393–400 (1941)
38. E. Kanai, On the quantization of the dissipative systems. Progr. Theor. Phys. **3**, 440–442 (1948)
39. T. Luo and Y. Guo, A Sort of Relation Between a Dissipative Mechanical System and Conservative Ones. arXiv:0803.2330v4 [math-ph] 13 pages (2010)
40. M.C. Baldiotti, R. Fresneda, D.M. Gitman, Quantization of the damped harmonic oscillator revisited. Phys. Lett. A **375**, 1630–1636 (2011)
41. C.M. Bender and M. Gianfreda, Twofold Transition in PT-Symmetric Coupled Oscillators. arXiv:1305.7107v1 [hep-th] 9 pages (2013)

42. W.E. Brittin, A note on the quantization of dissipative systems. Phys. Rev. **77**, 396 (1950)
43. J.R. Ray, Dissipation and quantum theory. Lett. Nuovo Cim. **25**, 47–50 (1979)
44. D.M. Greenberger, A critique of the major approaches to damping in quantum theory. J. Math. Phys. **20**, 762 (1979)
45. I.R. Svin'in, Quantum description of Brownian motion in an external field. Teor. Mat. Fiz. **27**, 270–278 (1976); translation I. R. Svin'in, Theoretical and Mathematical Physics **27**, 478–483 (1976)
46. J. Messer, Friction in quantum mechanics. Acta Phys. Austriaca **50**, 75–91 (1979)
47. P. Caldirola, Quantum theory of nonconservative systems. Nuovo Cim. **77**, 241–262 (1983)
48. N.A. Lemos, Canonical approach to the damped harmonic oscillator. Am. J. Phys. **47**, 857 (1979)
49. H. Gzyl, Quantization of the damped harmonic oscillator. Phys. Rev. A **27**, 2297 (1983)
50. I.A. Pedrosa, Canonical transformations and exact invariants for dissipative systems. J. Math. Phys. **28**, 2662–2664 (1987)
51. D. Schuch, A New Lagrange-Hamilton Formalism for Dissipative Systems. Int. J. Quant. Chem., Quant. Chem. Symp. **24**, 767–780 (1990)
52. E. Schrödinger, Quantisierung als Eigenwertproblem (Erste Mitteilung). Ann. Phys. **79**, 361 (1926)
53. G. Meinert, *Nichtkanonische Transformationen zur Behandlung klassischer, dissipativer Systeme und ihre quantenmechanischen Konsequenzen* (J.W. Goethe-University Frankfurt am Main, Diplom-thesis, 1996)
54. M.D. Kostin, On the Schrödinger-langevin equation. J. Chem. Phys. **57**, 3589 (1972)
55. B.K. Skagerstam, On the theory of quantized friction. Phys. Lett. B **58**, 21–25 (1975)
56. B.K. Skagerstam, Stochastic mechanics and dissipative forces. J. Math. Phys. **18**, 308 (1977)
57. K. Yasue, Quantization of dissipative dynamical systems. Phys. Lett. B 64 239–241 (1976); K. Yasue, A note on the derivation of the Schrödinger-Langevin equation. J. Stat. Phys. **16**, 113–116 (1977)
58. K. Yasue, Quantum mechanics of nonconservative systems. Ann. Phys. (N.Y.) **114**, 479–496 (1978)
59. K.K. Kan, J.J. Griffin, Quantized friction and the correspondence principle: single particle with friction. Phys. Lett. B **50**, 241–243 (1974)
60. J.D. Immele, K.K. Kan, J.J. Griffin, Special examples of quantized friction. Nucl. Phys. A **241**, 47–60 (1975)
61. J. Messer, Friction in quantum mechanics. Acta Phys. Austriaca **50**, 75–91 (1978)
62. B.R. Cho, Suhak Gwa Mulri (Korean) (Math. and Phys.) **3**, 37 (1980)
63. I. Bialynicki-Birula, J. Mycielskie, Nonlinear wave mechanics. Ann. Phys. (N.Y.) **100**, 62–93 (1976)
64. C.P. Enz, Beschreibung nicht-konservativer nicht-linearer Systeme (2. Teil). Physik in unserer Zeit **10**, 141–144 (1979)
65. G. Süssmann, unpublished, quoted in Albrecht, Ref. [66] and Hasse, Ref. [67]
66. K. Albrecht, A new class of Schrödinger operators for quantized friction. Phys. Lett. B **56**, 127–129 (1975)
67. R.W. Hasse, On the quantum mechanical treatment of dissipative systems. J. Math. Phys **16**, 2005 (1975)
68. N. Gisin, A simple nonlinear dissipative quantum evolution equation. J. Phys. A **14**, 2259–2267 (1981)
69. N. Gisin, Microscopic derivation of a class of non-linear dissipative Schrödinger-like equations. Physica A **111**, 364–370 (1982)
70. G.P. Beretta, Quantum thermodynamics of nonequilibrium. Onsager reciprocity and dispersion-dissipation relations. Found. Phys. **17**, 365–381 (1987)
71. G.P. Beretta, Nonlinear model dynamics for closed-system, constrained, maximal-entropy-generation relaxation by energy redistribution. Phys. Rev. E **73**, 026113 (2006)
72. G.P. Beretta, Maximum entropy production rate in quantum thermodynamics. J. Phys. Conf. Ser. **237**, 012004 (2010)

73. A.B. Nassar, Quantum trajectories and the Bohm time constant. Ann. Phys. (N.Y.) **331**, 317–322 (2013)
74. E. Madelung, *Die Mathematischen Hilfsmittel des Physikers* (Springer, Berlin, 1950), p. 432
75. B. Mrowka, Zur Darstellung der Quantenmechanik I. Z. Phys. **130**, 164–173 (1951)
76. D. Schuch, K.-M. Chung, H. Hartmann, Nonlinear Schrödinger-type field equation for the description of dissipative systems. I. Derivation of the nonlinear field equation and one-dimensional example. J. Math. Phys. **24**, 1652 (1983)
77. D. Schuch, K.-M. Chung, From macroscopic irreversibility to microscopic reversibility via a nonlinear Schrödinger-type field equation. Int. J. Quant. Chem. **29**, 1561–1573 (1986)
78. G. Grössing, J. Mesa Pasasio, H. Schwabl, A Classical Explanation of Quantization. Found. Phys. **41**, 1437–1453 (2011)
79. W.T. Coffey, Yu. P. Kalmykov and J.T. Waldron, *The Langevin Equation – With Applications in Physics, Chemistry and Electrical Engineering* (World Scientific, Singapore, 1996), Chap. 1.4
80. H.-D. Doebner, G.A. Goldin, On a general nonlinear Schrödinger equation admitting diffusion currents. Phys. Lett. A **162**, 397–401 (1992)
81. H.-D. Doebner, G.A. Goldin, Properties of nonlinear Schrödinger equations associated with diffeomorphism group representations. J. Phys. A: Math. Gen. **27**, 1771 (1994)
82. H.-D. Doebner and J. Tolar, Borel Quantization and Nonlinear Quantum Mechanics, in *Symmetries in Science XI*, ed. by: B.J. Gruber, G. Marmo and N. Yoshinaga (Kluwer Academic Publishers, Dortrecht, 2004), pp. 209–225
83. G.A. Goldin, in *Nonlinear, Deformed and Irreversible Quantum Systems*, ed. by: H.-D. Doebner, J.-D. Henning, W. Lücke, V.K. Dobrev and P. Nattermann (World Scientific, Singapore, 1995), p. 125
84. H.-D. Doebner, G.A. Goldin, Introducing nonlinear gauge transformations in a family of nonlinear Schrödinger equations. Phys. Rev. A **54**, 3764 (1996)
85. D. Schuch, Linearization of Nonlinear Schrödinger Equations - Non-unitary Transformations Compared to Nonlinear Gauge Transformations, in *Quantum Theory and Symmetries*, ed. by: H.-D. Doebner, J.-D. Henning, W. Lücke and V.K. Dobrev (World Scientific, Singapore, 2000), pp. 152–158
86. K. Przibram, *Briefe zur Wellenmechanik* (Springer, Wien, 1963), p. 52
87. D. Schuch, *Komplexe nichtlineare Zusammenhänge im Rahmen einer wellenmechanischen Beschreibung reversibler und irreversibler Dynamik (Habilitation-Thesis* (J.W. Goethe-University, Frankfurt am Main, 1991)
88. D. Schuch, On the complex relations between equations describing the dynamics of wave and particle aspects. Int. J. Quant. Chem. **42**, 663–683 (1992)
89. D. Schuch, Nonunitary connection between explicitly time-dependent and nonlinear approachesfor the description of dissipative quantum systems. Phys. Rev. A **55**, 935 (1997)
90. D. Schuch, Effective description of the dissipative interaction between simple model-systems and their environment. Int. J. Quant. Chem. **72**, 537–547 (1999)
91. D. Schuch, J. Guerrero, F.F. López-Ruiz, V. Aldaya, Interrelations between different canonical descriptions of dissipative systems. Phys. Scr. **90**, 045209 (2015)
92. J. Guerrero, F.F. López-Ruiz, V. Aldaya, F. Cossio, A round trip from Caldirola to Bateman systems. J. Phys. Conf. Ser. **284**, 012062 (2011)
93. H. Dekker, Classical and quantum mechanics of the damped harmonic oscillator. Phys. Rep. **80**, 1–112 (1981)

Chapter 5
Irreversible Dynamics and Dissipative Energetics of Gaussian Wave Packet Solutions

5.1 Direct Solution of the Riccati Equation, Ermakov Equation and Corresponding Invariant

In this section the same one-dimensional problems as in Chap. 2 will be considered, particularly the harmonic oscillator (HO) with constant frequency $\omega = \omega_0$ and (in the limit $\omega_0 \to 0$) the free motion, but now including, classically, a linear velocity dependent friction force. For the quantum mechanical description the effective nonlinear Schrödinger equation (NLSE) (4.115) with complex logarithmic nonlinearity,

$$\mathrm{i}\hbar\frac{\partial}{\partial t}\Psi_{\mathrm{NL}}(x,t) = \left\{-\frac{\hbar^2}{2m}\frac{\partial^2}{\partial x^2} + \frac{m}{2}\omega^2 x^2 + \gamma\frac{\hbar}{\mathrm{i}}\left(\ln\Psi_{\mathrm{NL}} - \langle\ln\Psi_{\mathrm{NL}}\rangle\right)\right\}\Psi_{\mathrm{NL}}(x,t), \quad (5.1)$$

is applied [1–5]. Its interrelation with other approaches for the description of open dissipative quantum systems has been shown in the last chapter. As in the non-dissipative[1] case, also the NLSE possesses analytic Gaussian wave packet (WP) solutions of the form

$$\Psi_{\mathrm{NL}}(x,t) = N_{\mathrm{NL}}(t)\exp\left\{\mathrm{i}\left[y_{\mathrm{NL}}(t)\tilde{x}^2 + \frac{\langle p\rangle_{\mathrm{NL}}}{\hbar}\tilde{x} + K_{\mathrm{NL}}(t)\right]\right\} \quad (5.2)$$

where again $\tilde{x} = x - \langle x\rangle_{\mathrm{NL}}$, but the mean values $\langle\ldots\rangle_{\mathrm{NL}}$ are now calculated using $\Psi_{\mathrm{NL}}(x,t)$. The dynamics of the system described by this WP is again completely

[1] Again, non-dissipative is used here instead of conservative because, for TD frequency $\omega = \omega(t)$, the corresponding Hamiltonian is also not a conserved quantity, though no dissipative force is present.

D. Schuch, *Quantum Theory from a Nonlinear Perspective*,
Fundamental Theories of Physics 191, https://doi.org/10.1007/978-3-319-65594-9_5

determined once the time-evolution of its maximum and width are known. Inserting (5.2) into (5.1) yields as equation of motion for the maximum

$$\ddot{\eta} + \gamma\dot{\eta} + \omega^2\eta = 0, \tag{5.3}$$

i.e., the averaged Langevin equation including the friction force.

The dynamics of the WP width can be obtained, for $\frac{2\hbar}{m} y_{\rm NL} = \mathcal{C}_{\rm NL}$, from the modified complex Riccati equation

$$\dot{\mathcal{C}}_{\rm NL} + \gamma\mathcal{C}_{\rm NL} + \mathcal{C}_{\rm NL}^2 + \omega^2 = 0 \tag{5.4}$$

with an additional term linear in $\mathcal{C}_{\rm NL}$. The WP width or position uncertainty is related with the imaginary part $\mathcal{C}_{\rm NL,I}$ in the same way as in the non-dissipative case,

$$\mathcal{C}_{\rm NL,I} = \frac{\hbar}{2m\langle\tilde{x}^2\rangle_{\rm NL}}. \tag{5.5}$$

It is also possible in this case to transform the inhomogeneous Riccati equation (5.4) into a homogeneous Bernoulli equation, providing a particular solution $\tilde{\mathcal{C}}_{\rm NL}$ is known. The resulting Bernoulli equation

$$\dot{\mathcal{V}} + 2\left(\tilde{\mathcal{C}}_{\rm NL} + \frac{\gamma}{2}\right)\mathcal{V} + \mathcal{V}^2 = 0 \tag{5.6}$$

differs from the one in the non-dissipative case (2.5), only in so far that $\tilde{\mathcal{C}}$ is replaced by $\tilde{\mathcal{C}}_{\rm NL} + \frac{\gamma}{2}$ in the coefficient of the linear term. The general solution is again $\mathcal{C}_{\rm NL}(t) = \tilde{\mathcal{C}}_{\rm NL} + \mathcal{V}(t)$.

Equation (5.6) can be linearized exactly like (2.5) via $\mathcal{V}(t) = \frac{1}{\kappa}$ leading to

$$\dot{\kappa} - 2\left(\tilde{\mathcal{C}}_{\rm NL} + \frac{\gamma}{2}\right)\kappa = 1. \tag{5.7}$$

In the case of a HO with constant frequency ω_0, the particular solution has the form

$$\tilde{\mathcal{C}}_{\pm} = -\frac{\gamma}{2} \pm \sqrt{\frac{\gamma^2}{4} - \omega_0^2}, \tag{5.8}$$

i.e., in general (apart from $\omega_0^2 = \frac{\gamma^2}{4}$), two (at least mathematical) solutions exist for $\tilde{\mathcal{C}}_{\rm NL}$.

The solution of the Bernoulli equation (5.6) can be written down immediately in analogy with the second term on the rhs of Eq. (2.7) as

$$\mathcal{V}_{\pm}(t) = \frac{1}{\kappa_{\pm}(t)} = \frac{{\rm e}^{-2(\tilde{\mathcal{C}}_{\pm}+\frac{\gamma}{2})t}}{\kappa_0 + \frac{1}{2(\tilde{\mathcal{C}}_{\pm}+\frac{\gamma}{2})}\left[1 - {\rm e}^{-2(\tilde{\mathcal{C}}_{\pm}+\frac{\gamma}{2})t}\right]}. \tag{5.9}$$

For TD particular solutions $\tilde{\mathcal{C}}_{\text{NL}}$, $\mathcal{V}(t)$ can be expressed in terms of the same integral $\mathcal{I}(t)$ as in (2.8), again only replacing $\tilde{\mathcal{C}}$ with $\tilde{\mathcal{C}}_{\text{NL}} + \frac{\gamma}{2}$.

Like in the non-dissipative case, the solutions of the Bernoulli (and hence the Riccati) equation are sensitive to the choice of the initial condition κ_0 which is realted to $\tilde{\mathcal{C}}_{\text{NL}}$ (for constant values) and the initial value $\mathcal{C}_{\text{NL}}(t=0) = \mathcal{C}_{\text{NL},0}$, via

$$\mathcal{V}_{\pm,0} = \frac{1}{\kappa_{\pm,0}} = \mathcal{C}_{\text{NL},0} - \tilde{\mathcal{C}}_{\pm,0}, \tag{5.10}$$

leading to expressions like (2.9).

The relations between the initial value $\mathcal{C}_{\text{NL},0}$ and the initial position and momentum uncertainties are specified below. For this purpose, it is also useful to rewrite the complex Riccati equation (5.4) as a real Ermakov equation.

Using the same definition of the imaginary part $\mathcal{C}_{\text{NL,I}}(t)$ in terms of the Ermakov variable $\alpha_{\text{NL}}(t)$, i.e.,

$$\mathcal{C}_{\text{NL,I}}(t) = \frac{1}{\alpha_{\text{NL}}^2(t)}, \tag{5.11}$$

the imaginary part of Eq. (5.4),

$$\dot{\mathcal{C}}_{\text{NL,I}} + \gamma\mathcal{C}_{\text{NL,I}} + 2\mathcal{C}_{\text{NL,I}}\mathcal{C}_{\text{NL,R}} = 0, \tag{5.12}$$

provides the modified real part as

$$\mathcal{C}_{\text{NL,R}}(t) = \frac{\dot{\alpha}_{\text{NL}}}{\alpha_{\text{NL}}} - \frac{\gamma}{2}. \tag{5.13}$$

Inserting both parts into the real part of Eq. (5.4),

$$\dot{\mathcal{C}}_{\text{NL,R}} + \gamma\mathcal{C}_{\text{NL,R}} + \mathcal{C}_{\text{NL,R}}^2 - \mathcal{C}_{\text{NL,I}}^2 + \omega^2 = 0, \tag{5.14}$$

finally yields the modified Ermakov equation

$$\ddot{\alpha}_{\text{NL}} + \left(\omega^2 - \frac{\gamma^2}{4}\right)\alpha_{\text{NL}} = \frac{1}{\alpha_{\text{NL}}^3} \tag{5.15}$$

that was already mentioned in Sect. 4.3 and has the same form as the Ermakov equation (4.66) of the description in expanding coordinates.

Together with the Newton-type equation (5.3) they also form an Ermakov system with the corresponding invariant

$$I_{\text{NL}} = \frac{1}{2}\mathrm{e}^{\gamma t}\left[\left(\dot{\eta}\alpha_{\text{NL}} - \left(\dot{\alpha}_{\text{NL}} - \frac{\gamma}{2}\alpha_{\text{NL}}\right)\eta\right)^2 + \left(\frac{\eta}{\alpha_{\text{NL}}}\right)^2\right] = \text{const.} \tag{5.16}$$

5.2 Position and Momentum Uncertainties in Terms of Ermakov and Riccati Variables

In the same way as in the non-dissipative case (see Sect. 2.3.1), the position and momentum-uncertainties as well as their correlation can be obtained by calculating the corresponding mean values, but now with the WP solution (5.2) of the NLSE (5.1). Expressed in terms of $\alpha_{\rm NL}(t)$ and $\dot{\alpha}_{\rm NL}(t)$, or $\mathcal{C}_{\rm NL,R}(t)$ and $\mathcal{C}_{\rm NL,I}(t)$, they attain the modified form

$$\langle\tilde{x}^2\rangle_{\rm NL}(t) = \frac{\hbar}{2m}\alpha_{\rm NL}^2 = \frac{\hbar}{2m}\frac{1}{\mathcal{C}_{\rm NL,I}}, \tag{5.17}$$

$$\langle\tilde{p}^2\rangle_{\rm NL}(t) = \frac{m\hbar}{2}\left[\left(\dot{\alpha}_{\rm NL} - \frac{\gamma}{2}\alpha_{\rm NL}\right)^2 + \frac{1}{\alpha_{\rm NL}^2}\right] = \frac{m\hbar}{2}\frac{\mathcal{C}_{\rm NL,R}^2 + \mathcal{C}_{\rm NL,I}^2}{\mathcal{C}_{\rm NL,I}}, \tag{5.18}$$

$$\langle[\tilde{x},\tilde{p}]_+\rangle_{\rm NL}(t) = \langle\tilde{x}\tilde{p} + \tilde{p}\tilde{x}\rangle_{\rm NL}(t) = \hbar\alpha_{\rm NL}\left[\dot{\alpha}_{\rm NL} - \frac{\gamma}{2}\alpha_{\rm NL}\right] = \hbar\frac{\mathcal{C}_{\rm NL,R}}{\mathcal{C}_{\rm NL,I}}, \tag{5.19}$$

where it is again obvious that these results minimize the Schrödinger–Robertson uncertainty relation [6, 7]

$$\langle\tilde{x}^2\rangle_{\rm NL}\langle\tilde{p}^2\rangle_{\rm NL} - \left(\frac{1}{2}\langle[\tilde{x},\tilde{p}]_+\rangle_{\rm NL}\right)^2 = \frac{\hbar^2}{4}. \tag{5.20}$$

Also in this dissipative case the explicit expressions for the uncertainties can be obtained by solving the Ermakov equation (5.15) for given $\alpha_{\rm NL}(t_0) = \alpha_{\rm NL,0}$ and $\dot{\alpha}_{\rm NL}(t_0) = \dot{\alpha}_{\rm NL,0}$ or the Riccati equation (5.4) for given κ_0.

In the non-dissipative case, we have seen that the solution of the Ermakov equation can also be obtained using the method of linear invariant operators, outlined in Appendix A, providing two linear independent solutions of the corresponding Newtonian equation are known. The same can be used in this case [8]. To make the formal similarity even clearer, a new (classical) variable $\hat{\xi}(t)$ is introduced that is related to the classical position variable $\eta(t) = \langle x\rangle_{\rm NL}$ that fulfils Eq. (5.3) via

$$\hat{\xi}(t) = \eta(t)e^{\frac{\gamma}{2}t}, \tag{5.21}$$

i.e., in the same way as the canonical variable $\hat{Q}$ in the expanding system is related with the physical variable x. Written in this new variable, the Newtonian equation (5.3) with the friction force formally turns into a Newtonian equation for an oscillator without friction, but shifted frequency $\Omega = \left(\omega^2 - \frac{\gamma^2}{4}\right)^{\frac{1}{2}}$,

$$\ddot{\hat{\xi}} + \left(\omega^2 - \frac{\gamma^2}{4}\right)\hat{\xi} = 0. \tag{5.22}$$

Together with the Ermakov equation (5.15) the corresponding invariant has the same form as in the non-dissipative case (and in the canonical expanding system)

$$I_{\rm NL} = \frac{1}{2}\left[\left(\dot{\hat{\xi}}\alpha_{\rm NL} - \hat{\xi}\dot{\alpha}_{\rm NL}\right)^2 + \left(\frac{\hat{\xi}}{\alpha_{\rm NL}}\right)^2\right] \tag{5.23}$$

(which is also useful later on in connection with the Wigner function), only η being replaced by $\hat{\xi}$ and α by $\alpha_{\rm NL}$.

In order to be consistent with the notation in Sect. 2.3.1 and Appendix A, where $\eta_i(t)$ had been replaced by $f_i(t)$ with $f_1(t) = -\frac{1}{v_0}\eta(t)$ (with $\eta = \langle x\rangle$), here $\hat{\xi}_i$ is replaced by $\xi_i = -\frac{1}{v_0}\hat{\xi}_i$ with $\xi_1 = -\frac{1}{v_0}\eta(t)e^{\frac{\gamma}{2}t}$.

The Ermakov variable $\alpha_{\rm NL\pm}(t)$ can now be obtained from

$$\alpha_{\rm NL\pm}(t) = \left[A\xi_1^2(t) + B\xi_2^2(t) \pm 2C\xi_1(t)\xi_2(t)\right]^{\frac{1}{2}}, \tag{5.24}$$

if two linear independent (but not necessarily orthogonal) solutions of Eq. (5.22) are known with the initial conditions

$$\xi_1(t_0) = 0, \quad \dot{\xi}_1(t_0) = -1, \quad \xi_2(t_0) = 1, \quad \dot{\xi}_2(t_0) = 0. \tag{5.25}$$

The subscript $\pm$ indicates that, due to the $\pm$-sign in front of the bilinear term, two solutions can be expected (at least from a mathematical point of view).

In the non-dissipative case, the coefficients A, B and C in the expression for $\alpha(t)$ are obtained by comparing the most general quadratic invariant with the Ermakov invariant (see Appendix A). Equations (5.15) and (5.22) differ from the Eqs. (2.16) and (2.3), leading to the invariant (2.21), only by the frequency $\Omega = \left(\omega^2 - \frac{\gamma^2}{4}\right)^{\frac{1}{2}}$ instead of ω. However, this frequency is eliminated anyway in the course of the derivation of the invariant. So $I_{\rm NL}$ in (5.23) has the identical form with invariant (2.21), only η is replaced by $\hat{\xi}$ and α by $\alpha_{\rm NL}$. Therefore, the coefficients A, B and C, expressed in terms of α_0 and $\dot{\alpha}_0$ also have an identical form, i.e.,

$$A = \dot{\alpha}_0^2 + \frac{1}{\alpha_0^2}, \quad B = \alpha_0^2, \quad C = \dot{\alpha}_0\alpha_0. \tag{5.26}$$

The major difference is that, in the non-dissipative case, the coefficients are directly related to either one of the initial position, momentum or correlation uncertainties via

$$A_{\rm L} = \dot{\alpha}_0^2 + \frac{1}{\alpha_0^2} = \frac{2}{m\hbar}\langle\tilde{p}^2\rangle_{\rm L,0}, \tag{5.27}$$

$$B_{\rm L} = \alpha_0^2 = \frac{2m}{\hbar}\langle\tilde{x}^2\rangle_{\rm L,0}, \tag{5.28}$$

$$C_{\rm L} = \dot{\alpha}_0\alpha_0 = \frac{1}{\hbar}\langle[\tilde{x},\tilde{p}]_+\rangle_{\rm L,0}, \tag{5.29}$$

whereas, in the dissipative case, these coefficients, expressed in terms of these initial uncertainties, are mixtures of them, i.e.,

$$A_{\rm NL} = \dot{\alpha}_0^2 + \frac{1}{\alpha_0^2} = \frac{2}{m\hbar}\left[\langle\tilde{p}^2\rangle_{\rm NL,0} + m\langle[\tilde{x},\tilde{p}]_+\rangle_{\rm NL,0} + m^2\frac{\gamma^2}{4}\langle\tilde{x}^2\rangle_{\rm NL,0}\right], \tag{5.30}$$

$$B_{\rm NL} = \alpha_0^2 = \frac{2m}{\hbar}\langle\tilde{x}^2\rangle_{\rm NL,0}, \tag{5.31}$$

$$C_{\rm NL} = \dot{\alpha}_0\alpha_0 = \frac{1}{\hbar}\left[\langle[\tilde{x},\tilde{p}]_+\rangle_{\rm NL,0} - m\gamma\langle\tilde{x}^2\rangle_{\rm NL,0}\right]. \tag{5.32}$$

The initial conditions α_0 and $\dot{\alpha}_0$ can also be expressed in terms of the initial uncertainties as

$$\alpha_{\rm NL,0} = \left(\frac{2m}{\hbar}\langle\tilde{x}^2\rangle_{\rm NL,0}\right)^{\frac{1}{2}},$$

$$\dot{\alpha}_{\rm NL,0} = \left(\frac{1}{2m\hbar\langle\tilde{x}^2\rangle_{\rm NL,0}}\right)^{\frac{1}{2}}\left[\langle[\tilde{x},\tilde{p}]_+\rangle_{\rm NL,0} - m\gamma\langle\tilde{x}^2\rangle_{\rm NL,0}\right]. \tag{5.33}$$

As shown in (5.10), for the determination of $\kappa_{\pm 0} = \frac{1}{\mathcal{V}_{\pm 0}}$, $\mathcal{C}_{\rm NL,0}$ must be known that can be expressed in terms of α_0 and $\dot{\alpha}_0$ as

$$\mathcal{C}_{\rm NL,0} = \frac{\dot{\alpha}_0}{\alpha_0} - \frac{\gamma}{2} + {\rm i}\frac{1}{\alpha_0^2}. \tag{5.34}$$

For the damped HO with constant frequency ω_0 this leads to

$$\begin{aligned}\mathcal{V}_{\pm 0} &= \left(\frac{\dot{\alpha}_0}{\alpha_0} - \frac{\gamma}{2}\right) + {\rm i}\frac{1}{\alpha_0^2} - \left[-\frac{\gamma}{2} \pm \sqrt{\frac{\gamma^2}{4} - \omega_0^2}\right]\\ &= \frac{\dot{\alpha}_0}{\alpha_0} + {\rm i}\left[\frac{1}{\alpha_0^2} \mp \sqrt{\omega_0^2 - \frac{\gamma^2}{4}}\right],\end{aligned} \tag{5.35}$$

where $\kappa_{\pm 0}$ is obtained by inverting this expression. For $\omega_0 \geq \frac{\gamma}{2}$, essentially ω_0 in the non-dissipative case is replaced by $\Omega = \left(\omega^2 - \frac{\gamma^2}{4}\right)^{\frac{1}{2}}$. However, for $\omega_0 < \frac{\gamma}{2}$ or $\omega_0 = 0$, additional real contributions (with different signs) occur that must be added to $\frac{\dot{\alpha}_0}{\alpha_0}$. These will be present even if $\dot{\alpha}_0 = 0$ or may compensate the contribution of $\frac{\dot{\alpha}_0}{\alpha_0} \neq 0$. So, depending on the choice of the initial parameters and the relation between friction coefficient γ and frequency ω_0, a variety of qualitatively-different time-dependencies of the uncertainties is possible.

Examples showing qualitatively new effects like bifurcations or resonance-type phenomena are given below. A more systematic discussion of the damped free motion and the damped HO for undercritical damping $\left(\omega_0 > \frac{\gamma}{2}\right)$, overdamping $\left(\omega_0 < \frac{\gamma}{2}\right)$ and the aperiodic limit $\left(\omega_0 = \frac{\gamma}{2}\right)$ as well as the corresponding uncertainties are given in Appendix B.

Before the examples are studied, the linearization of the Riccati equation (5.4) and consequences for a possible Lagrange/Hamilton formalism for the uncertainties in the dissipative case are considered next.

5.3 Linearization of the Riccati Equation and Dissipative Lagrange–Hamilton Formalism for Quantum Uncertainties

The complex NL Riccati equation (5.4) can also be linearized via a logarithmic derivative

$$\frac{2\hbar}{m} y_{\text{NL}} = \mathcal{C}_{\text{NL}} = \frac{\dot{\tilde{\lambda}}}{\tilde{\lambda}} = \frac{\dot{\alpha}}{\alpha} - \frac{\gamma}{2} + \mathrm{i}\dot{\varphi} = \frac{\dot{\lambda}}{\lambda} - \frac{\gamma}{2} \tag{5.36}$$

with $\tilde{\lambda} = \lambda \mathrm{e}^{-\frac{\gamma}{2}t} = \alpha_{\text{NL}} \mathrm{e}^{-\frac{\gamma}{2}t + \mathrm{i}\varphi}$. That is, the relation between $\dot{\varphi}$ and α_{NL}, $\dot{\varphi} = \frac{1}{\alpha_{\text{NL}}^2}$ remains unchanged as in the non-dissipative case (also the relation between $\mathcal{C}_{\text{I}}$ and α remains unchanged, $\mathcal{C}_{\text{NL,I}} = \frac{1}{\alpha_{\text{NL}}^2}$), only the amplitude α_{NL} is affected by the dissipation via the exponential damping factor $\mathrm{e}^{-\frac{\gamma}{2}t}$. Expressed in terms of $\tilde{\lambda}$, the linearized Riccati equation again takes the form of the corresponding (complex) Newtonian equation (now including the friction force),

$$\ddot{\tilde{\lambda}} + \gamma \dot{\tilde{\lambda}} + \omega^2 \tilde{\lambda} = 0. \tag{5.37}$$

In terms of the complex variable without tilde, $\lambda(t)$, a variable that expands exponentially when compared to $\tilde{\lambda}(t)$ (like $\hat{Q}(t)$ in comparison with $x(t)$), the equation of motion is

$$\ddot{\lambda} + \left(\omega^2 - \frac{\gamma^2}{4}\right) \lambda = 0, \tag{5.38}$$

i.e., again like in the expanding canonical system, an undamped oscillator with reduced frequency.

Writing $\tilde{\lambda}$ in polar coordinates as

$$\tilde{\lambda} = \tilde{u} + \mathrm{i}\,\tilde{z} = (u + \mathrm{i}\,z)\mathrm{e}^{-\frac{\gamma}{2}t} = \alpha_{\mathrm{NL}}\,\mathrm{e}^{-\frac{\gamma}{2}t}\cos\varphi + \mathrm{i}\,\alpha_{\mathrm{NL}}\,\mathrm{e}^{-\frac{\gamma}{2}t}\sin\varphi \tag{5.39}$$

shows that the conservation lawx $\dot{\varphi}\alpha_{\mathrm{NL}}^2 = \dot{z}u - \dot{u}z = 1$, in terms of the quantities with tilde, takes the form

$$\dot{\tilde{z}}\tilde{u} - \dot{\tilde{u}}\tilde{z} = \mathrm{e}^{-\gamma t}, \tag{5.40}$$

which will be important in connection with the time-dependent (TD) Green function discussed below.

In this dissipative case, the position and momentum uncertainties, and hence also their contribution(s) to the mean value of the energy, can again be expressed not only in terms of α and $\dot{\alpha}$ but also in terms of $\tilde{\lambda}$ and $\dot{\tilde{\lambda}}$. In particular, one obtains

$$\langle\tilde{x}^2\rangle_{\mathrm{NL}} = \frac{\hbar}{2m}\alpha_{\mathrm{NL}}^2 = \frac{\hbar}{2m}\mathrm{e}^{\gamma t}\left(\tilde{\lambda}\tilde{\lambda}^*\right), \tag{5.41}$$

$$\langle\tilde{p}^2\rangle_{\mathrm{NL}} = \frac{\hbar m}{2}\left[\left(\dot{\alpha}_{\mathrm{NL}} - \frac{\gamma}{2}\alpha_{\mathrm{NL}}\right)^2 + \frac{1}{\alpha_{\mathrm{NL}}^2}\right] = \frac{\hbar m}{2}\mathrm{e}^{\gamma t}(\dot{\tilde{\lambda}}\dot{\tilde{\lambda}}^*), \tag{5.42}$$

$$\langle[\tilde{x},\tilde{p}]_+\rangle_{\mathrm{NL}} = \hbar\left(\dot{\alpha}_{\mathrm{NL}}\alpha_{\mathrm{NL}} - \frac{\gamma}{2}\alpha_{\mathrm{NL}}^2\right) = \frac{\hbar}{2}\mathrm{e}^{\gamma t}\frac{\partial}{\partial t}(\tilde{\lambda}\tilde{\lambda}^*), \tag{5.43}$$

and the quantum mechanical energy contribution can be written as

$$\begin{aligned}\tilde{E}_{\mathrm{NL}} &= \frac{1}{2m}\langle\tilde{p}^2\rangle_{\mathrm{NL}} + \frac{m}{2}\omega^2\langle\tilde{x}^2\rangle_{\mathrm{NL}}\\ &= \frac{\hbar}{4}\left\{\left(\dot{\alpha}_{\mathrm{NL}} - \frac{\gamma}{2}\alpha_{\mathrm{NL}}\right)^2 + \dot{\varphi}^2\alpha_{\mathrm{NL}}^2 + \omega^2\alpha_{\mathrm{NL}}^2\right\}\\ &= \frac{\hbar}{4}e^{\gamma t}(\dot{\tilde{\lambda}}\dot{\tilde{\lambda}}^* + \omega^2\tilde{\lambda}\tilde{\lambda}^*),\end{aligned} \tag{5.44}$$

which is, in general, no longer constant. In particular, it can be shown that

$$\frac{d}{dt}\tilde{E}_{\mathrm{NL}} = -\frac{d}{dt}\left(\frac{\gamma}{4}\langle[\tilde{x},\tilde{p}]_+\rangle_{\mathrm{NL}}\right) \tag{5.45}$$

is valid, from which it follows that the quantity

$$\begin{aligned}\tilde{E}_0 &= \frac{1}{2m}\langle\tilde{p}^2\rangle_{\mathrm{NL}} + \frac{\gamma}{4}\langle[\tilde{x},\tilde{p}]_+\rangle_{\mathrm{NL}} + \frac{m}{2}\omega^2\langle\tilde{x}^2\rangle_{\mathrm{NL}}\\ &= \frac{\hbar}{4}e^{\gamma t}\left\{\dot{\tilde{\lambda}}\dot{\tilde{\lambda}}^* + \frac{\gamma}{2}\left(\dot{\tilde{\lambda}}\tilde{\lambda}^* + \dot{\tilde{\lambda}}^*\tilde{\lambda}\right) + \omega^2\tilde{\lambda}\tilde{\lambda}^*\right\}\\ &= \frac{\hbar}{4}\left\{\dot{\lambda}\dot{\lambda}^* + \left(\omega^2 - \frac{\gamma^2}{4}\right)\lambda\lambda^*\right\} = \text{const.}\end{aligned} \tag{5.46}$$

is a constant of motion for this dissipative system.

Based on this energetic invariant, that can also be written as

$$\tilde{E}_0 = \frac{\hbar}{4}\left(\dot{\alpha}_{\rm NL}^2 + \alpha_{\rm NL}^2\dot{\varphi}^2 + \Omega^2\alpha_{\rm NL}^2\right), \tag{5.47}$$

a Lagrangian/Hamiltonian formalism for the quantum uncertainties can be obtained in exactly the same way as described in Sect. 2.6, only ω must be replaced by $\Omega = \left(\omega^2 - \frac{\gamma^2}{4}\right)^{\frac{1}{2}}$ and the relevant variable $\lambda(t)$ is expanding exponentially compared to $\tilde{\lambda}$, i.e., $\lambda = \tilde{\lambda}e^{\frac{\gamma}{2}t}$.

In terms of real and imaginary parts of $\lambda = u + \mathrm{i}z$, the Ermakov invariant in the dissipative case can also be written as

$$\begin{aligned} I_{\rm NL} &= \frac{1}{2}\left(\frac{p_0\alpha_0}{m}\right)^2\left[\left(\frac{u}{\alpha_{\rm NL}}\right)^2 + \left(\frac{z}{\alpha_{\rm NL}}\right)^2\right] \\ &= \frac{1}{2}\left(\frac{p_0\alpha_0}{m}\right)^2\left[\sin^2\varphi + \cos^2\varphi\right] = \frac{1}{2}\left(\frac{p_0\alpha_0}{m}\right)^2. \end{aligned} \tag{5.48}$$

Comparison with (5.23) shows that $z = \tilde{z}\,e^{\frac{\gamma}{2}t} = \frac{m}{p_0\alpha_0}\hat{\xi} = \frac{m}{p_0\alpha_0}\eta\, e^{\frac{\gamma}{2}t}$, i.e., $\tilde{z} = \frac{m}{p_0\alpha_0}\eta(t)$, which is confirmed by the TD Green function shown below.

5.4 New Qualitative Quantum Effects Induced by a Dissipative Environment

5.4.1 Increase of Ground State Energy Due to Interaction with an Environment

Already the (undercritically) damped HO with constant frequency ω_0 makes it possible to show characteristic differences between the dissipative and non-dissipative case. Inserting a WP with constant width into NLSE (5.1) provides, as particular solution of the Riccati equation (5.4), the relation (5.8), $\tilde{C}_\pm = -\frac{\gamma}{2} \pm \mathrm{i}\Omega$ with $\Omega = \sqrt{\omega_0^2 - \frac{\gamma^2}{4}}$. Like in the case without dissipation, only the $+$-sign leads to normalizable Gaussian functions (for $\omega_0 > \frac{\gamma}{2}$). So, in agreement with $\dot{\alpha} = 0$, real and imaginary parts can be written as

$$\tilde{C}_{+\rm R} = -\frac{\gamma}{2}, \quad \tilde{C}_{+\rm I} = \Omega = \frac{\hbar}{2m\langle\tilde{x}^2\rangle_{\rm NL}} = \frac{1}{\alpha_0^2}. \tag{5.49}$$

The position uncertainty and the WP width can therefore be expressed in terms of

$$\langle\tilde{x}^2\rangle_{\rm NL} = \frac{\hbar}{2m\Omega}. \tag{5.50}$$

Expression (5.18) for the momentum uncertainty reduces to

$$\langle\tilde{p}^2\rangle_{\text{NL}} = \frac{m\hbar}{2}\left[\frac{\gamma^2}{4}\alpha_0^2 + \frac{1}{\alpha_0^2}\right] = \frac{m\hbar}{2}\frac{\omega_0^2}{\Omega}. \tag{5.51}$$

The quantum mechanical contribution to the energy, calculated as the mean value of H_{NL}, can therefore be written as [1]

$$\tilde{E}_{\text{NL}} = \frac{1}{2m}\langle\tilde{p}^2\rangle_{\text{NL}} + \frac{m}{2}\omega_0^2\langle\tilde{x}^2\rangle_{\text{NL}} = \frac{\hbar}{2}\frac{\omega_0^2}{\Omega} = \frac{\hbar}{2}\omega_0\left(\frac{\omega_0}{\Omega}\right) > \frac{\hbar}{2}\omega_0, \tag{5.52}$$

which is larger than in the undamped case! This is similar to the situation in classical Brownian motion described by the Langevin equation (including a stochastic fluctuating force). Even if the motion of the centre of mass of the observable system has (on an average) come to rest due to the action of the friction force, some erratic motion still remains around this final position due to the interaction with the surrounding via the fluctuating force. Even if this force vanishes on an average, this, for example, does not apply to quantities related to the square of this force. This supplies an additional contribution to the system's energy. Making the conventional assumption that the final state of the system's evolution is thermal equilibrium, via the equipartition theorem, this energy contribution can be set equal to $\frac{1}{2}\mathsf{k}_\mathsf{B}\mathsf{T}$ with k_B being Boltzmann's constant and T the temperature of the environment. In our dissipative quantum case, a similar situation appears to exist.

Comparison with Kostin's approach, where only the real part of the logarithmic nonlinearity enters (see (4.85)), shows that in his case the imaginary part of the particular solution is unchanged by the NL term, i.e., $\tilde{C}_{\text{K,I}} = \omega_0 = \frac{\hbar}{2m\langle\tilde{x}^2\rangle_{\text{K}}}$. Therefore, his energy contribution $\tilde{E}_{\text{K}} = \frac{\hbar}{2}\omega_0$ is the ground state energy of the undamped HO. Consequently, the increase of $\tilde{E}$, similar to the one caused by the stochastic force in the classical case, is due entirely to the imaginary part of the logarithmic nonlinearity and thus (via (4.113)) due to the diffusion term in the Smoluchowski equation (4.109).

This similarity can be taken even further, rewriting result (5.52) as

$$\tilde{E}_{\text{NL}} = \frac{\hbar}{2}\Omega\left(1 + \frac{\frac{\gamma^2}{4}}{\Omega}\right) = \frac{\hbar}{2}\Omega + m\frac{\gamma^2}{4}\langle\tilde{x}^2\rangle = \frac{\hbar}{2}\Omega + \frac{m}{2}\gamma D, \tag{5.53}$$

where relations (5.50) and (4.121), i.e., $D = \frac{\gamma}{2}\langle\tilde{x}^2\rangle$, have been used. With the same assumption of thermal equilibrium and replacing the diffusion coefficient according to (4.8) by the Einstein relation , $D = \frac{\mathsf{k}_\mathsf{B}\mathsf{T}}{m\gamma}$, Eq. (5.53) finally turns into

$$\tilde{E}_{\text{NL}} = \frac{\hbar}{2}\Omega + \frac{1}{2}\mathsf{k}_\mathsf{B}\mathsf{T}. \tag{5.54}$$

The effect of the imaginary logarithmic term is therefore twofold.

1. It changes the frequency of the ground state oscillation from ω_0 to the same damped frequency Ω that is the characteristic one for the system's classical oscillatory part.

2. It supplies an additional contribution to the energy from the interaction with the (non-observable) environment that, for the above choice of D, is equal to the thermal energy $\frac{1}{2}\mathrm{k_B T}$!

However, the choice of the diffusion coefficient D characterizes the specific environment that is considered. Therefore, different choices of D also lead to different expressions for the additional energy in (5.53).

One other interesting case occurs if, as already mentioned in Sect. 4.4.3, according to Grössing et al. [9] one identifies the Einstein diffusion coefficient with the quantum mechanical one (providing the SE is considered a diffusion equation with imaginary diffusion coefficient), i.e., $D = \frac{\mathrm{k_B T}}{m\gamma} = \frac{\hbar}{2m}$ (which would implicate $\frac{1}{2}\mathrm{k_B T} = \frac{\hbar}{2}\frac{\gamma}{2}$). Then the imaginary part $\tilde{W}_{\mathrm{I}}$ of the logarithmic nonlinearity (4.118) turns into

$$\tilde{W}_{\mathrm{I}} = -\mathrm{i T k_B} \left(\ln \varrho_{\mathrm{NL}} - \langle \ln \varrho_{\mathrm{NL}} \rangle\right) , \tag{5.55}$$

where $-\mathrm{k_B}\langle \ln \varrho \rangle = -\mathrm{k_B} \int_{-\infty}^{+\infty} dx\, \varrho \ln \varrho$ has a form like the statistical definition of entropy, $\mathcal{S}$. So, the mean value of the linear Hamiltonian, $\langle H_{\mathrm{L}} \rangle = E$, together with the second term of (5.55) would look like $E - \mathrm{i T}\mathcal{S}$, i.e., similar to the expression for the free energy, only here, again with the imaginary unit i appearing in the quantum mechanical context (see also [10]).

5.4.2 Bifurcation and Non-diverging Uncertainty Product

Next the damped free motion is considered and compared with the conservative case. Here, no WP solution with constant width exists (in both cases). The classical part of the dynamics, i.e., the motion of the WP-maximum, is affected by the dissipative environment as is expected. Whereas the isolated system moves with constant velocity $\langle v \rangle_{\mathrm{L}} = \dot{\eta}_{\mathrm{L}} = v_0 = \mathrm{const.}$, the damped system is slowed down by the interaction according to $\langle v \rangle_{\mathrm{NL}} = \dot{\eta}_{\mathrm{NL}} = v_0 e^{-\gamma t}$. Therefore, the position is no longer growing proportional to time t but approaches a finite value $\langle x \rangle_{\mathrm{NL}}(t \to \infty) = \frac{v_0}{\gamma}$, according to $\langle x \rangle_{\mathrm{NL}} = \eta_{\mathrm{NL}} = \frac{v_0}{\gamma}\left(1 - \mathrm{e}^{-\gamma t}\right)$.

Considering the time-dependence of the WP-width, the particular solution (5.8) for $\omega_0 = 0$ and $\gamma = 0$, i.e., for the isolated system, takes the value $\tilde{C}_{\pm} = \pm 0$. That means, for the particular solution there is no term quadratic in x in the exponent. This agrees with the statement in Sect. 5.1 that the limit of the WP solution of the HO with constant width, corresponding to a particular solution, in the limit $\omega_0 \to 0$, does not lead to the free motion Gaussian WP but to a plane wave, i.e., a function with no term quadratic in x but only linear in this variable.

Consequently in the corresponding Bernoulli equation, the linear term also vanishes. Therefore, one obtains only one solution that leads to the well-known position uncertainty that is growing proportional to t^2, an equally increasing uncertainty product due to a constant momentum uncertainty and, as a consequence of this, to a constant energy contribution, i.e.,

$$\langle \tilde{x}^2 \rangle_{\mathrm{L}} = \langle \tilde{x}^2 \rangle_0 \{1 + (\beta_0 t)^2\} = \frac{\hbar}{2m\beta_0}\{1 + (\beta_0 t)^2\}, \tag{5.56}$$

$$\langle \tilde{p}^2 \rangle_{\mathrm{L}} = \langle \tilde{p}^2 \rangle_0 = \frac{m\hbar}{2}\beta_0 = \text{const.}, \tag{5.57}$$

$$U_{\mathrm{L}} = \frac{\hbar^2}{4}\{1 + (\beta_0 t)^2\}, \tag{5.58}$$

$$\tilde{E}_{\mathrm{L}} = \frac{\hbar}{4}\beta_0, \tag{5.59}$$

where $\beta_0 = \frac{\hbar}{2m\langle \tilde{x}^2 \rangle_0}$ is still used as above.

For $t = 0$ the WP is a minimum uncertainty WP as $U_{\mathrm{L}}(0) = \frac{\hbar^2}{4}$; but then $U_{\mathrm{L}}(t)$ is spreading quadratically in time and diverging for $t \to \infty$.

In the dissipative case, i.e., $\omega_0 = 0$ but $\gamma \neq 0$, the particular solution (5.8) does not just vanish but can take two different (real) values, $\tilde{C}_+ = 0$, $\tilde{C}_- = -\gamma$. As a consequence, also the coefficient of the linear term in Bernoulli equation (5.6) can take two different values, $2(\tilde{C}_\pm + \frac{\gamma}{2}) = \pm\gamma$, leading to the same values with opposite sign in the exponents of the solutions (5.9).

However, the problem is not yet completely defined unless the initial condition κ_0, or equivalently α_0 and $\dot{\alpha}_0$, have been chosen. To make the resulting expressions as simple as possible and directly comparable with the ones given in Eqs. (5.56)–(5.59), a situation with no initial position-momentum-uncertainty correlations is chosen,[2] i.e., $\langle [\tilde{x}, \tilde{p}]_+ \rangle_{\mathrm{NL}}(0) = \hbar(\dot{\alpha}_{\mathrm{NL},0} - \frac{\gamma}{2}\alpha_{\mathrm{NL},0})\alpha_{\mathrm{NL},0} = 0$. This corresponds, in the conservative case, to the choice $\dot{\alpha}_0 = 0$ which avoids a term linear in t in the expression for $\langle \tilde{x}^2 \rangle_{\mathrm{L}}$.

The two different values $\pm\gamma$ lead to two solutions corresponding to two different states with different physical properties. In this sense one could say that a kind of degeneracy, existing in the conservative reversible case, has been removed by the breaking of time-reversal symmetry in the dissipative irreversible case. This splitting has close formal similarity with a Hopf-bifurcation where, depending on the value of a critical parameter, a system can either approach a fixed point or a limit cycle [11, 12]. This will be discussed again in Sect. 7.3 but there only for real Bernoulli equations whereas Eq. (5.6) is complex.

[2] The more general case is discussed in Appendix B.

From Eq. (5.9) follows for $+\gamma$:

$$\langle \tilde{x}^2 \rangle_+ = \langle \tilde{x}^2 \rangle_0 \left\{ \mathrm{e}^{\gamma t} + \left(\frac{\beta_0}{\frac{\gamma}{2}} \right)^2 \sinh^2 \frac{\gamma}{2} t \right\}, \tag{5.60}$$

$$\langle \tilde{p}^2 \rangle_+ = \langle \tilde{p}^2 \rangle_0 \mathrm{e}^{-\gamma t}, \tag{5.61}$$

$$U_+ = \frac{\hbar^2}{4} \left\{ 1 + \left[\left(\frac{\beta_0}{\gamma} \right) \left(1 - \mathrm{e}^{-\gamma t}\right) \right]^2 \right\}, \tag{5.62}$$

$$\tilde{E}_+ = \frac{\hbar}{4} \beta_0 \mathrm{e}^{-\gamma t}, \tag{5.63}$$

and for $-\gamma$:

$$\langle \tilde{x}^2 \rangle_- = \langle \tilde{x}^2 \rangle_0 \left\{ \mathrm{e}^{-\gamma t} + \left(\frac{\beta_0}{\frac{\gamma}{2}} \right)^2 \sinh^2 \frac{\gamma}{2} t \right\}, \tag{5.64}$$

$$\langle \tilde{p}^2 \rangle_- = \langle \tilde{p}^2 \rangle_0 \left\{ 1 + \left(\frac{\gamma}{\beta_0} \right)^2 \right\} \mathrm{e}^{-\gamma t}, \tag{5.65}$$

$$U_- = \frac{\hbar^2}{4} \left\{ 1 + \left[\left(\frac{\beta_0}{\gamma} \right) \left(1 - \mathrm{e}^{-\gamma t}\right) - \left(\frac{\gamma}{\beta_0} \right) \mathrm{e}^{-\gamma t} \right]^2 \right\}, \tag{5.66}$$

$$\tilde{E}_- = \frac{\hbar}{4} \beta_0 \left\{ 1 + \left(\frac{\gamma}{\beta_0} \right)^2 \right\} \mathrm{e}^{-\gamma t}. \tag{5.67}$$

In both cases, the corresponding WPs are spreading faster in time than in the reversible case (due to the $\sinh^2$-term even exponentially; an explanation for this being reasonable is given in the next sub-section and, more detailed, in [2]). Although the momentum uncertainties both decrease exponentially in time, the uncertainty products never violate Heisenberg's principle.

However, there are not only similarities but also distinct differences between the two WP solutions. While the $+\gamma$-WP has an initial value of the energy contribution that is identical to the one of the (constant) value of the reversible case, $\tilde{E}_+(0) = \frac{\hbar}{4}\beta_0 = \tilde{E}_{\mathrm{L}}$ the initial value of the $-\gamma$-WP has a larger contribution, given by

$$\tilde{E}_- (0) = \frac{\hbar}{4} \beta_0 \left\{ 1 + \left(\frac{\gamma}{\beta_0} \right)^2 \right\} > \tilde{E}_+ (0) \,. \tag{5.68}$$

What might be the physical interpretation of the energy difference between $\tilde{E}_+$ and $\tilde{E}_-$? Using the definition of β_0 as given above, the initial difference between the two energy values can be expressed as

$$\Delta\tilde{E}_0 = \tilde{E}_-(0) - \tilde{E}_+(0) = \frac{\hbar}{4}\frac{\gamma^2}{\beta_0} = \frac{m}{2}\gamma^2\langle\tilde{x}^2\rangle_0. \tag{5.69}$$

From this it follows that $\Delta\tilde{E}_0$ is independent of $\hbar$ but depends only on the parameters characterizing the environment. It vanishes for vanishing friction coefficient γ or, equivalently, for vanishing (initial value of the) diffusion coefficient $D_0 = \frac{\gamma}{2}\langle\tilde{x}^2\rangle_0$ in the Smoluchowski equation. Assuming again, as in the previous subsection, that D_0 could be identified with the Einstein relation $D = \frac{\mathsf{k_B T}}{m\gamma}$, would lead to

$$\Delta\tilde{E}_0 = \mathsf{k_B T}. \tag{5.70}$$

So, the two states differ initially by an energy that is related to the thermal energy of the environment in a way that only the energy $\tilde{E}_-$, not $\tilde{E}_+$, is raised by a corresponding amount before both exponentially decay (see also [13]).

Similar properties also apply to the initial values for the uncertainty product which, in the $+\gamma$-case, simply takes the minimum value $U_+(0) = \frac{\hbar^2}{4}$, whereas the value for the $-\gamma$-WP is larger,

$$U_-(0) = \frac{\hbar^2}{4}\left\{1 + \left(\frac{\gamma}{\beta_0}\right)^2\right\} > U_+(0). \tag{5.71}$$

The difference can be interpreted in the same way as the energy difference $\Delta\tilde{E}_0$. However, both uncertainty products do not diverge for $t \to \infty$, as in the conservative case, but asymptotically approach the *same finite* maximum value

$$U_{\pm\,\max} = U_\pm(t \to \infty) = \frac{\hbar^2}{4}\left[1 + \left(\frac{\beta_0}{\gamma}\right)^2\right]. \tag{5.72}$$

Some consequences:

1. For a system with large mass m, due to $\beta_0 = \frac{\hbar}{2m\langle\tilde{x}^2\rangle_L}$ the maximum value $U_{\pm\,\max}$ is small.
2. For larger γ, i.e., more frequent interaction between the system and the environment, the value of $U_{\pm\,\max}$ is getting smaller. This is similar to the quantum Zeno effect where frequent observation of (i.e., interaction with) the quantum system reduces the uncertainty of its state.
3. Even in the limit of permanent interaction (or observation), i.e., $\gamma \to \infty$, it is not possible to go below the lower bound of $\frac{\hbar^2}{4}$ for the uncertainty product.
4. Expressed in terms of the quantum mechanical diffusion coefficient $D_{qm} = \frac{\hbar}{2m}$ and the initial one of our Smoluchowski equation, $D_0 = \frac{\gamma}{2}\langle\tilde{x}^2\rangle_0$, the final uncertainty product can be expressed as $U_{\pm\,\max} = \frac{\hbar^2}{4}\left[1 + \left(\frac{1}{2}\frac{D_{qm}}{D_0}\right)^2\right]$ (for further details, see [2]).

5.4.3 Modified Plane Waves and Nonlinear Superposition

From the expressions for the position uncertainties of the damped free motion, Eqs. (5.60) and (5.64), it is obvious that in both cases (at least due to the $\sinh^2$-term) the corresponding WPs are spreading faster than in the conservative case. This is in contrast with some other models that predict a slower spreading or, to the contrary, shrinking of the WP width, sometimes even leading to an unphysical localisation with a position uncertainty shrinking to a delta function (while the momentum uncertainty stays finite, thus leading to a violation of the uncertainty principle like the one mentioned in Sect. 4.3.1). The suppression of the spreading is then interpreted in a way as if the WP would be a kind of matter density whose expansion is hindered by the collisions with the non-empty environment. As the WP should be considered a distribution of probability, not of matter, this interpretation is obviously wrong. To find arguments supporting the faster spreading of the damped WP, the reasons for the spreading in the conservative case are now firstly discussed in more detail and, afterwards, the modifications due to the interaction with the environment.

The Gaussian WP is not the only solution of the TDSE for the free motion, there are also (particular) solutions with the form of the plane waves

$$\Psi_{k,\mathrm{L}}(x,t) = \left(\frac{1}{2\pi}\right)^{\frac{1}{2}} \exp\{\mathrm{i}\,[kx - \omega_k t]\} \tag{5.73}$$

describing a wave travelling with the constant "group velocity" (= particle velocity)[3] $v_g = \frac{\hbar}{m}k$ (see also [6]) and constant energy $\hbar\omega_k$ with $\omega_k = \frac{\hbar}{2m}k^2$. The Gaussian WP solution of the same TDSE can be expressed as superposition of these plane waves for all possible k-values,

$$\begin{aligned} \Psi_{\mathrm{WP,L}}(x,t) &= \int_{-\infty}^{+\infty} dk A_{\mathrm{L}}(k)\Psi_{k,\mathrm{L}}(k,t) \\ &= \left(\frac{1}{2\pi}\right)^{\frac{1}{2}} \int_{-\infty}^{+\infty} dk A_{\mathrm{L}}(k) e^{\mathrm{i}[kx-\omega_k t]} \end{aligned} \tag{5.74}$$

where the subscripts L denotes quantities obeying the linear TDSE.

The expansion coefficients $A_{\mathrm{L}}(k)$ can be determined from

$$A_{\mathrm{L}}(k) = \int_{-\infty}^{+\infty} dx \Psi_{\mathrm{WP,L}}(x,t)\Psi^*_{k,\mathrm{L}}(x,t). \tag{5.75}$$

[3]The phase velocity v_p is related to v_g via $v_p = \frac{\omega_k}{k} = \frac{\hbar}{2m}k = \frac{1}{2}v_g$.

As these coefficients are not TD, it is practical to solve this integral for the most simple case, i.e., for $t = 0$,

$$A_{\mathrm{L}}(k) = \left(\frac{1}{2\pi}\right)^{\frac{1}{2}} \int_{-\infty}^{+\infty} dx \Psi_{\mathrm{WP,L}}(x, 0) \mathrm{e}^{-\mathrm{i}kx}, \tag{5.76}$$

which means that the Fourier transform of the initial WP $\Psi_{\mathrm{WP,L}}(x, 0)$ must be determined.

As shown in Sect. 2.7, the Fourier transform of the WP in position space leads to the corresponding one in momentum space. The width of this WP can be expressed in terms of the inverse of the quantity fulfilling the complex Riccati equation (2.4) in position space, i.e., $\left(\frac{am}{i\hbar}\right) = \left(\frac{2\hbar}{m}y\right)^{-1} = C^{-1}$ where particularly the real part of $a(t)$ is related to the momentum uncertainty via $\frac{a_{\mathrm{R}}m}{\hbar} = \frac{m\hbar}{2\langle\tilde{p}^2\rangle}$ (see also Eqs. (2.98–2.99)). Expressed with these two quantities, the coefficients $A_{\mathrm{L}}(k)$ can be written as

$$\begin{aligned} A_{\mathrm{L}}(k) &= \left(\frac{a_{\mathrm{R},0}}{\pi}\right)^{\frac{1}{4}} \exp\left\{-\frac{a_{\mathrm{R},0}}{2}\left(k_{\mathrm{WP}} - k\right)^2\right\} \\ &= \left(\frac{\hbar^2}{2\pi\langle\tilde{p}^2\rangle_0}\right)^{\frac{1}{4}} \exp\left\{-\frac{\hbar^2\left(k_{\mathrm{WP}} - k\right)^2}{4\langle\tilde{p}^2\rangle_0}\right\} \end{aligned} \tag{5.77}$$

with k_{WP} being the k-value of the WP related to the constant velocity of the classical free particle via $k_{\mathrm{WP}} = \frac{1}{\hbar}\langle p\rangle = \frac{m}{\hbar}\dot{\eta} = \frac{m}{\hbar}v_0$.

The reason for the spreading of the WP can be traced back to the fact that the plane waves that are superimposed in the WP have different k-values and thus different group and phase velocities, leading to a dephasing of the initially well-localized WP.

In order to find the modifications due to the interaction with the dissipative environment, a question must first be answered. Are there also wave-function solutions of the NLSE comparable with the plane waves of the linear problem?

Such functions $\Psi_{k,\mathrm{NL}}(x, t; k(t))$ can actually be determined and have the form [2]

$$\Psi_{k,\mathrm{NL}}(x, t; k(t)) = \left(\frac{1}{2\pi}\right)^{\frac{1}{2}} \exp\left\{\mathrm{i}\left[kx - \int_0^t \frac{\hbar k^2}{2m} dt' + \gamma \int_0^t k\langle x\rangle dt'\right]\right\}, \tag{5.78}$$

which is similar to plane waves and turns into these for $\gamma \to 0$. It can be shown that these functions are eigenfunctions of the momentum operator with TD eigenvalue $\hbar k(t)$ where

$$k(t) = k_0 \mathrm{e}^{-\gamma t} \tag{5.79}$$

and also eigenfunction of the kinetic energy operator, also with TD eigenvalue $\langle T\rangle_{\mathrm{NL}} = \frac{1}{2m}\hbar^2 k^2(t) = \frac{1}{2m}\hbar^2 k_0^2 \mathrm{e}^{-2\gamma t}$. However, due to the occurrence of $\langle x\rangle$ in the exponent, these functions are no eigenfunctions of the energy operator $i\hbar\frac{\partial}{\partial t}$ but still provide the correct mean value of the energy $\left\langle \mathrm{i}\hbar\frac{\partial}{\partial t}\right\rangle_{\mathrm{NL}} = \frac{1}{2m}\langle p\rangle_{\mathrm{NL}}^2$ with $\langle p\rangle_{\mathrm{NL}} = \hbar k(t)$ if $\dot{k} = -\gamma k$ is taken into account.

Like in the linear case, an expansion of the Gaussian WP solution of the NLSE for the damped free motion in terms of the functions $\Psi_{k,\mathrm{NL}}(x, t; k(t))$ is possible according to

$$\Psi_{\mathrm{WP,NL}}(x, t) = \int_{-\infty}^{+\infty} dk A_{\mathrm{NL}}(k) \Psi_{k,\mathrm{NL}}(x, t), \tag{5.80}$$

where the coefficients $A_{\mathrm{NL}}(k)$ can also be determined via

$$A_{\mathrm{NL}}(k) = \int_{-\infty}^{+\infty} dx \Psi_{\mathrm{WP,NL}}(x, t) \Psi^*_{k,\mathrm{NL}}(x, t). \tag{5.81}$$

However, in contrast with the linear problem, these coefficients A_{NL} are now TD due to the time-dependence of $k = k(t)$. Therefore, it is not sufficient to determine the coefficients for the most simple case $t = 0$ but one has to use the TD functions $\Psi_{\mathrm{WP,NL}}(x, t)$ and $\Psi_{k,\mathrm{NL}}(x, t)$ for the calculation. The coefficients obtained in this way are now complex functions where the imaginary term $a_{\mathrm{I}}(t)$ also occurs in the phase of these functions. This phase is irrelevant for the discussion of the spreading properties of the WP and will be ignored in the following (for further details, see [2]). Only the absolute value of the A_{NL} is considered which is given by

$$\begin{aligned} |A_{\mathrm{NL}}(k)| &= \left(\frac{a_{\mathrm{R}}(t)}{\pi}\right)^{\frac{1}{4}} \exp\left\{-\frac{a_{\mathrm{R}}(t)}{2}\tilde{k}^2\right\} \\ &= \left(\frac{\hbar^2}{2\pi\langle\tilde{p}^2\rangle_{\mathrm{NL}}(t)}\right)^{\frac{1}{4}} \exp\left\{-\frac{\hbar^2\tilde{k}^2}{4\langle\tilde{p}^2\rangle_{\mathrm{NL}}(t)}\right\} \end{aligned} \tag{5.82}$$

with $|\tilde{k}| = |k_{\mathrm{WP}} - k|$. This quantity depends on time (via $k(t)$) and on the difference between the k-value under consideration and the fixed value $k_{\mathrm{WP}}(t)$ determining the classical momentum $\langle p\rangle_{\mathrm{NL}} = \hbar k_{\mathrm{WP}}$ of the WP. Regarding the change in time, one finds[4]

$$\frac{d}{dt}|A_{\mathrm{NL}}(k)| = \gamma \left\{\frac{a_{\mathrm{R}}(t)}{2}\tilde{k}^2 + \frac{1}{4}\right\} \left(\frac{a_{\mathrm{R}}(t)}{\pi}\right)^{\frac{1}{4}} \exp\left\{-\frac{a_{\mathrm{R}}(t)}{2}\tilde{k}^2\right\}. \tag{5.83}$$

On the one hand, this shows that the change in time of $|A_{\mathrm{NL}}(k)|$ is proportional to the friction coefficient γ and vanishes for $\gamma \to 0$; on the other hand, it depends on $|A_{\mathrm{NL}}|$ itself and on $|\tilde{k}|$. The dependence on $|\tilde{k}|$ is of such that one term exists which is proportional to $\tilde{k}^2 e^{-\frac{a_{\mathrm{R}}}{2}\tilde{k}^2}$, therefore corresponding to a Maxwell–Boltzmann distribution, and a second term is proportional to $e^{-\frac{a_{\mathrm{R}}}{2}\tilde{k}^2}$, thus describing an exponential decay. This second term requires that also for $\tilde{k} = 0$, i.e., for $k = k_{\mathrm{WP}}$, the absolute value $|A_{\mathrm{NL}}|$ changes in time despite the fact that the first term vanishes in this case; i.e., *all* coefficients are TD.

[4]Note that $\frac{d}{dt}|A_{\mathrm{NL}}| \neq |\frac{d}{dt}A_{\mathrm{NL}}|$.

A rough qualitative draft of the change in time of $|A_{\mathrm{NL}}|$ as a function of $|\tilde{k}|$ shows a curve starting at $|\tilde{k}| = 0$ with the positive value $\frac{1}{4}\gamma\left(\frac{a_{\mathrm{R}}}{\pi}\right)^{\frac{1}{4}}$, running through a maximum at $|\tilde{k}_{\max}| = \left(\frac{3}{2a_R}\right)^{\frac{1}{2}}$ with the maximum value of $\mathrm{e}^{-\frac{3}{4}}\left[\gamma\left(\frac{a_{\mathrm{R}}}{\pi}\right)^{\frac{1}{4}}\right]$ and finally approaching the $|\tilde{k}|$-axis asymptotically (see Fig. 5.1).

The maximum value itself, as well as its position, i.e., $|\tilde{k}_{\max}|$, are TD whereby the position of the maximum approaches $|\tilde{k}| = 0$ for $t \to \infty$ and its value decreases to $\frac{1}{4}\gamma\left(\frac{a_{\mathrm{R}}}{\pi}\right)^{\frac{1}{4}}$ which itself approaches (due to $a_{\mathrm{R}}(t)$), zero in this limit.

The faster spreading of the free-motion WP solution of the logarithmic NLSE now becomes obvious and reasonable when compared with the linear case. In addition to the dephasing effect of the linear case based on different k-values, these values also vary in time in the NL case because $k(t) = k_0\mathrm{e}^{-\gamma t}$, wherefrom velocity differences result which are variable in time. Furthermore, due to the time-dependence of the coefficients $A_{\mathrm{NL}}(k)$, the contributions of the individual components $\Psi_{k,\mathrm{NL}}$ to the WP vary also in time. That means there is at least a twofold reason for the faster dephasing and, hence, spreading of the WP in comparison with the linear case.

Another interesting aspect of the results above is the fact that here an example for a NL differential equation obviously exists that possesses a solution with the form of a Gaussian function that can be expanded in terms of wave functions that are also themselves solutions of the *same* NL differential equation. This is usually valid only for linear differential equations. In this special case this might be due to the fact that

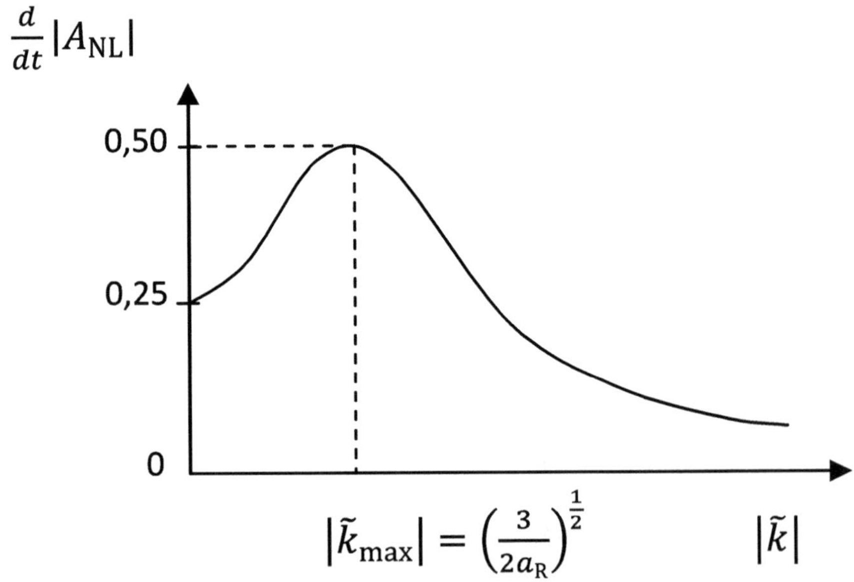

Fig. 5.1 Rough qualitative draft of $\frac{d}{dt}|A_{\mathrm{NL}}|$ as a function of $|\tilde{k}| = |k_{\mathrm{WP}} - k|$. Ordinate in units of $\gamma\left(\frac{a_{\mathrm{R}}}{\pi}\right)^{\frac{1}{4}}$

the absolute values of the plane-wave-type functions $\Psi_{k,\mathrm{NL}}$ are not depending on the position x. Therefore, the non-unitary transformation (4.135) described in Sect. 4.5 to get from a NL physical level to a linear canonical level essentially turns the exponentially-decaying k-values into constant ones but has no further consequences for the position-dependence of the individual functions $\Psi_{k,\mathrm{NL}}$. So the Gaussian WP can be expanded in terms of the usual plane waves as well as in terms of the modified ones (Eq. (5.78)) providing they also form a complete (orthonormal) basis set of functions at any moment in time.

In the case of the (undercritically) damped HO, the situation is different. The Gaussian WP solution (with constant width) can also be expanded in terms of the stationary solutions of the HO, i.e., a Gaussian (ground state) function multiplied by Hermitian polynomials where only ω_0 is replaced by $\Omega = \left(\omega_0^2 - \frac{\gamma^2}{4}\right)^{\frac{1}{2}}$. As these functions also provide a complete (orthonormal) basis this is not surprising. However, in this case, the individual (basis) functions are no longer solutions of the logarithmic NLSE because also the (real) Hermitian polynomials and their position dependence are affected by the above-mentioned non-unitary transformation.

5.4.4 Environmentally-Induced Tunnelling Currents and Resonant Energy Back-Transfer

For the damped HO three qualitatively different cases must be distinguished. As mentioned above, in classical mechanics a friction force with friction coefficient γ changes the frequency ω_0 of the undamped HO into $\Omega = \left(\omega_0^2 - \frac{\gamma^2}{4}\right)^{\frac{1}{2}}$.

For $\omega_0 > \frac{\gamma}{2}$, the *undercritical damping*, essentially ω_0 is replaced by the smaller frequency in the expression for the trajectory and the amplitude is damped by a factor $\mathrm{e}^{-\frac{\gamma}{2}t}$. This trajectory then also describes the motion of the maximum of the WP that is the solution of the logarithmic NLSE. In the Ermakov equation for $\alpha(t)$, only ω_0 is replaced by Ω and no additional damping term occurs. So, like in the undamped case, one obtains in this case WPs with constant widths (now depending on Ω instead of ω_0) and oscillating widths.

The quantity $\alpha(t)$ is also in the damped cases the one that determines the tunnelling currents or velocities. This becomes obvious when having another look at the Smoluchowski equation (4.109), written in the form

$$\frac{\partial}{\partial t}\varrho(x,t) + \frac{\partial}{\partial x}\left(\varrho v_- - D\frac{\partial}{\partial x}\varrho\right) = \frac{\partial}{\partial t}\varrho + \frac{\partial}{\partial x}\left(\varrho v_{\mathrm{T}}\right) = 0 \tag{5.84}$$

where the total velocity field v_{T}, again like in the continuity equation (2.43) has a convective term v_- (defined in (2.45)) but, in addition, a diffusive term v_{D}, i.e., $v_{\mathrm{T}} = v_- + v_{\mathrm{D}}$. Expressed in terms of α, $\dot{\alpha}$ and $\dot{\eta}$, v_- has now a slightly different form from the non-dissipative case, namely

$$v_- = \dot{\eta} + \left(\frac{\dot{\alpha}}{\alpha} - \frac{\gamma}{2}\right)\tilde{x}, \tag{5.85}$$

and, with the help of $D = \frac{\gamma}{2}\langle\tilde{x}^2\rangle$, v_{D} can be written as

$$v_{\text{D}} = -D\frac{\frac{\partial}{\partial x}\varrho}{\varrho} = \frac{\gamma}{2}\tilde{x}. \tag{5.86}$$

Therefore, also in this case the total velocity and hence the tunnelling currents depend on the relative change of the WP width, i.e.,

$$v_{\text{T}} = v_- + v_{\text{D}} = \dot{\eta} + \frac{\dot{\alpha}}{\alpha}\tilde{x}. \tag{5.87}$$

For $\omega_0 < \frac{\gamma}{2}$, the *overdamped* case, Ω obviously becomes purely imaginary, turning the trigonometric functions depending on it into hyperbolic ones. So the system makes, at most, one oscillation (depending on the initial conditions) before it approaches $x = 0$. In the Ermakov equation this changes the positive sign in front of the term depending on Ω^2 into a negative one which would correspond to a repulsive parabolic potential. The behaviour of these solutions is not discussed in detail here but a very similar situation arises when changing from a quantum mechanical system depending on (real) time t to a statistical mechanical system by replacing t with the *imaginary* "time" $\frac{\hbar}{k_B T}$ (with temperature T) which will be shown in Sect. 7.1.

With respect to the open quantum systems described by the logarithmic NLSE, the most interesting case is $\omega_0 = \frac{\gamma}{2}$, the *critical* damping or aperiodic limit. The solution of the classical equation of motion for $\eta(t)$, describing the motion of the WP maximum can be given as

$$\eta(t) = (c_1 t + c_2)\mathrm{e}^{-\frac{\gamma}{2}t} \tag{5.88}$$

where the constants c_1 and c_2 depend on the choice of the initial conditions. For $c_1 = v_0$ and $c_2 = 0$ for example, one obtains

$$\eta(t) = v_0 t e^{-\frac{\gamma}{2}t} \tag{5.89}$$

and

$$\dot{\eta}(t) = v_0\left(1 - \frac{\gamma}{2}t\right)\mathrm{e}^{-\frac{\gamma}{2}t} \tag{5.90}$$

with $\eta(0) = x_0 = 0$ and $\dot{\eta}(0) = v_0$, but also a choice $x_0 \neq 0$ and $v_0 = 0$ is possible.

The solution of the Ermakov equation is now trivial because, for $\omega = \frac{\gamma}{2}$, the frequency Ω vanishes and one obtains the Ermakov equation for the undamped free motion,

$$\ddot{\alpha} = \frac{1}{\alpha^3}, \tag{5.91}$$

with the well-known solution for the spreading free motion WP and, accordingly, a contribution to the current as shown in (2.46).

However, this means that if one starts with an undercritically-damped situation that can be described by a WP with constant width (and hence due to $\dot{\alpha} = 0$ no contribution to the overall tunnelling current) and changes frequency ω_0 or the "collision frequency" γ (or both) in a way that the resonance-type condition $\omega_0 = \frac{\gamma}{2}$ is met, suddenly a tunnelling current shows up. In an experimental setup, a change of frequency ω_0 could be achieved by considering a similar problem of the two-dimensional motion in a magnetic field with a damping environment. For this system, also a description by a logarithmic NLSE is possible (for details, see [15–18]). As the (cyclotron) frequency ω_c is connected with the magnetic field via $\omega_c = \frac{e}{mc}B$ (with e = electric elementary charge, B = magnetic field,[5] c = (vacuum) speed of light), it can be easily changed by varying B. A variation of γ might be more difficult but not impossible, assuming that it is connected with the temperature of the environment.

An even more surprising result is obtained when considering the energy of this aperiodic limit. The classical contribution for the corresponding WP solution behaves as expected

$$E_{cl,ap} = \frac{1}{2m}\langle p\rangle_{ap}^2 + \frac{m}{2}\frac{\gamma^2}{4}\langle x\rangle_{ap}^2 = E_0\left[\left(1-\frac{\gamma}{2}t\right)^2 + \left(\frac{\gamma}{2}t\right)^2\right]\mathrm{e}^{-\gamma t} \tag{5.92}$$

with $E_0 = \frac{m}{2}v_0^2$, if $\langle x\rangle_{ap} = \eta = (5.89)$ and $\langle p\rangle_{ap} = m\dot{\eta}$ with $\dot{\eta} = (5.90)$ is taken. So, a quadratic growth in time of the energy is overcompensated by an exponential decay, leading to vanishing energy for $t \to \infty$.

However, the *quantum mechanical* contribution $\tilde{E}_{ap}$ now fulfils

$$\tilde{E}_{ap} = \frac{1}{2m}\langle\tilde{p}^2\rangle + \frac{m}{2}\frac{\gamma^2}{4}\langle\tilde{x}^2\rangle = \frac{\hbar}{4}\beta_0\left[2\left(\frac{\frac{\gamma}{2}}{\beta_0}\right)^2 + \left(1-\frac{\gamma}{2}t\right)^2 + \left(\frac{\gamma}{2}t\right)^2\right]. \tag{5.93}$$

Apart from a constant contribution, it also contains the two terms that are growing quadratically in time, just as in the classical energy, but here without the exponential damping factor. The initial ground-state energy $\tilde{E}_{ap,0} = \frac{\hbar}{4}\beta_0\left[1 + 2\left(\frac{\frac{\gamma}{2}}{\beta_0}\right)^2\right]$ drops after $t_{\min} = \frac{1}{\gamma}$ to its minimum value $\tilde{E}_{ap,\min} = \frac{\hbar}{4}\beta_0\left[\frac{1}{2} + 2\left(\frac{\frac{\gamma}{2}}{\beta_0}\right)^2\right]$ and, afterwards, grows quadratically in time. Where does the energy gained by the quantum system come from? Considering the term describing the correlations of position and momentum-fluctuations,

$$\frac{\gamma}{4}\langle[\tilde{x},\tilde{p}]_+\rangle_{ap} = -\frac{\hbar}{4}\beta_0\left[\left(1-\frac{\gamma}{2}t\right)^2 + \left(\frac{\gamma}{2}t\right)^2\right] + \frac{\hbar}{4}\beta_0\left[1 - 2\left(\frac{\frac{\gamma}{2}}{\beta_0}\right)^2\right], \tag{5.94}$$

---

[5] More precisely, magnetic induction, but the difference does not matter in this context (see, e.g., [19]).

that represents the effect of the environment, one finds the same terms that are growing quadratically in time, as in $\tilde{E}_{ap}$, only with a negative sign; and the sum of the quantum contributions of the system and the environment,

$$\tilde{E}_{ap} + \frac{\gamma}{4}\langle[\tilde{x}, \tilde{p}]_+\rangle_{ap} = \frac{\hbar}{4}\beta_0 = \text{const.} \tag{5.95}$$

is always a constant.

Considering the environment as a heat bath (eventually infinite), this might at first sight appear to be a violation of the laws of thermodynamics. However, this is not the case as the energy from the heat bath is not transferred into the classical degrees of freedom of the system. In other words, the WP maximum does not start accelerating or even oscillating with increasing amplitude in this resonance-like situation, but is still damped exponentially. Only the quantum part of the systems energy, $\tilde{E}$, absorbs energy from the surrounding. So, thermal energy is transferred into quantum mechanical energy and, moreover, in a way that the sum of the energies of the system and surrounding always remains constant. Somehow this has similarities with thermal chemical reactions but it would be desirable to find experiments that could also detect this resonance-like behaviour of the energy and of the current. Macroscopic quantum effects (like superconducting or quantum-Hall currents) might be candidates but require further investigation (see also [14]).

5.5 Time-Dependent Green Function for the Dissipative Case

As has been shown in the non-dissipative case, the Gaussian WP solution of the TDSE could also be obtained from an initial Gaussian WP $\Psi_{\text{WP}}(x', t')$ with the help of a TD Green function or Feynman kernel $G(x, x', t, t')$. The same is assumed to be possible also for the logarithmic NLSE. In the linear case and for at most quadratic Hamiltonians, this Green function could be expressed in terms of the real and imaginary parts (and their derivatives) of the complex quantity $\lambda = u + iz$ that linearizes the complex Riccati equation (2.4) for $\mathcal{C}(t)$ to a linear complex Newtonian equation for $\lambda(t)$, Eq. (2.49). It was also possible to express the Ermakov invariant in terms of these quantities according to (see Eqs. (2.21) and (2.70))

$$I_{\text{L}} = \frac{1}{2}\left[\left(\dot{\eta}\alpha_{\text{L}} - \eta\dot{\alpha}_L\right)^2 + \left(\frac{\eta}{\alpha_{\text{L}}}\right)^2\right] = \frac{1}{2}\left(\frac{\alpha_0 p_0}{m}\right)\left[\left(\frac{u}{\alpha_{\text{L}}}\right)^2 + \left(\frac{z}{\alpha_{\text{L}}}\right)^2\right] = \text{const.}$$

In the NL case, this invariant has the form (see Eq. (5.16))

$$I_{\text{NL}} = \frac{1}{2}e^{\gamma t}\left[\left(\dot{\eta}\alpha_{\text{NL}} - \left(\dot{\alpha}_{\text{NL}} - \frac{\gamma}{2}\alpha_{\text{NL}}\right)\eta\right)^2 + \left(\frac{\eta}{\alpha_{\text{NL}}}\right)^2\right] = \text{const.}$$

Comparing these two invariants and taking into account that $\eta(t)$ and $\dot{\eta}(t)$ have the same physical meaning in both cases (namely classical position and velocity), suggests that in the transition from α_0 to $\alpha_{\rm NL}(t)$, an exponential scaling factor might be involved. In order to take this into account, the terms containing α_0 in the initial WP $\Psi_{\rm NL}(x', t' = 0)$, as well as in the Feynman kernel or time propagator $G(x, x', t, t' = 0)$, are multiplied by purely TD functions $f_i(t)$ that must fulfil the condition $f_i(t' = 0) = 1$ so that the definition of the initial WP is identical to that in the linear case. Further specifications of $f_i(t)$ are given later.

Therefore, the WP solution of the NLSE can be written as

$$\Psi_{\rm NL}(x,t) = \int_{-\infty}^{+\infty} dx' G_{\rm NL}(x,x',t,t')\Psi_{\rm NL}(x',t') \tag{5.96}$$

with the initial WP[6]

$$\Psi_{\rm NL}(x', t' = 0) = \left(\frac{m\beta_0}{\pi\hbar}\right)^{\frac{1}{4}} \exp\left\{\frac{im}{2\hbar}\left[\mathrm{i}f_2(t')\left(\frac{x'}{\alpha_0}\right)^2 + 2f_1(t')\frac{p_0}{m}\left(\frac{x'}{\alpha_0}\right)\right]\right\} \tag{5.97}$$

and the integral kernel takes the form

$$G_{\rm NL}(x,x',t,0) = F(t)\left(\frac{mf_2(t)\mathrm{e}^{-\frac{\gamma}{2}t}}{2\pi\mathrm{i}\hbar\alpha_0\tilde{z}}\right)^{\frac{1}{2}} \exp\left\{\frac{im}{2\hbar}\left[\frac{\dot{\tilde{z}}}{\tilde{z}}x^2 - 2\frac{x}{\tilde{z}}f_1(t)\left(\frac{x'}{\alpha_0}\right) + \frac{\tilde{u}}{\tilde{z}}f_2(t)\left(\frac{x'}{\alpha_0}\right)^2\right]\right\} \tag{5.98}$$

where a possibly TD function $F(t)$ has been introduced to eventually absorb some phase factors that do not occur in the linear case with this kind of ansatz. Furthermore, for comparison with the linear case discussed in Chap. 2, the nomenclature with the TD parameters u and z has been kept. The meaning of the tilde above the parameters, i.e., $\tilde{u}$ and $\tilde{z}$, will turn out to be the same as in the linearized complex Newtonian equation (5.37), as shown below, and does *not* indicate a shift of the variable by its mean value, like in $\tilde{x} = x - \langle x \rangle$.

Inserting (5.97) and (5.98) into (5.96), the Gaussian WP at time t is obtained in the form

$$\Psi_{\rm NL}(x,t) = \left(\frac{m}{\pi\hbar}\right)^{\frac{1}{4}} \left(\frac{\mathrm{e}^{-\frac{\gamma}{2}t}}{\tilde{\lambda}}\right)^{\frac{1}{2}} F(t) \exp\left\{\frac{im}{2\hbar}\left[\frac{\dot{\tilde{z}}}{\tilde{z}}x^2 - \frac{(x - \frac{p_0\alpha_0}{m}\tilde{z})^2}{\tilde{\lambda}\tilde{z}}\frac{f_1^2}{f_2}\right]\right\} \tag{5.99}$$

with the complex quantity $\tilde{\lambda} = \tilde{\hat{u}} + \mathrm{i}\tilde{\hat{z}}$.

---

[6]Because for $t \neq t' = 0$ the functions $f_i(t)$ can be different from 1, this also can lead to a time-dependence of the terms multiplied by f_i in the initial WP for $t > t'$. Therefore, it is no longer the case that $G_{\rm NL}(x, x'.t, t')$ itself has also to fulfil the NLSE, like $G_{\rm L}$ has to fulfil the SE in the linear case.

Inserting G_{NL} into the NLSE (5.1), one obtains terms proportional to powers[7] of x and x'. From the terms proportional to x^2, it follows that $\tilde{z}$ must fulfil the classical equation of motion for the system, i.e.,

$$\ddot{\tilde{z}} + \gamma\dot{\tilde{z}} + \omega^2\tilde{z} = 0. \tag{5.100}$$

As in the linear case, $\tilde{z}$ and $\tilde{u}$ are not independent of each other but must fulfil a relation that is obtained from the terms proportional to x'^2, namely,

$$\dot{\tilde{z}}\tilde{u} - \dot{\tilde{u}}\tilde{z} = \frac{f_1^2}{f_2}. \tag{5.101}$$

However, from Eq. (5.40) we already know the value on the rhs, i.e.,

$$\frac{f_1^2}{f_2} = \mathrm{e}^{-\gamma t}. \tag{5.102}$$

The WP as given in (5.99) must be identical to WP (5.2). Comparison therefore shows the following relations:

$$\tilde{z} = \frac{m}{\alpha_0 p_0}\eta(t) \tag{5.103}$$

and

$$\frac{2\hbar}{m}y_{NL} = \mathcal{C}_{NL} = \frac{\dot{\tilde{z}}}{\tilde{z}} - \frac{1}{\tilde{z}\tilde{\lambda}}\mathrm{e}^{-\gamma t} = \frac{\dot{\tilde{\lambda}}}{\tilde{\lambda}}, \tag{5.104}$$

where (5.101) and (5.102) have been used to obtain the very rhs of (5.104).

Using $\tilde{\lambda} = \alpha_{NL}\mathrm{e}^{\mathrm{i}\varphi - \frac{\gamma}{2}t}$, and hence $\mathcal{C}_{NL} = \frac{\dot{\alpha}_{NL}}{\alpha_{NL}} - \frac{\gamma}{2} + \mathrm{i}\frac{1}{\alpha_{NL}^2}$, the WP solution (5.99) can finally be written in the form

$$\Psi_{WP,NL}(x,t) = \left(\frac{m}{\pi\hbar}\right)^{\frac{1}{4}}\left(\frac{\mathrm{e}^{-\frac{\gamma}{2}t}}{\tilde{\lambda}}\right)^{\frac{1}{2}} F(t)\exp\left\{\frac{im}{2\hbar}\frac{\dot{\tilde{\lambda}}}{\tilde{\lambda}}\tilde{x}^2 + \frac{\mathrm{i}}{\hbar}\langle p\rangle_{NL}\tilde{x} + \frac{im}{2\hbar}\dot{\eta}\eta\right\} \tag{5.105}$$

with

$$F(t) = \exp\left\{\frac{\mathrm{i}}{\hbar}\int_0^t \left(\frac{\gamma}{4}\langle[\tilde{x},\tilde{p}]_+\rangle_{NL} + \frac{\gamma}{2}m\dot{\eta}\eta\right)dt'\right\}. \tag{5.106}$$

As in the linear case, also the Ermakov invariant I_{NL} (5.16) can be expressed in terms of the parameters that enter the propagator G_{NL}. In particular, with

$$\eta(t) = \frac{\alpha_0 p_0}{m}\tilde{z} = \frac{\alpha_0 p_0}{m}z\mathrm{e}^{-\frac{\gamma}{2}t} = \hat{\xi}\mathrm{e}^{-\frac{\gamma}{2}t}, \tag{5.107}$$

[7]The terms independent of x and x' do not necessarily all cancel due to what was mentioned in the last footnote.

the invariant can be written as

$$
\begin{aligned}
I_{\mathrm{NL}} &= \frac{1}{2}\left(\frac{\alpha_0 p_0}{m}\right)^2 \left[\mathrm{e}^{\gamma t}\left(\dot{\tilde{z}}\alpha_{\mathrm{NL}} - \left(\dot{\alpha}_{\mathrm{NL}} - \frac{\gamma}{2}\alpha_{\mathrm{NL}}\right)\tilde{z}\right)^2 + \mathrm{e}^{\gamma t}\left(\frac{\tilde{z}}{\alpha_{\mathrm{NL}}}\right)^2\right] \\
&= \frac{1}{2}\left(\frac{\alpha_0 p_0}{m}\right)^2 \left[\left(\frac{u}{\alpha_{\mathrm{NL}}}\right)^2 + \left(\frac{z}{\alpha_{\mathrm{NL}}}\right)^2\right] = \text{const.}
\end{aligned}
\tag{5.108}
$$

where, again, $e^{\gamma t}\left(\frac{\tilde{z}}{\alpha_{\mathrm{NL}}}\right)^2 = \left(\frac{z}{\alpha_{\mathrm{NL}}}\right)^2 = \sin^2\varphi$ has been used from which it follows that the first term in the square bracket must be $\left(\frac{u}{\alpha_{\mathrm{NL}}}\right)^2 = \cos^2\varphi$ in order to obtain a constant value for I_{NL}.

The two orthogonal solutions of the complex linear Eq. (5.38) are again $u(t)$ and $z(t)$ where this time $z(t)$ is not directly proportional to $\eta(t)$ but to $\hat{\xi}(t)$ fulfilling Eq. (5.22) and related to $\eta(t)$ via (5.21) or (5.107).

However, u and z, or $\tilde{u}$ and $\tilde{z}$, respectively, can again be expressed in terms of η and $\dot{\eta}$. Up to a $\pm$-sign, one obtains

$$
\tilde{u} = \mathrm{e}^{-\frac{\gamma}{2}t}u = \alpha_{\mathrm{NL}}^2\dot{\tilde{z}} - \left(\dot{\alpha}_{\mathrm{NL}}\alpha_{\mathrm{NL}} - \frac{\gamma}{2}\alpha_{\mathrm{NL}}^2\right)\tilde{z}
\tag{5.109}
$$

or

$$
\frac{\tilde{u}}{\tilde{z}} = \alpha_{\mathrm{NL}}^2\left(\frac{\dot{\tilde{z}}}{\tilde{z}} - \left(\frac{\dot{\alpha}_{\mathrm{NL}}}{\alpha_{\mathrm{NL}}} - \frac{\gamma}{2}\right)\right)
\tag{5.110}
$$

or

$$
\frac{\dot{\tilde{z}}}{\tilde{z}} = \frac{\dot{\eta}}{\eta} = \frac{1}{\alpha_{\mathrm{NL}}^2}\frac{\tilde{u}}{\tilde{z}} + \left(\frac{\dot{\alpha}_{\mathrm{NL}}}{\alpha_{\mathrm{NL}}} - \frac{\gamma}{2}\right).
\tag{5.111}
$$

These relations are true for all times t; they also hold for $t = 0$. In this case (and for $\left(\frac{\dot{\alpha}_{\mathrm{NL}}}{\alpha_{\mathrm{NL}}}\right)_0 = 0$, as assumed also in the linear case, though other choices are possible),[8] $\tilde{u}$ turns into

$$
\tilde{u} = \alpha_0^2\dot{\tilde{z}} + \frac{\gamma}{2}\alpha_0^2\tilde{z}.
\tag{5.112}
$$

Together with (5.107), the expression for α_{NL}^2 becomes

$$
\alpha_{\mathrm{NL}}^2(t) = u^2 + z^2 = \mathrm{e}^{\gamma t}\left(\tilde{u}^2 + \tilde{z}^2\right) = \mathrm{e}^{\gamma t}\left(\frac{m}{\alpha_0 p_0}\right)^2\left[\alpha_0^4\left(\dot{\eta}^2 + \gamma\eta\dot{\eta} + \frac{\gamma^2}{4}\eta^2\right) + \eta^2\right]
\tag{5.113}
$$

---

[8]The choice $\dot{\alpha}_{\mathrm{NL},0} = 0$ is different from the choice $\langle[\tilde{x}, \tilde{p}]_+\rangle_{\mathrm{NL}}(0) = (\dot{\alpha}_{\mathrm{NL},0} - \frac{\gamma}{2}\alpha_{\mathrm{NL},0})\alpha_{\mathrm{NL},0} = 0$ used in Sect. 5.4.2. Therefore the analytical expression for $\alpha_{\mathrm{NL}}^2(t) = \frac{\hbar}{2m}\langle\tilde{x}^2\rangle_{\mathrm{NL}}$ below (for the damped free motion) differs from the two expressions $\langle\tilde{x}^2\rangle_\pm$ in Eqs. (5.60) and (5.64), showing again the influence of the initial conditions. For $\gamma = 0$, $\dot{\alpha}_0 = 0$ is equivalent to $\langle[\tilde{x}, \tilde{p}]_+\rangle_{\mathrm{L}}(0) = \dot{\alpha}_{\mathrm{L},0}\alpha_{\mathrm{L},0} = 0$.

or

$$\alpha^2_{\mathrm{NL}}(t) = \alpha_0^2 \frac{\mathrm{e}^{\gamma t}}{v_0^2} \left[\dot{\eta}^2 + \gamma \eta \dot{\eta} + \left(\beta_0^2 + \frac{\gamma^2}{4} \right) \eta^2 \right] = \frac{2m}{\hbar} \langle \tilde{x}^2 \rangle_{\mathrm{NL}} . \tag{5.114}$$

Also in the dissipative case, the influence of the initial uncertainty on the time-evolution of the position uncertainty is shown explicitly, particularly for the damped HO (with the limiting cases $\omega_0 \to 0$, i.e., damped free motion, and $\gamma \to 0$).

For this purpose, the expressions for $\eta(t)$ and $\dot{\eta}(t)$ of the damped HO are inserted into Eq. (5.114), i.e., with $\eta(t) = \left(\frac{v_0}{\Omega}\right) \mathrm{e}^{-\frac{\gamma}{2}t} \sin \Omega t$ (for $\eta(0) = 0$) and $\dot{\eta}(t) = v_0 \mathrm{e}^{-\frac{\gamma}{2}t} \cos \Omega t - \frac{\gamma}{2} \left(\frac{v_0}{\Omega}\right) \mathrm{e}^{-\frac{\gamma}{2}t} \sin \Omega t$ (for $\dot{\eta}(0) = v_0$), one obtains

$$\alpha^2_{\mathrm{NL,HO}} = \alpha_0^2 \left\{ \cos^2 \Omega t + \left(\frac{\beta_0}{\Omega} \right)^2 \sin^2 \Omega t \right\} = \alpha_0^2 \left\{ 1 + \left(\frac{\beta_0^2 - \Omega^2}{\Omega^2} \right) \sin^2 \Omega t \right\} , \tag{5.115}$$

i.e., the same expression as in the conservative case, only that the frequency ω_0 is now replaced by the reduced frequency $\Omega = \left(\omega_0^2 - \frac{\gamma}{4}\right)^{\frac{1}{2}}$. In analogy with the linear SE, also for the NLSE, a constant width is only obtained for $\beta_0 = \Omega$. This is the case when the system is in the state of lowest possible energy (see also Ref. [20]); otherwise, the WP width oscillates. In the case of disappearing friction, i.e., $\gamma \to 0$ and thus $\Omega \to \omega_0$, the result of the linear SE is obviously regained. More interesting is the limit $\omega_0 \to 0$, i.e., the transition to the damped free motion. In this case, $\lim_{\omega_0 \to 0} \left(\omega_0^2 - \frac{\gamma^2}{4} \right)^{\frac{1}{2}} = \pm \mathrm{i} \frac{\gamma}{2}$ is purely imaginary. Using the relations between the trigonometric and the hyperbolic functions, i.e., $\cos(\mathrm{i}x) = \cosh x$ and $\sin(\mathrm{i}x) = \mathrm{i} \sinh x$, one obtains (with $\cosh^2 x = 1 + \sinh^2 x$)

$$\alpha^2_{\mathrm{NL},fr} = \alpha_0^2 \left\{ \cosh^2 \frac{\gamma}{2} t + \left(\frac{\beta_0}{\frac{\gamma}{2}} \right)^2 \sinh^2 \frac{\gamma}{2} t \right\} = \alpha_0^2 \left\{ 1 + \left(\frac{\beta_0^2 + \frac{\gamma^2}{4}}{\frac{\gamma^2}{4}} \right) \sinh^2 \frac{\gamma}{2} t \right\} , \tag{5.116}$$

which, in the limit $\gamma \to 0$, indeed turns into the correct expression for the spreading free WP

$$\lim_{\gamma \to 0} \alpha^2_{\mathrm{NL},fr}(t) = \alpha_0^2 [1 + (\beta_0 t)^2] = \alpha^2_{\mathrm{L},fr}(t). \tag{5.117}$$

Also for α^2_{NL} given by (5.116), due to the exponential factor $\mathrm{e}^{\frac{\gamma}{2}t}$ in $\sinh \frac{\gamma}{2} t$, the WP width for the damped free motion is spreading much faster (exponentially) in comparison with the undamped case (quadratically in time); for further details, see also Sect. 5.4.3 and [12].

The same result (5.116) for the damped free motion can also be obtained directly by inserting $\dot{\eta}(t) = v_0 e^{\gamma t}$ (for $\dot{\eta}(0) = v_0$) and $\eta(t) = (\frac{v_0}{\gamma})(1 - e^{-\gamma t})$ (with $\eta(0) = 0$) into (5.114).

That the time-dependence of α_{NL} (now again for the damped HO), essentially depends on the difference between the initial state's position uncertainty and its ground state uncertainty $\langle \tilde{x}^2 \rangle_{GS}$ (as in the conservative case) becomes even more obvious regarding again the time-derivative of α_{NL}^2

$$\dot{\alpha}_{NL} \alpha_{NL} = \frac{\alpha_0^2}{v_0^2} e^{\gamma t} \left[\left(\beta_0^2 - \Omega^2\right) \eta \dot{\eta} + \frac{\gamma}{2} \left(\beta_0^2 - \Omega^2\right) \eta^2 \right], \tag{5.118}$$

which only vanishes for $\beta_0 = \Omega$ (which corresponds to $\langle \tilde{x}^2 \rangle_{NL,0} = \langle \tilde{x}^2 \rangle_{GS}$) although a second term appears that is not present in the non-dissipative case.

Finally, a comparison is made between G_{NL} and a Feynman kernel G_F for the damped HO that is frequently quoted and used in the literature (see Grosche and Steiner [21], p. 186, 6.2.1.19 and references therein). Their Feynman kernel derived from the classical Lagrangian L_{CK} (4.32) of Caldirola and Kanai reads in our notation

$$\begin{aligned} G_F(x, x', t, t' = 0) = \left(\frac{m\Omega}{2\pi i \hbar \sin \Omega t} \right)^{\frac{1}{2}} \exp \left\{ \frac{im\Omega}{2\hbar \sin \Omega t} \left[(x^2 e^{\gamma t} + x'^2 e^{\gamma t'}) \cos \Omega t \right. \right. \\ \left. \left. - \; 2xx' e^{\frac{\gamma}{2}(t+t')} \right] - \frac{im}{2\hbar} \frac{\gamma}{2} \left(x^2 e^{\gamma t} - x'^2 e^{\gamma t'} \right) + \gamma \frac{t + t'}{4} \right\}. \end{aligned} \tag{5.119}$$

In comparison, our propagator for the damped HO is explicitly given by

$$\begin{aligned} G_{NL,HO}(x, x', t, 0) = & F(t) \left(\frac{m\Omega}{2\pi i \hbar \sin \Omega t} \right)^{\frac{1}{2}} \\ & \times \exp \left\{ \frac{im\Omega}{2\hbar \sin \Omega t} \left[(x^2 + x'^2) \cos \Omega t - 2xx' \right] - \frac{im}{4\hbar} \gamma x^2 \right\}. \end{aligned} \tag{5.120}$$

The essential difference is that, in the exponent of G_F, all terms depending on x or x' have a factor growing exponentially in time. This factor also ends up in the resulting WP (and for normalization purposes must be compensated for by the last term in the exponent of (5.119)). Therefore, this WP is a canonical WP (which can be expected as it is based on $\hat{L}_{CK}$) and, as discussed in detail above, leads to unphysical results if interpreted as a physical WP. So, if the Feynman kernel of Grosche and Steiner is used, afterwards, the transformation (4.135) with subsequent normalization is necessary if the WP is also to be used within the context of the usual quantum mechanical operators to avoid unphysical results (such as violation of the uncertainty principle, etc.).

5.6 Dissipative Schrödinger Equation in Momentum Space

5.6.1 Friction Term in Momentum Space

In order to include the observable effects of irreversibility and dissipation, the modification of the TDSE in position space was obtained by adding a time-reversal-symmetry breaking diffusion term $-D\Delta\varrho$ to the continuity equation for $\varrho = \Psi^* \Psi$, turning it into the Smoluchowski equation (4.109) and separating this equation into the NLSE (4.115) and its complex conjugate via the separation condition (see Eq. (4.113))

$$-D\frac{\Delta\varrho}{\varrho} = \gamma\left(\ln\varrho - \langle\ln\varrho\rangle\right).$$

Due to this condition, the modified SE contains an additional complex logarithmic nonlinearity (see Eq. (4.116))

$$\tilde{W}_{\rm SCH} = \gamma\frac{\hbar}{\rm i}\left(\ln\Psi - \langle\ln\Psi\rangle\right)$$

when written as $\mathrm{i}\hbar\frac{\partial}{\partial t}\Psi_{\rm NL}(\mathbf{r},t) = \left\{-\frac{\hbar^2}{2m}\Delta + V(\mathbf{r}) + \tilde{W}_{\rm SCH}\right\}\Psi_{\rm NL}(\mathbf{r},t)$.

This additional term leads to a linear velocity dependent friction force in the averaged equation of motion (4.117) and a corresponding decrease of energy proportional to 2γ times the kinetic energy of the system (at least for the classical contribution).

However, the particular logarithmic form of the friction term $\tilde{W}_{\rm SCH}$ is connected with the definition of the momentum- or velocity-operator in position space, using Schrödinger's definition of the wave function $\Psi(\mathbf{r},t)$ via the action function S as $S_c = \frac{\hbar}{\rm i}\ln\Psi(\mathbf{r},t)$ and the momentum definition in the Hamilton-Jacobi theory, $\mathbf{p} = \nabla S$, as detailed in Sect. 4.5.

This leads to

$$\mathbf{p}_c = \frac{\hbar}{\rm i}\nabla\ln\Psi(\mathbf{r},t) = \frac{\hbar}{\rm i}\frac{\nabla\Psi}{\Psi} = m\mathbf{v}_c \tag{5.121}$$

with mean value

$$\langle\mathbf{p}_c\rangle = \int d\mathbf{r}\Psi^*\frac{\hbar}{\rm i}\frac{\nabla\Psi}{\Psi}\Psi = \int d\mathbf{r}\Psi^*\frac{\hbar}{\rm i}\nabla\Psi = \left\langle\frac{\hbar}{\rm i}\nabla\right\rangle = \langle\mathbf{p}_{\rm op}\rangle, \tag{5.122}$$

i.e., the same mean value as the quantum mechanical momentum operator $\mathbf{p}_{\rm op} = \frac{\hbar}{i}\nabla$ in position space. In this case, the negative gradient of $\tilde{W}_{\rm SCH}$ has the form of a friction force linearly proportional to the (complex) velocity $\mathbf{v}_c$, $\nabla\tilde{W}_{\rm SCH} = -m\gamma\mathbf{v}_c$.

On the contrary, in momentum-space, the momentum operator is simply a c-number, $\mathbf{p}_{\rm op} = \mathbf{p}$, whereas now the position operator is connected to the Nabla-operator in momentum space, ∇_p, via $\mathbf{r}_{\rm op} = -\frac{\hbar}{\rm i}\nabla_p$. Therefore, if $\tilde{W}_{\rm SCH}$ is used in the same logarithmic form as in position space, the negative momentum-space gradient of $\tilde{W}_{\text{SCH}}$, $-\nabla_p \tilde{W}_{\text{SCH}}$, no longer yields a friction force linearly proportional to velocity (with a negative sign), but an accelerating force linearly proportional to position (with a positive sign).

So in momentum space, this form of the NLSE is no longer correct to describe the same physical situation. This can also be seen from the fact that the Fourier-transform of the WP solution (5.2) in position space (in contrast with the linear theory) is no longer a solution of the NLSE in momentum space if the form of $\tilde{W}_{\text{SCH}}$ remains unchanged.

In order to find a form of the NL friction term which is valid in position as well as in momentum space, results obtained in the NL description in position space can be used [22].

Recalling the situation in the linear case, the complex Riccati equations (Eq. (2.4) in position space and (2.101) in momentum space) that we are most interested in could both be linearized to the same complex Newtonian equation (2.49) for $\lambda(t)$ (see also Fig. 2.4). Assuming that the same should also be valid in the NL case, the complex equation (5.37) for $\tilde{\lambda}(t)$,

$$\ddot{\tilde{\lambda}} + \gamma\dot{\tilde{\lambda}} + \omega^2\tilde{\lambda} = 0,$$

should play a central role in the reformulation of the friction term.

As mentioned above in connection with the rate of energy dissipation, the form of the friction term should not explicitly depend on the potential of the problem but somehow be connected with the kinetic energy of the system.

In the derivation of the friction term in position space, the comparison of the diffusion term proportional to $\Delta\varrho$, with the logarithm $\ln\varrho$ was essential. Now the procedure on the level of the complex function Ψ can be reversed. The logarithm of $\Psi(\mathbf{r}, t)$, $\ln\Psi(\mathbf{r}, t)$, is now expressed in terms of $\Delta\Psi(\mathbf{r}, t)$ or, more precisely, $\tilde{W}_{\text{SCH}}$ in terms of the kinetic energy term, which shall be written as $\tilde{W}_{\text{D}}$. Requiring that the condition

$$\tilde{W}_{\text{SCH}}\Psi_{\text{WP}} = \tilde{W}_{\text{D}}\Psi_{\text{WP}} \tag{5.123}$$

must be fulfilled,[9] it is possible to express $\tilde{W}_{\text{SCH}}$, with the help of $\tilde{\lambda}$, in terms of the kinetic energy (and its mean value),

$$\tilde{W}_{\text{D}} = \gamma\left(\frac{\tilde{\lambda}}{\dot{\tilde{\lambda}}}\right)\frac{1}{2m}\left(\mathbf{p}_{\text{op}}^2 - \langle\mathbf{p}_{\text{op}}^2\rangle\right). \tag{5.124}$$

This form of the friction term turns out to be valid in position and momentum space. It obviously has the same property as the logarithmic nonlinearity $\tilde{W}_{\text{SCH}}$, namely a disappearing mean value $\langle\tilde{W}_{\text{D}}\rangle = 0$. Like in position space $\tilde{W}_{\text{D}}$ influences the

[9]Hasse's form of the friction term can also be regained by comparing it with $\tilde{W}_{\text{SCH}}$ according to $\tilde{W}_{\text{SCH}} = \tilde{W}_{\text{Has}} - \langle\tilde{W}_{\text{Has}}\rangle$. More details are below in Chap. 6.

dynamics of the observable system by altering the wave function that is used to calculate the (TD) mean values but is itself not observable. This is again in agreement with a situation where the motion of a Brownian particle is affected by the interaction with the environment although the motion of the surrounding molecules cannot be observed directly.

As the coefficient of the kinetic energy operator in (5.124), $\left(\frac{\tilde{\lambda}}{\tilde{\tilde{\lambda}}}\right)$, is complex, this form of the dissipative term shows similarities with other approaches mentioned before due to the imaginary contribution.

In Gisin's approach, described in Sect. 4.4.2, a non-Hermitian term (in the particular form $\frac{\hbar}{\mathrm{i}}B = \frac{\hbar}{\mathrm{i}}kH_{\mathrm{L}}$) is considered that is proportional to the linear Hamiltonian and thus also to the kinetic energy with imaginary coefficient. However, this assumption (including the potential V) leads to a wrong energy dissipation.

Also the additional imaginary term of Doebner and Goldin (mentioned at the end of Sect. 4.4.3) can be expressed in terms of the Laplacian operator and thus the kinetic energy operator. But as mentioned before, this term by itself does not describe dissipation but essentially introduces irreversibility due to the diffusion term in the Smoluchowski equation.

As in Chap. 2 or Sects. 5.1–5.5, the following discussion is restricted to the one-dimensional case: Prior to the WP solutions of the dissipative case in momentum space, a short discussion is given of the friction term $\tilde{W}_D$ in position space for the case of the free motion, i.e., $V = 0$. This has similarities with other equations, such as the diffusion equation, if $\tilde{W}_{\mathrm{D}}$ is written as

$$\tilde{W}_{\mathrm{D}} = \gamma\left(\frac{\tilde{\lambda}}{\tilde{\tilde{\lambda}}}\right)\left(-\frac{\hbar^2}{2m}\frac{\partial^2}{\partial x^2} - \left\langle -\frac{\hbar^2}{2m}\frac{\partial^2}{\partial x^2}\right\rangle\right). \tag{5.125}$$

The second derivative with complex coefficient can be combined with the operator of kinetic energy allowing the NLSE to be written in position space in the form

$$\frac{\partial}{\partial t}\Psi(x,t) = \left\{D(t)\frac{\partial^2}{\partial x^2} + \frac{1}{\mathrm{i}\hbar}V(x) - g(t)\right\}\Psi(t) \tag{5.126}$$

with

$$D(t) = \left(1 + \gamma\frac{\tilde{\lambda}}{\tilde{\tilde{\lambda}}}\right)\mathrm{i}\frac{\hbar}{2m} : \text{complex "diffusion coefficient"}, \tag{5.127}$$

$$g(t) = \frac{\gamma}{\mathrm{i}\hbar}\frac{\tilde{\lambda}}{\tilde{\tilde{\lambda}}}\left\langle\frac{p_{op}^2}{2m}\right\rangle : \text{nonlinearity}. \tag{5.128}$$

Formally, $g(t)$ is just a (complex) TD function. However, in order to actually calculate the mean value, the solution Ψ of the NLSE must be known.[10]

[10] It might be possible to apply iterative techniques as in the Hartree–Fock method.

Now considering the case $V = 0$, i.e., free motion with damping, and treating $g(t)$ formally as a mere TD function, $g(t)$ can be removed from the equation via the transformation

$$\Psi(x,t) = \tilde{\Psi}(x,t)\exp\left\{-\int_0^t dt' g(t')\right\}. \tag{5.129}$$

The equation for $\tilde{\Psi}(x,t)$,

$$\frac{\partial}{\partial t}\tilde{\Psi}(x,t) - D(t)\frac{\partial^2}{\partial x^2}\tilde{\Psi}(x,t) = 0 \tag{5.130}$$

has the form of a diffusion equation with a *complex* TD diffusion coefficient $D(t)$. Keeping in mind that the SE for the free motion has the form of a diffusion equation with a purely imaginary diffusion coefficient, $\mathrm{i}\frac{\hbar}{2m}$, Eq. (5.130) is a combination of a reversible linear SE and an irreversible (also linear) diffusion equation. For both of these equations, the superposition principle holds. This might explain why our NLSE displays a kind of superposition property for the damped free motion in position space, as plane-wave-type solutions exist that can be superimposed to form a Gaussian WP-type solution of the NLSE (see Sect. 5.4.3).

5.6.2 *Wave Packet Solutions in Momentum Space*

In the following, the new form (5.124) of our dissipative friction term, $\tilde{W}_{\mathrm{D}}$, is considered in momentum space and the solutions of the corresponding NLSE (particularly again for the HO in one dimension) are analyzed. Analogous to the linear theory, it is assumed that the solution in momentum space in our NL case can also simply be obtained by Fourier transformation of the WP solution (5.2),

$$\Psi_{\mathrm{NL}}(p,t) = \left(\frac{1}{2\pi\hbar}\right)^{\frac{1}{2}}\int_{-\infty}^{+\infty} dx\,\Psi_{\mathrm{NL}}(x,t)\mathrm{e}^{-\frac{\mathrm{i}}{\hbar}px}, \tag{5.131}$$

leading to the Gaussian WP in momentum space

$$\Psi_{\mathrm{WP}}(p,t) = \left(\frac{a}{\hbar}\right)^{\frac{1}{2}} N(t)\exp\left\{\mathrm{i}\left[\frac{ia(t)}{2\hbar^2}\tilde{p}^2 - \frac{1}{\hbar}\langle x\rangle_{\mathrm{NL}}\tilde{p} + K(t) - \frac{\langle x\rangle_{\mathrm{NL}}\langle p\rangle_{\mathrm{NL}}}{\hbar}\right]\right\} \tag{5.132}$$

with $\tilde{p} = p - \langle p\rangle_{\mathrm{NL}}$ and the complex TD quantity

$$\frac{am}{\mathrm{i}\hbar} = \frac{\tilde{\lambda}}{\dot{\tilde{\lambda}}} = C_{\mathrm{NL}}^{-1}. \tag{5.133}$$

The particle aspect is again expressed by the fact that the maximum of the WP coincides with the classical momentum $\langle p \rangle_{\rm NL} = m\dot{\eta}$ where the classical trajectory $\eta(t) = \langle x \rangle_{\rm NL}$ is determined by the Newtonian equation (5.3).

The wave aspect, i.e., the momentum uncertainty $\langle \tilde{p}^2 \rangle = \langle p^2 \rangle - \langle p \rangle^2$ or the WP width in momentum space, is connected to the real part of the complex coefficient $a(t)$ as in the linear case via

$$\frac{a_{\rm R} m}{\hbar} = \frac{m\hbar}{2\langle \tilde{p}^2 \rangle}. \tag{5.134}$$

The time-evolution of the WP width in momentum space is also governed by a complex Riccati equation for $\left(\frac{am}{i\hbar}\right)$, as in the linear case, but now with an additional linear term with coefficient γ (as in position space), i.e.,

$$-\frac{\dot{a}m}{i\hbar} + \gamma \frac{am}{i\hbar} + \omega^2 \left(\frac{am}{i\hbar}\right)^2 + 1 = 0. \tag{5.135}$$

With the definition (5.133) this Riccati equation can be linearized to the same complex Newtonian equation (5.37), $\ddot{\tilde{\lambda}} + \gamma \dot{\tilde{\lambda}} + \omega^2 \tilde{\lambda} = 0$, as in position space. Therefore, the transition between position and momentum space in the NL case is possible in exactly the same way as in the linear case by considering inverse quantities instead of necessitating Fourier transformations. Figure 2.4 showing this connection in the linear case can be taken over almost unchanged; only quantities without tilde must be replaced by those with tilde and the Riccati equations must be supplemented with the linear terms; the details are shown in Fig. 5.2.

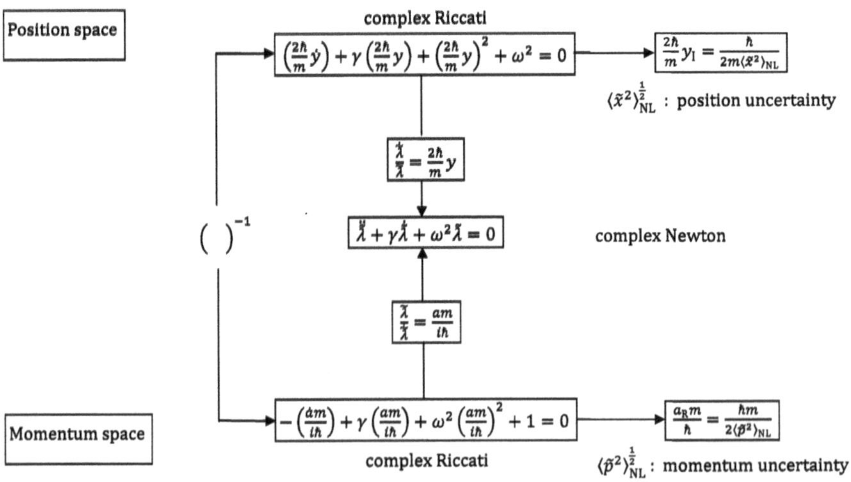

Fig. 5.2 Connections amongst the equations determining the time-dependence of position and momentum uncertainties. The complex variable entering the nonlinear Riccati equation for the momentum uncertainty is the inverse of the corresponding variable of the Riccati equation for the position uncertainty and both are linked via the variable $\tilde{\lambda}(t)$, fulfilling the complex linear Newtonian equation

In order to turn the complex Riccati equation (5.135) into a real Ermakov equation, (also in the dissipative case) a variable $\epsilon_{\rm NL}(t)$ is introduced in analogy with (2.105) that is related to $\frac{a_{\rm R} m}{\hbar}$, and thus to the momentum uncertainty, in the same way as in the linear case

$$\frac{a_{\rm R} m}{\hbar} = \frac{m\hbar}{2\langle \tilde{p}^2 \rangle_{\rm NL}} = \frac{1}{\epsilon_{\rm NL}^2}. \tag{5.136}$$

It should be mentioned that the situation in position space differs somewhat. Whereas α^2, expressed in terms of λ and λ^*, has the same form in the linear and the nonlinear cases,

$$\alpha_{\rm L}^2 = \lambda\lambda^*, \tag{5.137}$$

$$\alpha_{\rm NL}^2 = {\rm e}^{\gamma t}\left(\tilde{\lambda}\tilde{\lambda}^*\right) = \lambda\lambda^*, \tag{5.138}$$

ϵ^2 expressed in terms of $\dot{\lambda}$ and $\dot{\lambda}^*$, or equivalently in terms of α and $\dot{\alpha}$, differs in the two cases,

$$\epsilon_{\rm L}^2 = \dot{\lambda}\dot{\lambda}^* = \left(\dot{\alpha}_{\rm L}^2 + \frac{1}{\alpha_{\rm L}^2}\right), \tag{5.139}$$

$$\epsilon_{\rm NL}^2 = {\rm e}^{\gamma t}\left(\dot{\tilde{\lambda}}\dot{\tilde{\lambda}}^*\right) = \left[\left(\dot{\alpha}_{\rm NL} - \frac{\gamma}{2}\alpha_{\rm NL}\right)^2 + \frac{1}{\alpha_{\rm NL}^2}\right], \tag{5.140}$$

reflecting the difference in the expressions for $\langle \tilde{p}^2 \rangle_{\rm L}$ (Eq. (2.23)) and $\langle \tilde{p}^2 \rangle_{\rm NL}$ (Eq. (5.42)).

With this definition of $\epsilon_{\rm NL}$, the imaginary part of $\frac{am}{i\hbar}$, in comparison with (2.106), now takes the modified form

$$\frac{a_{\rm I} m}{\hbar} = \frac{1}{2\omega^2}\left(\frac{\dot{a}_{\rm R}}{a_{\rm R}} - \gamma\right) = -\frac{1}{\omega^2}\left(\frac{\dot{\epsilon}_{\rm NL}}{\epsilon_{\rm NL}} + \frac{\gamma}{2}\right). \tag{5.141}$$

Inserting this and the real part (5.136) into Riccati equation (5.135), allows this equation to be rewritten as an Ermakov equation of the form

$$\ddot{\epsilon}_{\rm NL} + \left(\omega^2 - \frac{\gamma^2}{4}\right)\epsilon_{\rm NL} = \frac{\omega^4}{\epsilon_{\rm NL}^3}, \tag{5.142}$$

that looks like the Ermakov equation (5.15) in position space (apart from the factor ω^4 on the rhs). Changing to a new variable $R_{\rm NL}$ like in the linear case (Eq. (2.109)) according to

$$R_{\rm NL} = \frac{\epsilon_{\rm NL}}{\omega}, \tag{5.143}$$

Eq. (5.142) can actually be rewritten in exactly the same form as in position space as

$$\ddot{R}_{\mathrm{NL}} + \left(\omega^2 - \frac{\gamma^2}{4}\right) R_{\mathrm{NL}} = \frac{1}{R_{\mathrm{NL}}^3}. \tag{5.144}$$

Using again the variable $\Pi = m\dot{\eta}$ for the classical momentum, where now $\dot{\Pi} = -m\omega^2\eta - m\gamma\dot{\eta}$ and thus

$$\ddot{\Pi} + \gamma\dot{\Pi} + \omega^2\Pi = 0 \tag{5.145}$$

is valid, the last two equations can be combined to yield the Ermakov invariant in momentum space for the NL case in the form

$$I_{\mathrm{NL,p}} = \frac{1}{2}\mathrm{e}^{\gamma t}\left[\left(\dot{\Pi}R_{\mathrm{NL}} - \left(\dot{R}_{\mathrm{NL}} - \frac{\gamma}{2}R_{\mathrm{NL}}\right)\Pi\right)^2 + \left(\frac{\Pi}{R_{\mathrm{NL}}}\right)^2\right], \tag{5.146}$$

having exactly the same structure as (5.16) in position space.

5.6.3 Time-Dependent Green Function in Momentum Space

Again like in position space, the dissipative WP solution, also in momentum space, can be obtained from an initial WP $\Psi_{\mathrm{NL}}(p', 0)$ with the help of a TD Green function via

$$\Psi_{\mathrm{NL}}(p,t) = \int_{-\infty}^{+\infty} dp' G_{\mathrm{NL}}(p,p',t,t'=0)\Psi_{\mathrm{NL}}(p',0) \tag{5.147}$$

with

$$\Psi_{\mathrm{NL}}(p',0) = \left(\frac{1}{\pi\hbar m\epsilon_0}\right)^{\frac{1}{4}} \exp\left\{\frac{im}{2\hbar}\left[\mathrm{i}f_2(t')\left(\frac{p'}{m\epsilon_0}\right)^2 - 2\mathrm{i}f_1(t')\frac{p_0 p'}{m^2\epsilon_0^2}\right]\right\}$$
$$\times \exp\left\{-\frac{m}{2\hbar}\left(\frac{p_0}{m\epsilon_0}\right)^2\right\} \tag{5.148}$$

where the last exponential factor that is independent of p' and t' can be compensated by the normalization factor of the WP and $\epsilon_0 = \epsilon_{\mathrm{NL}}(t=0)$ implies that the initial state is a minimum uncertainty WP (see also Sect. 2.7). The integral kernel now takes the form

$$G_{\mathrm{NL}}(p,p',t,0) = F(t)\left(\frac{-f_2(t)\mathrm{e}^{-\frac{\gamma}{2}t}}{2\pi\hbar m\epsilon_0\dot{\tilde{z}}}\right)^{\frac{1}{2}}$$
$$\times \exp\left\{-\frac{im}{2\hbar}\left[\frac{\tilde{z}}{\dot{\tilde{z}}}\frac{p^2}{m^2} - 2\mathrm{i}f_1(t)\left(\frac{p}{m\dot{\tilde{z}}}\right)\left(\frac{p'}{m\epsilon_0}\right) + f_2(t)\left(\frac{\dot{\tilde{u}}}{\dot{\tilde{z}}} + 2\mathrm{i}\right)\left(\frac{p'}{m\epsilon_0}\right)^2\right]\right\}. \tag{5.149}$$

The purely TD function $F(t)$ is the same as in position space (see (5.106)) and also here the exponential scaling factor discussed at the beginning of Sect. 5.5 has been taken into account by the TD functions $f_1(t)$ and $f_2(t)$ with $\frac{f_1^2}{f_2} = e^{-\gamma t}$.

Performing the integration in (5.147) using (5.148) and (5.149) yields the Gaussian WP

$$\Psi_{\text{WP,NL}}(p,t) = \left(\frac{1}{\pi\hbar m}\right)^{\frac{1}{4}} \left(\frac{\mathrm{i}e^{-\frac{\gamma}{2}t}}{\dot{\tilde{\lambda}}}\right)^{\frac{1}{2}} \exp\left\{-\frac{\mathrm{i}m}{2\hbar}\left[\frac{\tilde{z}}{\dot{\tilde{z}}}\frac{p^2}{m^2} + \frac{e^{-\gamma t}}{\tilde{z}\dot{\tilde{\lambda}}}\frac{\tilde{p}^2}{m^2}\right]\right\}. \tag{5.150}$$

Using the relations between $\tilde{\lambda}$, $\tilde{u}$, $\tilde{z}$ and their time-derivatives, this WP can finally be written in a form like the corresponding position space WP (5.105), namely [23]

$$\Psi_{\text{WP,NL}}(p,t) = \left(\frac{1}{\pi\hbar m}\right)^{\frac{1}{4}} \left(\frac{\mathrm{i}e^{-\frac{\gamma}{2}t}}{\dot{\tilde{\lambda}}}\right)^{\frac{1}{2}} F(t) \exp\left\{-\frac{\mathrm{i}m}{2\hbar}\frac{\tilde{\lambda}}{\dot{\tilde{\lambda}}}\frac{\tilde{p}^2}{m^2} - \frac{\mathrm{i}}{\hbar}\langle x\rangle_{\text{NL}}\tilde{p} - \frac{\mathrm{i}}{\hbar}\frac{\langle x\rangle_{\text{NL}}\langle p\rangle_{\text{NL}}}{2}\right\}. \tag{5.151}$$

Comparison with the dissipative WP in position space written either as in (5.99),

$$\Psi_{\text{WP,NL}}(x,t) = \left(\frac{m}{\pi\hbar}\right)^{\frac{1}{4}} \left(\frac{e^{-\frac{\gamma}{2}t}}{\tilde{\lambda}}\right)^{\frac{1}{2}} F(t) \exp\left\{\frac{\mathrm{i}m}{2\hbar}\left[\frac{\dot{\tilde{z}}}{\tilde{z}}x^2 - \frac{e^{-\gamma t}}{\tilde{z}\tilde{\lambda}}\tilde{x}^2\right]\right\}, \tag{5.152}$$

or, as in (5.105) as

$$\Psi_{\text{WP,NL}}(x,t) = \left(\frac{m}{\pi\hbar}\right)^{\frac{1}{4}} \left(\frac{e^{-\frac{\gamma}{2}t}}{\tilde{\lambda}}\right)^{\frac{1}{2}} F(t) \exp\left\{\frac{\mathrm{i}m}{2\hbar}\frac{\dot{\tilde{\lambda}}}{\tilde{\lambda}}\tilde{x}^2 + \frac{\mathrm{i}}{\hbar}\langle p\rangle_{\text{NL}}\tilde{x} + \frac{\mathrm{i}}{\hbar}\frac{\langle p\rangle_{\text{NL}}\langle x\rangle_{\text{NL}}}{2}\right\} \tag{5.153}$$

shows that the *symmetry* between the WP solutions in position and momentum space in the linear case also applies in the NL case, only z, u and λ must be replaced by the quantities with tilde. So, also here, the transition between position and momentum space essentially can be achieved if the following substitutions are applied: $x \leftrightarrow \frac{p}{m}$, $+ \leftrightarrow -$ and $\tilde{z}$, $\tilde{u}$ or $\tilde{\lambda} \leftrightarrow \dot{\tilde{z}}$, $\dot{\tilde{u}}$ or $\dot{\tilde{\lambda}}$.

In Table 5.1 the Green functions for position and momentum space in the linear and NL cases are presented for comparison. Similarly, Table 5.2 shows the corresponding Gaussian WPs and Table 5.3 displays the interrelations between the coefficients of the quadratic terms in the exponent of the Gaussians, i.e., the terms that fulfil the Riccati equations, also expressed in terms of λ and $\tilde{\lambda}$ and their components.

Table 5.1 Green functions for the Gaussian WP solutions in position and momentum space for the linear and nonlinear case. Terms independent of $x', p', t' = 0$ that can be absorbed in the normalization coefficient are omitted. For $t' = 0$ a minimum uncertainty WP is assumed, thus $\epsilon_0 = \frac{1}{\alpha_0}$

$$G_L(x,x',t,t'=0) = \left(\frac{m}{2\pi i\hbar\alpha_0 z}\right)^{\frac{1}{2}} \exp\left\{\frac{im}{2\hbar}\left[\frac{\dot{z}}{z}x^2 - 2\frac{x}{z}\left(\frac{x'}{\alpha_0}\right) + \frac{u}{z}\left(\frac{x'}{\alpha_0}\right)^2\right]\right\}$$

$$G_{NL}(x,x',t,t'=0) = F(t)\left(\frac{m f_2(t) e^{-\frac{\gamma}{2}t}}{2\pi i\hbar\alpha_0 \tilde{z}}\right)^{\frac{1}{2}} \exp\left\{\frac{im}{2\hbar}\left[\frac{\dot{\tilde{z}}}{\tilde{z}}x^2 - 2\frac{x}{\tilde{z}}f_1(t)\left(\frac{x'}{\alpha_0}\right) + \frac{\tilde{u}}{\tilde{z}}f_2(t)\left(\frac{x'}{\alpha_0}\right)^2\right]\right\}$$

$$G_L(p,p',t,t'=0) = \left(\frac{-1}{2\pi\hbar m\varepsilon_0 \dot{z}}\right)^{\frac{1}{2}} \exp\left\{-\frac{im}{2\hbar}\left[\frac{z}{\dot{z}}\frac{p^2}{m^2} - 2i\frac{p}{m\dot{z}}\left(\frac{p'}{m\varepsilon_0}\right) + \left(\frac{\dot{u}}{\dot{z}} + 2i\right)\left(\frac{p'}{m\varepsilon_0}\right)^2\right]\right\}$$

$$G_{NL}(p,p',t,t'=0) = \left(\frac{-f_2(t) e^{-\frac{\gamma}{2}t}}{2\pi\hbar m\varepsilon_0 \dot{\tilde{z}}}\right)^{\frac{1}{2}} \exp\left\{-\frac{im}{2\hbar}\left[\frac{\tilde{z}}{\dot{\tilde{z}}}\frac{p^2}{m^2} - 2i\frac{p}{m\dot{\tilde{z}}}f_1(t)\left(\frac{p'}{m\varepsilon_0}\right) + \left(\frac{\dot{\tilde{u}}}{\dot{\tilde{z}}} + 2i\right)f_2(t)\left(\frac{p'}{m\varepsilon_0}\right)^2\right]\right\}$$

Table 5.2 Gaussian WP solutions in position and momentum space for the linear and nonlinear case – expressed in terms of λ and $\tilde{\lambda}$

$$\Psi_{WP,L}(x,t) = \left(\frac{m}{\pi\hbar}\right)^{\frac{1}{4}}\left(\frac{1}{\lambda}\right)^{\frac{1}{2}} \exp\left\{\frac{im}{2\hbar}\left(\frac{\dot{\lambda}}{\lambda}\right)\tilde{x}^2 + \frac{i}{\hbar}\langle p\rangle\tilde{x} + \frac{i}{\hbar}\frac{\langle p\rangle\langle x\rangle}{2}\right\}$$

$$\Psi_{WP,NL}(x,t) = \left(\frac{m}{\pi\hbar}\right)^{\frac{1}{4}}\left(\frac{e^{-\frac{\gamma}{2}t}}{\tilde{\lambda}}\right)^{\frac{1}{2}} F(t)\exp\left\{\frac{im}{2\hbar}\left(\frac{\dot{\tilde{\lambda}}}{\tilde{\lambda}}\right)\tilde{x}^2 + \frac{i}{\hbar}\langle p\rangle\tilde{x} + \frac{i}{\hbar}\frac{\langle p\rangle\langle x\rangle}{2}\right\}$$

$$\Psi_{WP,L}(p,t) = \left(\frac{1}{\pi\hbar m}\right)^{\frac{1}{4}}\left(\frac{i}{\dot{\lambda}}\right)^{\frac{1}{2}} \exp\left\{-\frac{im}{2\hbar}\left(\frac{\lambda}{\dot{\lambda}}\right)\frac{\tilde{p}^2}{m^2} - \frac{i}{\hbar}\langle x\rangle\tilde{p} - \frac{i}{\hbar}\frac{\langle x\rangle\langle p\rangle}{2}\right\}$$

$$\Psi_{WP,NL}(p,t) = \left(\frac{1}{\pi\hbar m}\right)^{\frac{1}{4}}\left(\frac{i\, e^{-\frac{\gamma}{2}t}}{\dot{\tilde{\lambda}}}\right)^{\frac{1}{2}} F(t)\exp\left\{-\frac{im}{2\hbar}\left(\frac{\tilde{\lambda}}{\dot{\tilde{\lambda}}}\right)\frac{\tilde{p}^2}{m^2} - \frac{i}{\hbar}\langle x\rangle\tilde{p} - \frac{i}{\hbar}\frac{\langle x\rangle\langle p\rangle}{2}\right\}$$

Table 5.3 Interrelations between the coefficients of the quadratic terms in the exponent of the Gaussians, i.e., the terms that fulfil the Riccati equations, also expressed in terms of λ and $\tilde{\lambda}$ fulfilling the linearized complex Newtonian equations

	x	p
$\gamma = 0$	$\left(\frac{2\hbar}{m}y_L\right) = \mathcal{C}_L = \frac{\dot{\lambda}}{\lambda} = \frac{\dot{z}\lambda - 1}{z\lambda}$	$\left(\frac{a_L m}{i\hbar}\right) = \mathcal{C}_L^{-1} = \frac{\lambda}{\dot{\lambda}} = \frac{z\dot{\lambda} + 1}{\dot{z}\dot{\lambda}}$
$\gamma \neq 0$	$\left(\frac{2\hbar}{m}y_{NL}\right) = \mathcal{C}_{NL} = \frac{\dot{\tilde{\lambda}}}{\tilde{\lambda}} = \frac{\dot{\tilde{z}}\tilde{\lambda} - e^{-\gamma t}}{\tilde{z}\tilde{\lambda}}$	$\left(\frac{a_{NL} m}{i\hbar}\right) = \mathcal{C}_{NL}^{-1} = \frac{\tilde{\lambda}}{\dot{\tilde{\lambda}}} = \frac{\tilde{z}\dot{\tilde{\lambda}} + e^{-\gamma t}}{\dot{\tilde{z}}\dot{\tilde{\lambda}}}$

Comparison of the Green functions in Table 5.1 shows that the transition from position to momentum space (in the TDSE and NLSE) mainly requires the replacement[11] of u and z without "dot" (for time-derivative) by those with dot and vice versa.

Transition from the TDSE to the NLSE (in position as well as in momentum space) is achieved by replacing u and z with the corresponding quantities with tilde, $\tilde{u}$ and $\tilde{z}$, which means multiplying by $e^{-\frac{\gamma}{2}t}$ and introducing the TD functions $F(t)$, $f_1(t)$ and $f_2(t)$.

Very similar relations apply to the WP solutions expressed in terms of the parameters z and λ, their time-derivatives or their counterparts with tilde. The time-dependence of these parameters can be obtained from the corresponding Riccati equations or the linearized complex Newtonian form.

5.7 Wigner Function and Ermakov Invariant for the Dissipative Case

In Sect. 2.8 the close relation between the Wigner function and the Ermakov invariant has been shown. Whether this still holds in the dissipative case or, if not, the modifications necessary are discussed next.

For this purpose, I_{L} in expression (2.136) for $W(0, 0, t)$ is replaced by I_{NL}, i.e.,

$$W_{dis}(0, 0; t) = \frac{1}{\pi\hbar} \exp\left\{-\frac{2m}{\hbar} I_{\mathrm{NL}}\right\} \tag{5.154}$$

and also η and $\dot{\eta}$ in I_{NL} by $\tilde{x}$ and $\frac{\tilde{p}}{m}$, as well as α, $\dot{\alpha}$ etc. by the position and momentum uncertainties $\langle \tilde{x}^2 \rangle_{\mathrm{NL}}$ and $\langle \tilde{p}^2 \rangle_{\mathrm{NL}}$ and their correlation $\langle [\tilde{x}, \tilde{p}]_+ \rangle_{\mathrm{NL}}$. In doing so, one arrives at a form for the dissipative Wigner function

$$W_{dis}(x, p; t) = \frac{e^{-\gamma t}}{\pi\hbar} \exp\left\{-\frac{2}{\hbar^2} e^{\gamma t} \left[\langle \tilde{p}^2 \rangle_{\mathrm{NL}} \tilde{x}^2 - \langle [\tilde{x}, \tilde{p}]_+ \rangle_{\mathrm{NL}} \tilde{x}\tilde{p} + \langle \tilde{x}^2 \rangle_{\mathrm{NL}} \tilde{p}^2\right]\right\}, \tag{5.155}$$

which looks exactly like the one in the conservative case, only $\langle \ldots \rangle_{\mathrm{L}}$ being replaced by $\langle \ldots \rangle_{\mathrm{NL}}$ and two exponential factors occurring, one in the exponent and one in front of it, whereby the latter is a consequence of the other due to the normalization requirement of W_{dis}.

As has been shown above, the same invariant, but with different physical interpretation, exists on the formal canonical level (related to the CK approach, see Eq. (4.45)) and on the physical level (related to the logarithmic NLSE, see Eq. (5.16)).

---

[11] This is concerning the TD quantities; in addition, for the transition from position to momentum space, α_0 must also be replaced by $\epsilon_0 = \frac{1}{\alpha_0}$ (for an initial minimum-uncertainty WP) and a factor i appears in the coefficient of the term depending on p'.

In order to clarify the meaning of the form (5.155) of the dissipative Wigner function, the equation of motion that it fulfils is considered. It can be shown straightforwardly that this is a Fokker–Planck equation in phase space,

$$\frac{\partial}{\partial t} W_{dis} + \frac{p}{m}\frac{\partial}{\partial x} W_{dis} - \frac{\partial V}{\partial x}\frac{\partial}{\partial p} W_{dis} - \gamma p \frac{\partial}{\partial p} W_{dis} - \gamma W_{dis} = 0, \tag{5.156}$$

or

$$\frac{\partial}{\partial t} W_{dis} + \frac{\partial}{\partial x}(\dot{x} W_{dis}) + \frac{\partial}{\partial p}(\dot{p} W_{dis}) = 0, \tag{5.157}$$

with $\dot{p} = -\frac{\partial}{\partial x} V - \gamma p$ (for V being at most quadratic in x) but without diffusion terms, i.e., without second derivatives.

As in classical statistical mechanics, the diffusion coefficients of these second-derivative terms are usually proportional to the temperature of the system. This would correspond to a situation at temperature $\mathsf{T} = 0$ which is one of the assumptions Sun and Yu [24, 25] made in their derivation of the CK Hamiltonian. The notion that W_{dis} corresponds to the canonical level is further confirmed if one integrates W_{dis} over the momentum p in order to obtain the distribution function in position space that should be equal to $\varrho_{\mathrm{NL}}(x, t)$, the solution of the Smoluchowski equation (4.109). In doing so, however, one obtains

$$\int_{-\infty}^{+\infty} dp W_{dis}(x, p; t) = \sqrt{\frac{\mathrm{e}^{-\gamma t}}{2\pi\langle\tilde{x}^2\rangle_{\mathrm{NL}}}} \exp\left\{-\frac{\mathrm{e}^{\gamma t}\tilde{x}^2}{2\langle\tilde{x}^2\rangle_{\mathrm{NL}}}\right\} \neq \varrho_{\mathrm{NL}}(x, t) = \Psi_{\mathrm{NL}}^*(x, t)\Psi_{\mathrm{NL}}(x, t). \tag{5.158}$$

So the proper solution of the Smoluchowski equation is obtained when the transformation from the canonical to the physical level is performed according to Eq. (4.135), $\ln\Psi_{\mathrm{NL}} = e^{-\gamma t}\ln\hat{\Psi}$. This essentially requires multiplying the exponent by $e^{-\gamma t}$ and normalizing the resulting density function. As a consequence, in the equation of motion for the density, the addition terms $-\gamma(\ln\varrho(x, t) - \langle\ln\varrho(x, t)\rangle)\varrho(x, t)$, originating from $\frac{\partial}{\partial t}\varrho(x, t)$ and the normalization, must be compensated for in order to fulfil the conservation law for $\varrho(x, t)$. This compensation term for the Gaussian WP that can be written in the form

$$\gamma\left(\ln\varrho_{\mathrm{NL}} - \langle\ln\varrho_{\mathrm{NL}}\rangle\right)\varrho_{\mathrm{NL}} = -D\frac{\partial^2}{\partial x^2}\varrho_{\mathrm{NL}}, \tag{5.159}$$

with $D = \frac{\gamma}{2}\langle\tilde{x}^2\rangle_{\mathrm{NL}}$, is nothing but the separation condition (4.113) used in the derivation of the logarithmic NLSE in Sect. 4.4.3. So, the transformation from the formal canonical level to the physical dissipative level introduces the missing diffusion term that is necessary to turn the reversible continuity equation for the distribution function in position space into the irreversible Smoluchowski equation.

Therefore, the Wigner function W_{dis} in Eq. (5.155), obtained with the help of the Ermakov invariant I_{NL}, is the one on the canonical level and must be treated accordingly.

5.8 Algebraic Derivation of the Dissipative Ermakov Invariant

In Sect. 2.10 it has been shown how the dynamical Ermakov invariant can be obtained in an algebraic way, making use of the canonical Hamiltonian formalism, particularly exploiting the fact that the time-derivative of a dynamical quantity can be obtained with the help of the Poisson brackets. A Lie algebra which is closed in relation to these brackets finally allowed to obtain the desired invariant.

In the dissipative case, we have seen in Sect. 4.3 that there is also a description possible using the established canonical formalism. However, not to be overlooked is the fact that the canonical variables are connected with the physical ones via a non-canonical transformation (in the classical case, corresponding to a non-unitary transformation in the quantum mechanical one). The canonical description that showed closest formal similarity with the conventional non-dissipative one was the description in terms of exponentially-expanding variables (see Sect. 4.3.2). This is used in the following as a basis for the construction of the dynamical invariant for the dissipative system.[12]

For this purpose we recall that the equation of motion for the canonical variable (see Eq. (4.49)),

$$\hat{Q} = x e^{\frac{\gamma}{2}t},$$

is given by (see Eq. (4.52))

$$\ddot{\hat{Q}} + \Omega^2 \hat{Q} = 0$$

and the time-evolution of any phase-space function $\hat{F}(\hat{Q}, \hat{P}, t)$ on the canonical level is given by (see Eqs. (4.56) and (4.51))

$$\frac{d}{dt}\hat{F} = \{\hat{F}, \hat{H}_{exp}\}_{(Q,P)-} + \frac{\partial}{\partial t}\hat{F}$$

with

$$\hat{H}_{exp} = \frac{1}{2m}\hat{P}^2 + \frac{m}{2}\left(\omega^2 - \frac{\gamma^2}{4}\right)\hat{Q}^2.$$

---

[12] An algebraic derivation of a dissipative Ermakov invariant based on the canonical Caldirola–Kanai approach has been shown by Korsch [26], see also [27].

Expressed in terms of the corresponding physical variables x and p, one obtains (see Eq. (4.57))

$$\frac{d}{dt}F(x,p,t) = \{F,H\}_- + \{F,\frac{\gamma}{2}xp\}_- - \{F,\frac{\gamma}{2}xp\}_+ + \frac{\partial}{\partial t}F$$

where $H = \frac{1}{2m}p^2 + \frac{m}{2}\omega^2 x^2$; i.e., on the physical level, additional Poisson brackets $\{,\}_-$ and anti-Poisson brackets $\{,\}_+$ occur.

Expressing the dynamical invariant for the dissipative case in terms of the same phase-space functions[13] $\Gamma_n (n = 1-3)$ as defined in (2.184), i.e., $\Gamma_1 = \frac{p^2}{2m}$, $\Gamma_2 = xp$, $\Gamma_3 = \frac{mx^2}{2}$ it can be written as

$$I_{exp} = \sum_n \hat{\kappa}_n(t)\Gamma_n \tag{5.160}$$

where the $\hat{\kappa}_n$ are the expansion coefficients on the *canonical* level. The invariant must again fulfil the condition $\frac{d}{dt}I = 0$. As we are interested in the result on the *physical* level, the time-derivative in the form (4.57), i.e., in terms of physical position and momentum is used, leading to

$$\begin{aligned} \frac{d}{dt}I_{exp} &= \{I_{exp},H\}_- + \{I_{exp},\frac{\gamma}{2}xp\}_- - \{I_{exp},\frac{\gamma}{2}xp\}_+ + \frac{\partial}{\partial t}I_{exp} \\ &= \{I_{exp},H\}_- + \frac{\gamma}{2}\{I_{exp},\Gamma_2\}_- - \frac{\gamma}{2}\{I_{exp},\Gamma_2\}_+ + \frac{\partial}{\partial t}I_{exp} = 0. \end{aligned} \tag{5.161}$$

In comparison with the non-dissipative case, there are *additional terms* from $\{\Gamma_n,\Gamma_2\}_-$ and $\{\Gamma_n,\Gamma_2\}_+$. From these, only

$$\{\Gamma_1,\Gamma_2\}_- - \{\Gamma_1,\Gamma_2\}_+ = -4\Gamma_1 \tag{5.162}$$

and

$$\{\Gamma_2,\Gamma_2\}_+ = 2\Gamma_2 \tag{5.163}$$

contribute.

At this point it should be kept in mind that I_{exp} has the dimension of an action and, like $\hat{H}_{exp}$, is an invariant on the canonical level.[14] As has been shown in Sect. 4.5, the connection between the Hamiltonians and actions on the canonical and the physical level (in the classical and in the quantum case) is given via

$$\hat{S}_c = e^{\gamma t}S_c \quad \text{and} \quad \hat{H} = e^{\gamma t}H_L.$$

[13]On the canonical level, x and p can be expressed in terms of $\hat{Q}$ and $\hat{P}$.

[14]This also agrees with our finding in connection with the dissipative Wigner function in the previous subsection and the dissipative creation-/annihilation operators discussed in the following subsection.

Therefore, the expansion coefficients $\kappa_{n,\mathrm{NL}}$ on the physical level are related to the canonical coefficients $\hat{\kappa}_n$ via $\kappa_{n,\mathrm{NL}} = \mathrm{e}^{-\gamma t}\hat{\kappa}_n$.

Taking all this into account, one obtains a modified set of coupled evolution equations for $\kappa_{n,\mathrm{NL}}$,

$$\dot{\kappa}_{1,\mathrm{NL}} = \gamma\kappa_{1,\mathrm{NL}} - 2\kappa_{2,\mathrm{NL}}, \tag{5.164}$$

$$\dot{\kappa}_{2,\mathrm{NL}} = \omega^2\kappa_{1,\mathrm{NL}} - \kappa_{3,\mathrm{NL}}, \tag{5.165}$$

$$\dot{\kappa}_{3,\mathrm{NL}} = 2\omega^2\kappa_{2,\mathrm{NL}} - \gamma\kappa_{3,\mathrm{NL}}. \tag{5.166}$$

With $\kappa_{1,\mathrm{NL}} = \frac{1}{m}\alpha_{\mathrm{NL}}^2$, in analogy with the non-dissipative case, this set of coupled equations can be condensed into the *modified Ermakov equation*

$$\ddot{\alpha}_{\mathrm{NL}} + \left(\omega^2 - \frac{\gamma^2}{4}\right)\alpha_{\mathrm{NL}} = \frac{1}{\alpha_{\mathrm{NL}}^3}, \tag{5.167}$$

which is identical to the Ermakov equation (5.15) of the logarithmic NLSE (5.1) (hence the subscript NL).

As in the non-dissipative case, it is again assumed that the relations between α_{NL} and $\dot{\alpha}_{\mathrm{NL}}$ on the one side and the quantum uncertainties on the other, are the same as found in (5.17)–(5.19) but now for the dissipative case. The $\kappa_{n,\mathrm{NL}}$ can then be expressed as

$$\kappa_{1,\mathrm{NL}} = \frac{1}{m}\alpha_{\mathrm{NL}}^2 = \frac{2}{\hbar}\langle\tilde{x}^2\rangle_{\mathrm{NL}}, \tag{5.168}$$

$$\kappa_{2,\mathrm{NL}} = -\frac{1}{m}\alpha_{\mathrm{NL}}\left(\dot{\alpha}_{\mathrm{NL}} - \frac{\gamma}{2}\alpha_{\mathrm{NL}}\right) = -\frac{1}{\hbar m}\langle[\tilde{x},\tilde{p}]_+\rangle_{\mathrm{NL}}, \tag{5.169}$$

$$\kappa_{3,\mathrm{NL}} = \frac{1}{m}\left[\left(\dot{\alpha}_{\mathrm{NL}} - \frac{\gamma}{2}\alpha_{\mathrm{NL}}\right)^2 + \frac{1}{\alpha_{\mathrm{NL}}^2}\right] = \frac{2}{\hbar m}\langle\tilde{p}^2\rangle_{\mathrm{NL}}. \tag{5.170}$$

Equations (5.164–5.166) can also be written as evolution equations for the quantum uncertainties in the form

$$\frac{\partial}{\partial t}\langle\tilde{x}^2\rangle_{\mathrm{NL}} = \frac{1}{m}\langle[\tilde{x},\tilde{p}]_+\rangle_{\mathrm{NL}} + \gamma\langle\tilde{x}^2\rangle_{\mathrm{NL}}, \tag{5.171}$$

$$\frac{\partial}{\partial t}\langle[\tilde{x},\tilde{p}]_+\rangle_{\mathrm{NL}} = 4\left(\frac{1}{2m}\langle\tilde{p}^2\rangle_{\mathrm{NL}} - \frac{m}{2}\omega^2\langle\tilde{x}^2\rangle_{\mathrm{NL}}\right), \tag{5.172}$$

$$\frac{\partial}{\partial t}\langle\tilde{p}^2\rangle_{\mathrm{NL}} = -m\omega^2\langle[\tilde{x},\tilde{p}]_+\rangle_{\mathrm{NL}} - \gamma\langle\tilde{p}^2\rangle_{\mathrm{NL}}, \tag{5.173}$$

similar to Eqs. (2.138)–(2.140) in the non-dissipative case.

Finally, the dissipative Ermakov invariant on the physical level can be written in terms of x and α_{NL}^2 as

$$I_{\rm NL} = \frac{1}{2}\mathrm{e}^{\gamma t}\left[\left(\dot{x}\alpha_{\rm NL} - \left(\dot{\alpha}_{\rm NL} - \frac{\gamma}{2}\alpha_{\rm NL}\right)x\right)^2 + \left(\frac{x}{\alpha_{\rm NL}}\right)^2\right] = \text{const.} \tag{5.174}$$

which is identical to the invariant (5.16) of the logarithmic NLSE for $\eta = x$ (for further details, see [28]).

Expressing x and $\dot{x}$ in terms of the expanding coordinate $\hat{Q}$ and velocity $\dot{\hat{Q}}$, the invariant can be rewritten in the form

$$\hat{I}_{exp} = \frac{1}{2}\left[\left(\dot{\hat{Q}}\alpha_{\rm NL} - \hat{Q}\dot{\alpha}_{\rm NL}\right)^2 + \left(\frac{\hat{Q}}{\alpha_{\rm NL}}\right)^2\right] = \text{const}, \tag{5.175}$$

in agreement with Eqs. (4.52) and (5.167), reproducing Eq. (4.66) for $\hat{\alpha}_{exp} = \alpha_{\rm NL}$, thus confirming the results from Sects. 4.3.2 and 4.5.

This is again the form showing that the invariant is not only independent of ω, i.e., also existing for $\omega = \omega(t)$, but also independent of γ, i.e., also existing for $\gamma = \gamma(t)$!

5.9 Generalized Creation and Annihilation Operators and Coherent States for the Dissipative Case

In the non-dissipative case, we found that the creation and annihilation operators, and particularly their generalizations, are related to the Ermakov invariant via (see Eqs. (2.218) and (2.217))

$$I_{{\rm L},op} = \frac{\hbar}{m}\left[a^+(t)a(t) + \frac{1}{2}\right]$$

with

$$I_{\rm L} = \frac{\hbar}{m}ww^*,$$

where w is the complex eigenvalue of the coherent state $|w\rangle$ that is an eigenstate of the annihilation operator $a(t)$. That is, $a(t)$ and $a^+(t)$ can be obtained by the factorization of the operator $I_{{\rm L},op}$ corresponding to the invariant $I_{\rm L}$.

Attempting to do the same with $I_{\rm NL} = \frac{\hbar}{m}\hat{w}_{\rm NL}\hat{w}^*_{\rm NL}$ (see Eq. (5.16)) leads to

$$\hat{w}_{\rm NL} = \sqrt{\frac{m}{2\hbar}}\left[\left(\frac{\eta}{\alpha_{\rm NL}}\right) + \mathrm{i}\left(\dot{\eta}\alpha_{\rm NL} - \left(\dot{\alpha}_{\rm NL} - \frac{\gamma}{2}\alpha_{\rm NL}\right)\eta\right)\right]\mathrm{e}^{\frac{\gamma}{2}t}, \tag{5.176}$$

$$\hat{w}^*_{\rm NL} = \sqrt{\frac{m}{2\hbar}}\left[\left(\frac{\eta}{\alpha_{\rm NL}}\right) - \mathrm{i}\left(\dot{\eta}\alpha_{\rm NL} - \left(\dot{\alpha}_{\rm NL} - \frac{\gamma}{2}\alpha_{\rm NL}\right)\eta\right)\right]\mathrm{e}^{\frac{\gamma}{2}t}, \tag{5.177}$$

and, after replacing $\dot{\eta}$ by $\frac{p_{\mathrm{op}}}{m}$ (with $p_{\mathrm{op}} = \frac{\hbar}{\mathrm{i}}\frac{\partial}{\partial x}$) and η by x, to

$$\hat{a}_{\mathrm{NL}}(t) = \mathrm{i}\sqrt{\frac{m}{2\hbar}}\alpha_{\mathrm{NL}}\left(\frac{p_{\mathrm{op}}}{m} - C_{\mathrm{NL}}x\right)\mathrm{e}^{\frac{\gamma}{2}t}, \tag{5.178}$$

$$\hat{a}^{+}_{\mathrm{NL}}(t) = -\mathrm{i}\sqrt{\frac{m}{2\hbar}}\alpha_{\mathrm{NL}}\left(\frac{p_{op}}{m} - C^{*}_{\mathrm{NL}}x\right)\mathrm{e}^{\frac{\gamma}{2}t}. \tag{5.179}$$

From (5.178) and (5.179) it is clear that, apart from the exponential factor $e^{\frac{\gamma}{2}t}$, the form of the (generalized) creation and annihilation operators is identical to the ones defined in (2.202) and (2.203) in the non-dissipative case. Only the solutions of Riccati equation (2.4) must be replaced by the ones of Eq. (5.4) belonging to the logarithmic NLSE (5.1).

Still the meaning of the exponential factor $\mathrm{e}^{\frac{\gamma}{2}t}$ needs to be explained. As I_{NL} is a constant of motion for the dissipative system (like $\hat{H}_{exp}$ on the canonical level) but the NL Hamiltonian and its mean value are not, the invariant should correspond to the canonical level, in agreement with the remarks in the previous sub-section (and the results for the dissipative Wigner function).

In the classical case, the connection between the canonical and the physical level was given by (4.133); in the quantum mechanical case this corresponds to the non-unitary connection between the canonical and physical wave functions according to (4.135), i.e.

$$\ln\hat{\Psi} = \mathrm{e}^{\gamma t}\ln\Psi_{\mathrm{NL}}.$$

Therefore, the factor $e^{\gamma t}$ in I_{NL} should be *omitted* if it is used for the definition of the creation and annihilation operators that supply the wave functions and coherent states which are solutions of the Hamiltonian operator on the *physical* level.

In this dissipative case, the action-type quantity $mI_{\mathrm{NL}}\mathrm{e}^{-\gamma t}$ on the physical level can be expressed in the same way in terms of uncertainties multiplied by the conjugate mean values as in the non-dissipative case (with $\gamma = 0$, i.e., $\mathrm{e}^{-\gamma t} = 1$), only the mean values are now calculated with $\Psi_{\mathrm{NL}}(x,t)$, i.e.,

$$mI_{\mathrm{NL}}e^{-\gamma t} = \frac{1}{\hbar}\left[\langle\tilde{x}^2\rangle_{\mathrm{NL}}\langle p\rangle^2_{\mathrm{NL}} - \langle[\tilde{x},\tilde{p}]_{+}\rangle_{\mathrm{NL}}\langle x\rangle_{\mathrm{NL}}\langle p\rangle_{\mathrm{NL}} + \langle\tilde{p}^2\rangle_{\mathrm{NL}}\langle x\rangle^2_{\mathrm{NL}}\right]. \tag{5.180}$$

The equivalent relation between the invariant I_{NL} and the Wigner function for the dissipative case has already been mentioned above.

So finally the generalized creation and annihilation operators for the dissipative systems can also be written in the form

$$a_{\mathrm{NL}}(t) = \sqrt{\frac{m}{2\hbar}}\alpha_{\mathrm{NL}}(t)\left(\frac{\hbar}{m}\frac{\partial}{\partial x} - \mathrm{i}\,C_{\mathrm{NL}}x\right) = \hat{a}_{\mathrm{NL}}\,\mathrm{e}^{-\frac{\gamma}{2}t}, \tag{5.181}$$

$$a^{+}_{\mathrm{NL}}(t) = \sqrt{\frac{m}{2\hbar}}\alpha_{\mathrm{NL}}(t)\left(-\frac{\hbar}{m}\frac{\partial}{\partial x} + \mathrm{i}\,C^{*}_{\mathrm{NL}}x\right) = \hat{a}^{+}_{\mathrm{NL}}\,\mathrm{e}^{-\frac{\gamma}{2}t}. \tag{5.182}$$

The corresponding coherent states can again be obtained via application of $a_{\mathrm{NL}}(t)$ as eigenstates of this operator with eigenvalue $w_{\mathrm{NL}} = \hat{w}_{\mathrm{NL}}\mathrm{e}^{-\frac{\gamma}{2}t}$ or via the displacement operator, leading to

$$|w\rangle_{\mathrm{NL}} = \mathrm{e}^{-\frac{1}{2}|w_{\mathrm{NL}}|^2}\sum_{n=0}^{\infty}\frac{w^{n}_{\mathrm{NL}}(a^{+}_{\mathrm{NL}})^{n}}{n!}|0\rangle \tag{5.183}$$

with

$$w_{\mathrm{NL}} = \sqrt{\frac{m}{2\hbar}}\alpha_{\mathrm{NL}}\left[\mathcal{C}_{\mathrm{NL,I}}\eta + \mathrm{i}\,(\dot{\eta} - \mathcal{C}_{\mathrm{NL,R}}\eta)\right], \tag{5.184}$$

where the coherent states $|w\rangle_{\mathrm{NL}}$ are identical to the exact WP solutions $\Psi_{\mathrm{NL}}(x,t)$ (5.2) of the log NLSE (5.1) discussed in this section and related to the canonical dissipative WPs via the non-unitary transformations given in Sect. 4.5 (for further details, see [29, 30]).

References

1. D. Schuch, K.-M. Chung, H. Hartmann, Nonlinear Schrödinger-type field equation for the description of dissipative systems. I. Derivation of the nonlinear field equation and one-dimensional example. J. Math. Phys. **24**, 1652–1660 (1983)
2. D. Schuch, K.-M. Chung, H. Hartmann, Nonlinear Schrödinger-type field equation for the description of dissipative systems. III. Frictionally damped free motion as an example for an aperiodic motion. J. Math. Phys. **25**, 3086–3092 (1984)
3. D. Schuch, K.-M. Chung, From macroscopic irreversibility to microscopic reversibility via a nonlinear Schrödinger-type field equation. Int. J. Quant. Chem. **29**, 1561–1573 (1986)
4. D. Schuch, Connections between Newton- and Schrödinger-type equations in the description of reversible and irreversible dynamics. Int. J. Quant. Chem., Quant. Chem. Symp. **36**(S23), 57–72 (1989)
5. D. Schuch, On the complex relations between equations describing the dynamics of wave and particle aspects. Int. J. Quant. Chem. **42**, 663–683 (1992)
6. E. Schrödinger, Sitzungsber. Preuss. Akad. Wiss. Math.-Phys. Kl., XIX (Berlin, 1930), pp. 296–303
7. H.P. Robertson, The uncertainty principle. Phys. Rev. **34**, 163 (1929); ibid. **35**, 667 (1930); ibid. **46**, 794 (1934)
8. H. Cruz, D. Schuch, O. Castaños, O. Rosas-Ortiz, Time-evolution of quantum systems via a complex nonlinear Riccati equation. II. Dissipative systems. Ann. Phys. **373**, 609–630 (2016)
9. G. Grössing, J. Mesa Pasasio, H. Schwabl, A classical explanation of quantization. Found. Phys. **41**, 1437–1453 (2011)
10. D. Schuch, Nonlinear quantum mechanics, complex classical mechanics and conservation laws for closed and open systems. J. Phys. Conf Ser. **361**, 012020 (2012)
11. H.G. Schuster, *Deterministic Chaos; An Introduction* (Physik Verlag, Weinheim, 1984), p. 112
12. D. Schuch, Is quantum theory intrinsically nonlinear? Phys. Scr. **87**, 038117, 10 pp (2013) and Highlights 2013

13. D. Schuch, Can the breaking of time-reversal symmetry remove degeneracies in quantum mechanics? Phys. Lett. A **294**, 31–36 (2002)
14. D. Schuch, New energetic and dynamic quantum effects originating from the breaking of time-reversal symmetry. J. Phys. A: Math. Gen. **35**, 8615–8629 (2002)
15. D. Schuch, K.-M. Chung, H. Hartmann, Nonlinear Schrödinger-type field equation for the description of dissipative systems. II. Frictionally damped motion in a magnetic field. Int. J. Quant. Chem. **25**, 391–410 (1984)
16. D. Schuch, in *New Challenges in Computational Quantum Chemistry*, ed. by R. Broer, P.J.C. Aerts, P.S. Bagus (University of Groningen, 1994), pp. 255–269
17. D. Schuch, M. Moshinsky, Coherent states and dissipation for the motion of a charged particle in a constant magnetic field. J. Phys. A: Math. Gen. **36**, 6571–6586 (2003)
18. M. Nuñez, P.O. Hess, D. Schuch, Quantum mechanics in dissipative systems with a strong magnetic field. Phys. Rev. A **70**, 032103, 5 pp (2004)
19. J. Schwinger, L.L. De Raad Jr., K.A. Milton, Wu-yang Tsai, *Classical Electrodynamics* (Perseus Books, Reading Massachusetts, 1998), p. 42
20. D. Schuch, Connection between quantum mechanical and classical time evolution of certain dissipative systems via a dynamical invariant. J. Math. Phys. **48**, 122701 (2007)
21. C. Grosche, F. Steiner, *Handbook of Feynman Path Integrals, Springer Tracts in Modern Physics*, vol. 145 (Springer, Heidelberg, 1998)
22. D. Schuch, On a form of nonlinear dissipative wave mechanics valid in position- and momentum-space. Int. J. Quant. Chem., Quant. Chem. Symp. **28**, 251–259 (1994)
23. D. Schuch, Green's function for dissipative quantum systems and its relation to nonlinear evolution equations, in *Group 24—Physical Applications and Mathematical Aspects of Geometry, Groups and Algebra*, ed. by J.P. Gazeau, R. Kerner (IOP Conference Series, Bristol, 2003), pp 741–745
24. L.H. Yu, C.P. Sun, Evolution of the wave function in a dissipative system. Phys. Rev. A **49**, 592 (1994)
25. C.P. Sun, L.H. Yu, Exact dynamics of a quantum dissipative system in a constant external field. Phys. Rev. A **51**, 1845 (1995)
26. H.J. Korsch, Dynamical invariants and time-dependent harmonic systems. Phys. Lett. A **74**, 294–296 (1979)
27. R.S. Kaushal, H.J. Korsch, Dynamical Noether invariants for time-dependent nonlinear systems. J. Math. Phys. **22**, 1904 (1981)
28. D. Schuch, R.S. Kaushal, Some remarks on dissipative Ermakov systems and damping in Bose–Einstein condensates. J. Phys. Conf. Ser. **306**, 012032, 10 pp (2011)
29. O. Castaños, D. Schuch, O. Rosas-Ortiz, Generalized coherent states for time-dependent and nonlinear Hamiltonian operators via complex Riccati equations. J. Phys. A: Math. Theor. **46**, 075304, 20 pp (2013)
30. D. Schuch, O. Castaños, O. Rosas-Ortiz, Generalized creation and annihilation operators via complex nonlinear Riccati equations. J. Phys. Conf. Ser. **442**, 012058, 10 pp (2013)

Chapter 6
Dissipative Version of Time-Independent Nonlinear Quantum Mechanics

In the time-dependent (TD) non-dissipative case, the information on the time-evolution of the wave packet (WP) width, and thus on the quantum mechanical aspect, was contained in the nonlinear (NL) complex Riccati equation for the quantity $\frac{2\hbar}{m} y(t) = \mathcal{C} = \frac{\dot{\lambda}(t)}{\lambda(t)}$,

$$\frac{d}{dt}\left(\frac{\dot{\lambda}}{\lambda}\right) + \left(\frac{\dot{\lambda}}{\lambda}\right)^2 + \omega^2 = 0 \,, \tag{6.1}$$

that can be linearized to the complex Newtonian equation for $\lambda(t) = u + \mathrm{i}z = \alpha\,\mathrm{e}^{\mathrm{i}\varphi}$ (where $z(t)$ is proportional to the classical trajectory $\eta(t) = \langle x \rangle$),

$$\ddot{\lambda} + \omega^2 \lambda = 0 \,. \tag{6.2}$$

Taking into account a linear velocity dependent friction force, an additional term linear in $\mathcal{C}_{\text{NL}} = \frac{\dot{\tilde{\lambda}}}{\tilde{\lambda}}$ enters the Riccati equation (with $\tilde{\lambda} = \lambda\,\mathrm{e}^{-\frac{\gamma}{2}t}$),

$$\frac{d}{dt}\left(\frac{\dot{\tilde{\lambda}}}{\tilde{\lambda}}\right) + \gamma\left(\frac{\dot{\tilde{\lambda}}}{\tilde{\lambda}}\right) + \left(\frac{\dot{\tilde{\lambda}}}{\tilde{\lambda}}\right)^2 + \omega^2 = 0 \,, \tag{6.3}$$

which, after linearization, causes an additional linear term with first derivative in the second-order differential equation for $\tilde{\lambda}(t)$,

$$\ddot{\tilde{\lambda}} + \gamma\dot{\tilde{\lambda}} + \omega^2 \tilde{\lambda} = 0 \,. \tag{6.4}$$

In both cases, the additional terms depend on the coefficient γ of the friction force.

D. Schuch, *Quantum Theory from a Nonlinear Perspective*,
Fundamental Theories of Physics 191, https://doi.org/10.1007/978-3-319-65594-9_6

In the NL formulation of time-independent (TI) quantum mechanics (discussed in Sect. 3.2), the complex Riccati equation

$$\nabla \left(\frac{\nabla\Psi}{\Psi}\right) + \left(\frac{\nabla\Psi}{\Psi}\right)^2 + \frac{2m}{\hbar^2}(E - V) = 0 \tag{6.5}$$

in the non-dissipative case could be linearized to the usual (complex) TI Schrödinger equation (SE), written in the form

$$\Delta\Psi + \frac{2m}{\hbar^2}(E - V)\,\Psi = 0\,. \tag{6.6}$$

The question now arising is, what are the modifications to this Riccati equation and its linearized form caused by the friction force in the dissipative case? Comparison with the TD situation suggest a linear term $\Gamma\left(\frac{\nabla\Psi}{\Psi}\right)$ should be added to the Riccati Eq. (6.5), i.e.,

$$\nabla \left(\frac{\nabla\Psi}{\Psi}\right) + \Gamma\left(\frac{\nabla\Psi}{\Psi}\right) + \left(\frac{\nabla\Psi}{\Psi}\right)^2 + \frac{2m}{\hbar^2}(E - V) = 0\,, \tag{6.7}$$

and a corresponding first-derivative term $\Gamma\nabla\Psi$ to Eq. (6.6), i.e.,

$$\Delta\Psi + \Gamma\nabla\Psi + \frac{2m}{\hbar^2}(E - V)\,\Psi = 0\,. \tag{6.8}$$

The coefficient Γ should somehow be related to the friction force and its parameter γ but, in general, could be a complex function of $\mathbf{r}$ (or in one dimension x) and t, $\Gamma(\mathbf{r}, t)$.

As Ψ is complex, a contribution from the imaginary part of the additional term $\Gamma\nabla\Psi$ should show up in the equation for $\varrho = \Psi^*\Psi$, thus modifying the continuity equation (3.30) written in the notation in Sect. 3.2 with $\Psi = a\,\mathrm{e}^{-\frac{i}{\hbar}S}$ (and $\varrho = a^2$) as

$$\frac{\partial}{\partial t}a^2 + \frac{1}{m}\,\nabla\left(a^2\nabla S\right) = 0\,.$$

What should be the specific form of this additional term and, particularly, of $\Gamma(\mathbf{r}, t)$? To answer this question the following assumptions are made:

1. The SE can be obtained from the continuity equation (3.30) according to Madelung and Mrowka [1, 2] (see Sect. 4.4.3) via separation. The log NLSE (4.115) can be obtained from the Smoluchowski equation

$$\frac{\partial}{\partial t}a^2 + \frac{1}{m}\,\nabla\left(a^2\nabla S\right) - D\Delta a^2 = 0 \tag{6.9}$$

with the additional diffusion term $-D\Delta a^2$ and the condition

$$-D\Delta a^2 = \gamma(\ln a^2 - \langle \ln a^2 \rangle)a^2 \tag{6.10}$$

also via separation [3, 4]. Therefore, the additional term $\Gamma \nabla \Psi$ in the modified TISE should also originate from an additional term in the continuity equation, plus separation. That is, a term

$$\Gamma \nabla a^2 = \Gamma \left(\Psi^* \nabla \Psi + \Psi \nabla \Psi^*\right) \tag{6.11}$$

should occur in the evolution equation for $a^2 = \varrho$. The contribution from this term would enter the modified SE with an *imaginary* coefficient.

2. The additional term in the equation for $a^2 = \varrho$, at least for cases with Gaussian WP solutions (i.e., $V = 0$ and HO) discussed before in this context, should have the same effect as the diffusion term or the $\ln \rho$-term, i.e.,

$$-D\Delta_x \varrho = \gamma(\ln \varrho - \langle \ln \varrho \rangle)\varrho = \frac{\gamma}{2}\left(1 - \frac{\tilde{x}^2}{\langle \tilde{x}^2 \rangle}\right)\varrho\,. \tag{6.12}$$

Comparison shows that

$$\nabla_x \left(\frac{\gamma}{2}\tilde{x}\varrho\right) = \frac{\gamma}{2}\left(1 - \frac{\tilde{x}^2}{\langle \tilde{x}^2 \rangle}\right)\varrho \tag{6.13}$$

yields the desired result! (In our one-dimensional case that is discussed in detail, ∇_x means $\frac{\partial}{\partial x}$ and the subscript x will be omitted in the following as the results apply to problems in one, two and three dimensions.)

After separation, this contributes an additional term

$$W_{\mathrm{TI}} = \frac{\gamma}{2}\frac{\hbar}{\mathrm{i}}\left(\frac{1}{2} + \tilde{x}\frac{\nabla \Psi}{\Psi}\right) \tag{6.14}$$

that already looks familiar when rewritten as

$$W_{\mathrm{TI}}\,\Psi = \left(\frac{\gamma}{4}\frac{\hbar}{\mathrm{i}} + \frac{\gamma}{2}\tilde{x}\frac{\hbar}{\mathrm{i}}\nabla\right)\Psi = \frac{\gamma}{4}\,[\tilde{x}, p_{\mathrm{op}}]_+\,\Psi = \frac{1}{2}W_{\mathrm{S\ddot{u}s}}\,\Psi\,, \tag{6.15}$$

where $W_{\mathrm{S\ddot{u}s}}$ is identical to Süssman's [5] approach (4.89) that was discussed in Sect. 4.4.1.

However:

(a) its mean value $\langle W_{\mathrm{TI}} \rangle = \frac{\gamma}{4}\langle [\tilde{x}, \tilde{p}]_+ \rangle = \langle W_{\mathrm{Has}} \rangle \neq 0$ does not vanish (where W_{Has} is Hasse's friction term [6], also discussed in Sect. 4.4.1) but this could be compensated for by a normalization factor; and

(b) the corresponding mean value of the friction force is

$$\langle -\nabla W_{\mathrm{TI}} \rangle = -\frac{\gamma}{2}\langle p \rangle\,,$$

i.e., half of the friction force is missing.

Considering for instance the TISE for the HO with real wave functions, this would not actually cause a problem as, in this case, $<p> = 0$. However, if one would also take into account cases where the phase of the wave function matters, a further real contribution

$$\frac{\gamma}{2}\tilde{x}\langle p\rangle = \frac{1}{2}W_{\text{Al}} \tag{6.16}$$

with Albrecht's [7] friction term W_{Al} (4.90) would be needed, in addition to W_{TI}.

Where could the missing terms originate from? To answer this question we need to revisit Madelung's hydrodynamical formulation. From the continuity equation, rewritten in the notation with $\varrho = a^2$, see Eq. (3.30) above, in the TI case with $\frac{\partial}{\partial t}a^2 = 0$ the conservation law

$$\nabla S = \frac{C}{a^2}$$

was derived. Taking into account the additional term (6.13), this is obviously no longer valid. However, introducing a modified action function S' via

$$\begin{aligned} &\frac{\partial}{\partial t}a^2 + \frac{1}{m}\,\nabla\left(a^2\left(\nabla S + m\frac{\gamma}{2}\tilde{x}\right)\right) = \\ &\frac{\partial}{\partial t}a^2 + \frac{1}{m}\,\nabla\left(a^2\nabla S'\right) = 0 \end{aligned} \tag{6.17}$$

with [1]

$$S' = S + m\frac{\gamma}{4}\tilde{x}^2 + f(t) \tag{6.18}$$

allows us to find, for $\frac{\partial}{\partial t}a^2 = 0$, again a conservation law of the form

$$\nabla S' = \frac{C}{a^2}\,. \tag{6.19}$$

Separation of (6.17) provided the modified TISE with the additional term W_{TI}. However, interpreting the term $m\frac{\gamma}{2}\tilde{x}$ as part of the gradient of a modified (real) action function S' means that also the second equation of Madelungs formulation describing the time-evolution of the action function S via

$$\frac{\partial}{\partial t}S + \frac{1}{2m}(\nabla S)^2 + V - \frac{\hbar^2}{2m}\frac{\Delta a}{a} = 0$$

must be modified to a corresponding equation for S'.

[1] Note that the phase factor $\frac{im}{\hbar}\frac{\gamma}{4}\tilde{x}^2$ (apart from the factor $e^{\gamma t}$ linking the physical with canonical level) corresponds to the unitary transformation between the Caldirola–Kanai approach and the one in expanding variables, as given in Eq. (4.70).

In a first step, from the SE plus the term W_{TI}, one obtains the Hamilton–Jacobi-type equation (3.31) plus an additional term from W_{TI} (the last on the rhs) but still for the action S,

$$\frac{\partial}{\partial t}S + \frac{1}{2m}(\nabla S)^2 + V - \frac{\hbar^2}{2m}\frac{\Delta a}{a} + \frac{\gamma}{2}\tilde{x}(\nabla S) = 0\,. \tag{6.20}$$

In a second step to obtain a form that is consistent with the modified action in the second line of Eq. (6.17), S is replaced by S' according to (6.18). Specifically, the (explicit) time-derivative of the action supplies the additional terms

$$\frac{\partial}{\partial t}S = \frac{\partial}{\partial t}S' + \frac{\gamma}{2}\langle p\rangle\tilde{x} - \frac{\partial}{\partial t}f(t)\,, \tag{6.21}$$

leading to Eq. (6.20) expressed in terms of S' as

$$\frac{\partial}{\partial t}S' + \frac{1}{2m}(\nabla S')^2 + \left[V - \frac{m}{2}\frac{\gamma}{4}\tilde{x}^2 + \frac{\gamma}{2}\langle p\rangle\tilde{x} - \frac{\partial}{\partial t}f\right] - \frac{\hbar^2}{2m}\frac{\Delta a}{a} = 0\,. \tag{6.22}$$

With $\frac{\partial}{\partial t}S' = -E'$ and $\frac{\partial}{\partial t}f = \frac{\gamma}{4}\langle[\tilde{x},\tilde{p}]_+\rangle$ and applying the new conservation law (6.19), Eq. (6.22) can be rewritten as Ermakov equation

$$\Delta a + \frac{2m}{\hbar^2}\left[\,\hat{E}'_{\mathrm{R}} - \hat{V}\,\right]a = (\frac{1}{\hbar}\nabla S')^2\,a = \left(\frac{C}{\hbar}\right)^2\frac{1}{a^3} \tag{6.23}$$

with $\hat{E}'_{\mathrm{R}} = E' - \frac{\gamma}{2}\langle p\rangle\tilde{x} + \frac{\gamma}{4}\langle[\tilde{x},\tilde{p}]_+\rangle$ and[2] $\hat{V} = V - \frac{m}{2}\frac{\gamma}{4}\tilde{x}^2$.

Including the two additional terms originating from the time-dependence of S expressed in terms of S', as shown in (6.21) with $\frac{\partial}{\partial t}f = \frac{\gamma}{4}\langle[\tilde{x},\tilde{p}]_+\rangle$, the full irreversible dissipative term $\tilde{W}_{\mathrm{TI}}$ now attains the form

$$\tilde{W}_{\mathrm{TI}} = \frac{\gamma}{2}\,[\tilde{x}, p_{\mathrm{op}}]_+ + \frac{\gamma}{2}\langle p\rangle\tilde{x} - \frac{\gamma}{4}\,\langle[\tilde{x},\tilde{p}]_+\rangle = W_{\mathrm{Has}} - \langle W_{\mathrm{Has}}\rangle = \tilde{W}_{\mathrm{SCH}}\,, \tag{6.24}$$

with $\tilde{W}_{\mathrm{SCH}}$ being the logarithmic nonlinearity as given in (4.118).

The first additional (real) term $\frac{\gamma}{2}\langle p\rangle\tilde{x}$ simply provides the missing contribution $\frac{1}{2}W_{\mathrm{AI}}$ that is necessary to obtain the correct friction force in the equation of motion for the mean values and the second term is equal to the mean value of Hasse's friction term and must be subtracted so that $\langle\tilde{W}_{\mathrm{TI}}\rangle = 0$. For Gaussian WPs the equality $\tilde{W}_{\mathrm{TI}}\Psi_{\mathrm{WP}} = \tilde{W}_{\mathrm{SCH}}\Psi_{\mathrm{WP}}$ is valid.

---

[2]The replacement $V = \frac{m}{2}\omega^2x^2 \to \hat{V} = \frac{m}{2}\left(\omega^2 - \frac{\gamma^2}{4}\right)x^2$ in this position-dependent problem corresponds to the replacement $\omega^2 \to \left(\omega^2 - \frac{\gamma^2}{4}\right)$ in the TD problem Eq. (5.15) and the description in expanding variables, Eq. (4.66).

The modified TISE including these two additional terms can be rewritten as

$$E'\Psi = \left\{ -\frac{\hbar^2}{2m}\Delta_x + \frac{\gamma}{2}\tilde{x}\,\frac{\hbar}{\mathrm{i}}\nabla_x + \left(V + \frac{\gamma}{2}\,\tilde{x}\langle p\rangle - \frac{\gamma}{4}\,\langle[\tilde{x},\tilde{p}]_+\rangle + \mathrm{i}\hbar\frac{\gamma}{4}\right)\right\}\Psi\,, \tag{6.25}$$

or in the form of the corresponding modified complex Riccati equation

$$\nabla_x\left(\frac{\nabla_x\Psi}{\Psi}\right) + \mathrm{i}\gamma\frac{m}{\hbar}\,\tilde{x}\left(\frac{\nabla_x\Psi}{\Psi}\right) + \left(\frac{\nabla_x\Psi}{\Psi}\right)^2 + \frac{2m}{\hbar^2}(\hat{E} - V) = 0 \tag{6.26}$$

with the complex quantity

$$\hat{E} = E' - \frac{\gamma}{2}\langle p\rangle\tilde{x} + \frac{\gamma}{4}\,\langle[\tilde{x},\tilde{p}]_+\rangle - \mathrm{i}\hbar\frac{\gamma}{4}\,. \tag{6.27}$$

In Fig. 6.1, the TD and space-dependent cases are again compared but now including the environmental effect. Whereas in the TD situation the coefficient of the additional linear term in the Riccati equation and the first-derivative term in its linearized form do not explicitly depend on time t (only an implicit time-dependence via $\gamma = \gamma(t)$ is possible), in the space-dependent case the coefficient depends linearly on the spatial variable and is, in contrast with the TD case, purely imaginary. In the TD case, the conservation law remains unchanged while in the space-dependent case the phase is changed in a way that corresponds to a unitary transformation of the wave function. On the other hand, the amplitude of the complex variable in the

	Time	Space
NL complex	$\frac{d}{dt}\left(\frac{\dot{\tilde\lambda}}{\tilde\lambda}\right) + \gamma\left(\frac{\dot{\tilde\lambda}}{\tilde\lambda}\right) + \left(\frac{\dot{\tilde\lambda}}{\tilde\lambda}\right)^2 + \omega^2 = 0$	$\nabla\left(\frac{\nabla\Psi}{\Psi}\right) + i\gamma\frac{m}{\hbar}\tilde{x}\left(\frac{\nabla\Psi}{\Psi}\right) + \left(\frac{\nabla\Psi}{\Psi}\right)^2 + \frac{2m}{\hbar^2}\left(\hat{E}_R - V + i\hbar\frac{\gamma}{4}\right) = 0$
linearized	$\ddot{\tilde\lambda} + \gamma\dot{\tilde\lambda} + \omega^2\tilde\lambda = 0$	$\Delta\Psi + i\gamma\frac{m}{\hbar}\tilde{x}(\nabla\Psi) + \frac{2m}{\hbar^2}\left(\hat{E}_R - V + i\hbar\frac{\gamma}{4}\right)\Psi = 0$
complex variable	$\tilde\lambda = \alpha e^{i\varphi - \frac{\gamma}{2}t}$	$\Psi = a e^{iS/\hbar}$
conservation law	$\dot\varphi = \frac{1}{\alpha^2}$	$\nabla S' = \nabla\left(S + m\frac{\gamma}{4}\tilde{x}^2\right) = \frac{k}{a^2}$
NL real	$\ddot\alpha + \left(\omega^2 - \frac{\gamma^2}{4}\right) = \frac{1}{\alpha^3}$	$\Delta a + \frac{2m}{\hbar^2}\left[\hat{E}'_R - \left(V - \frac{m}{2}\frac{\gamma^2}{4}\tilde{x}^2\right)\right]a = \frac{(k/\hbar)^2}{a^3}$
Riccati variables	$\frac{\dot{\tilde\lambda}}{\tilde\lambda} = \frac{\dot\alpha}{\alpha} - \frac{\gamma}{2} + i\frac{1}{\alpha^2}$	$\frac{\nabla\Psi}{\Psi} = \frac{\nabla a}{a} + i\frac{1}{\hbar}\left(\frac{k}{a^2} - m\frac{\gamma}{2}\tilde{x}\right)$

Fig. 6.1 Comparison between time-dependent and space-dependent Riccati equations, their linearized forms and the corresponding Ermakov equations for the dissipative case

space-dependent situation remains unchanged whereas in the TD case, the amplitude is multiplied by an exponential damping factor. Nevertheless, the resulting Ermakov equations have the same structure in both cases. So it seems that a change of the phase, corresponding to a unitary transformation, can have the same effect as a change in the amplitude of a complex variable which would correspond to a non-unitary transformation (for further details, see also [8]).

References

1. E. Madelung, *Die Mathematischen Hilfsmittel des Physikers* (Springer, Berlin, 1950), p. 432
2. B. Mrowka, Zur Darstellung der Quantenmechanik I. Z. Phys. **130**, 164–173 (1951)
3. D. Schuch, K.-M. Chung, H. Hartmann, Nonlinear Schrödinger-type field equation for the description of dissipative systems. I. Derivation of the nonlinear field equation and one-dimensional example. J. Math. Phys. **24**, 1652 (1983)
4. D. Schuch, K.-M. Chung, From macroscopic irreversibility to microscopic reversibility via a nonlinear Schrödinger-type field equation. Int. J. Quant. Chem. **29**, 1561–1573 (1986)
5. G. Süssmann, unpublished, quoted in Hasse, Ref. [6] and Albrecht, Ref. [7]
6. R.W. Hasse, On the quantum mechanical treatment of dissipative systems. J. Math. Phys. **16**, 2005–2011 (1975)
7. K. Albrecht, A new class of Schrödinger operators for quantized friction. Phys. Lett. B **56**, 127–129 (1975)
8. D. Schuch, Complex Riccati equations as a link between different approaches for the description of dissipative and irreversible systems, in J. Phys. Conf. Ser. **380**, p. 012009 (2012) (22 pp)

Chapter 7
Nonlinear Riccati Equations in Other Fields of Physics

In Chap. 2 it has been shown that, for systems where the time-dependent Schrödinger equation (TDSE) possesses analytic solutions (i.e., at most quadratic Hamiltonians), the information about the dynamics of the classical as well as quantum mechanical aspects of the system can equally well be obtained from a quadratically nonlinear (NL) complex Riccati equation. The different ways of treating this equation also make it possible to connect it to other established formulations of this quantum mechanical problem, like the ones using a TD Green function or Feynman kernel, a description in momentum space or via the Wigner function in phase space. Also, well-defined relations can be shown to algebraic and group-theoretical descriptions as well as to those using generalized creation/annihilation operators. Not only was it possible to rewrite all these formulations of quantum mechanics in terms of the Riccati variable (or variables derived from it), but additional information is also gained that is not obvious in the linear formulation, like sensitivity to the initial quantum uncertainties or additional dynamical invariants.

In the time-independent (TI) case (discussed in Chap. 3), the Riccati formulation could even be generalized to any position-dependent potential and a unique relation could be established between phase and amplitude of the wave function (that is not obvious in the linear formulation, but immediately clear from the quadratic nonlinearity of the complex Riccati equation). Comparison with supersymmetric (SUSY) quantum mechanics even allowed the construction of new complex (non-Hermitian) potentials with real energy eigenvalues.

Chapters 4, 5 and 6 showed how the entire formalism could be extended consistently to include dissipative velocity-dependent friction forces. The relations between different approaches to reach this goal were shown and, in conclusion, all could be traced back to simple modifications of the Riccati equations introduced in Chaps. 2 and 3.

D. Schuch, *Quantum Theory from a Nonlinear Perspective*,
Fundamental Theories of Physics 191, https://doi.org/10.1007/978-3-319-65594-9_7

In this Chapter, the advantage of a formulation in terms of Riccati equations is demonstrated by rewriting problems from various fields of physics in terms of (real or complex) Riccati equations. This enables direct comparison with the corresponding formulation of the above-mentioned quantum systems and therefore creates a unified formulation of all these fields of physics (and more, as the following examples are only a small selection of a much larger variety).

7.1 Riccati Equations in Statistical Thermodynamics

In Sect. 2.2 it has been shown that the solution $\mathcal{C}(t)$ of the complex Riccati equation (2.4) that determines the time-dependence of the wave packet (WP) width, and thus the position uncertainty can be obtained, once a particular solution $\tilde{\mathcal{C}}$ and the solution of the homogeneous Bernoulli equation (2.5) for the complex quantity $\mathcal{V}(t)$ are known, in the form $\mathcal{C}(t) = \tilde{\mathcal{C}} + \mathcal{V}(t)$. Particularly for constant $\tilde{\mathcal{C}}$, the general solution of the Riccati equation could be written down explicitly in the form (2.7), rewritten now for the following purpose as

$$\mathcal{C}(t) = \tilde{\mathcal{C}} + \frac{2\tilde{\mathcal{C}}\,\mathrm{e}^{-2\tilde{\mathcal{C}}t}}{\kappa_0\, 2\tilde{\mathcal{C}} + \left(1 - \mathrm{e}^{-2\tilde{\mathcal{C}}\,t}\right)} = \tilde{\mathcal{C}} + \frac{2\tilde{\mathcal{C}}}{\kappa_0\, 2\tilde{\mathcal{C}}\,\mathrm{e}^{2\tilde{\mathcal{C}}t} + \left(\mathrm{e}^{2\tilde{\mathcal{C}}t} - 1\right)} . \tag{7.1}$$

As shown also in Sect. 2.2, the particular solution $\tilde{\mathcal{C}}$ for the HO is purely imaginary, turning the exponential functions into trigonometric ones.

In Sect. 2.5.1. it has been shown why trigonometric functions are good candidates for fulfilling Riccati equations. At the same time, however, it was also shown that switching to an imaginary argument ($\varphi \to \mathrm{i}\varphi$), which turns the trigonometric functions into hyperbolic ones, also leads to functions that fulfil Riccati equations but to those that would correspond to a repulsive quadratic potential.

In Eq. (7.1) this would mean that $2\tilde{\mathcal{C}}t = \mathrm{i}2\omega_0 t$ in the exponent must be replaced by a real quantity that can be attained by replacing the time-variable "t" by an "imaginary time" τ, e.g., $\tau = \mathrm{i}\frac{\hbar}{k_B T}$, with k_B = Boltzmanns constant and T = temperature, thus turning the quantum mechanical problem into one of statistical thermodynamics. This trick of replacing the real physical time with an imaginary time (thus turning the TDSE into a kind of diffusion equation) is also used, e.g., in quantum chemistry for computational purposes to avoid problems with oscillatory solutions. The concept of an imaginary time is applied also in cosmology [1].

In our case, the formal similarity is used to show that the Riccati formalism established in the context of TD quantum mechanics can also be found in other fields of physics, in this case in statistical thermodynamics. To show that this is not just pure mathematical formalism, but that it has also well-known physical implications, the replacements $2\tilde{C} \rightarrow \hbar\omega$ and[1] $t \rightarrow \frac{1}{\mathrm{K_B T}} = \beta$ are considered and the initial condition κ_0 is chosen to be zero, $\kappa_0 = 0$.

Relation (7.1) then turns into

$$C\left(\frac{1}{\mathrm{K_B T}}\right) = \frac{\hbar}{2}\omega_0 + \frac{\hbar\omega_0}{\mathrm{e}^{\frac{\hbar\omega_0}{\mathrm{K_B T}}} - 1} = \frac{\hbar}{2}\,\omega_0 \,\coth\left(\frac{\hbar\omega_0}{2\mathrm{K_B T}}\right) = \langle E\rangle_{th} \tag{7.2}$$

which is the well-known expression [2] for the average energy $\langle E\rangle_{th}$ of a single oscillator in thermal equilibrium. The first term on the rhs is just the ground state energy of the harmonic oscillator (HO), the second is equal to Planck's distribution function for the black body radiation. This type of relation between Eq. (7.2) and the Riccati equation has also been found by Rosu et al. [3, 4].

In the TD quantum mechanical context, the Riccati variable could be written as a logarithmic derivative indicating that only relative changes matter. The same is also possible in this case. Using the familiar expression of the partition function Z as

$$Z = \sum_n \mathrm{e}^{-n\hbar\omega_0\beta} = \frac{1}{1 - \mathrm{e}^{\hbar\omega_0\beta}}\,, \tag{7.3}$$

Equation (7.2) can be rewritten as

$$\langle E\rangle_{th} = \frac{\hbar}{2}\,\omega_0 + \frac{\frac{\partial}{\partial\beta}Z^{-1}}{Z^{-1}} = \frac{\hbar}{2}\,\omega_0 - \frac{\partial}{\partial\beta}\,\ln\,Z\,. \tag{7.4}$$

It is worth mentioning that this energy, particularly obvious in the first part on the rhs of Eq. (7.2), is the sum of the ground state energy of the HO and a second term that is simply a Bose–Einstein distribution function.

Knowing this solution and the formal structure of the Riccati systems developed in the case of the TDSE, the corresponding Riccati equation can be written straightforwardly as

$$C' + C^2 - \tilde{C}^2 = 0 \tag{7.5}$$

where prime denotes derivative with respect to $\beta = \frac{1}{\mathrm{K_B T}}$, $' = \frac{d}{d\beta}$.

As shown in Sect. 2.7 (where the momentum-space representation of the TDSE was discussed), if C fulfils a Riccati equation like (2.4) the quantity $\mathcal{K}(t) = -\tilde{C}^2 C^{-1}$ also fulfils a Riccati equation, namely $-\mathcal{K}' + \mathcal{K}^2 + \omega_0^2 = 0$ (see (2.103)), leading in our case to

$$-\mathcal{K}' + \mathcal{K}^2 - \tilde{C}^2 = 0\,, \tag{7.6}$$

[1] As we are interested in quantities C that are related to energy, $\hbar$ has been taken away from the definition of complex time and included in $\tilde{C}$. This is simply a formal matter of definition. The quantity $\beta = \frac{1}{\mathrm{K_B T}}$ should not be confused with $\beta(t) = \frac{\hbar}{2m\langle\tilde{x}^2\rangle(t)}$ that was used in the context of the TDSE!

i.e., the same Riccati equation but now with a negative sign for the derivative term. Therefore, the solution is proportion to the inverse of $\mathcal{C}$, i.e.,

$$\mathcal{K}\left(\frac{1}{\mathrm{k_B T}}\right) = \frac{\hbar}{2}\omega_0 - \frac{\hbar\omega_0}{\mathrm{e}^{\frac{\hbar\omega_0}{\mathrm{k_B T}}}+1} = \frac{\hbar}{2}\omega_0 \tanh\left(\frac{\hbar\omega_0}{2\mathrm{k_B T}}\right) \tag{7.7}$$

where the contribution beside the ground-state energy is now a Fermi–Dirac distribution.

This is strikingly similar to the situation in SUSY quantum mechanics as mentioned in Sect. 3.1, where the Riccati equations (3.6) and (3.7) determining the potentials for the bosonic and fermionic Hamiltonians also only differ by the sign of the derivative term.

So, the quantum statistical properties distinguishing bosons and fermions by the sign of a derivative in a Riccati equation in SUSY quantum mechanics have a counterpart in classical statistical thermodynamics, providing this is also formulated in terms of Riccati equations.

7.2 The Logistic or Verhulst Equation

A slight modification of Eq. (7.7) leads to the solution of another Riccati/Bernoulli-type equation, the logistic equation proposed by the Belgian mathematician Pierre Verhulst in 1838 [5].

For this purpose, Eq. (7.7) can be written in terms of "neutral" variables, i.e., $\hbar\omega \to 1$ and $\frac{1}{\mathrm{k_B T}} \to x$, thus $\mathcal{K}\left(\frac{1}{\mathrm{k_B T}}\right) \to g(x)$, as $g(x) = \frac{1}{2}\tanh\left(\frac{x}{2}\right)$, fulfilling

$$g' + g^2 - \frac{1}{4} = 0 \tag{7.8}$$

with $' = \frac{d}{dx}$. Adding the constant $\frac{1}{2}$ to $g(x)$ leads to

$$f(x) = g(x) + \frac{1}{2} = \frac{1}{2}\left(\tanh\left(\frac{x}{2}\right) + 1\right) = \frac{1}{1+\mathrm{e}^{-x}}\,, \tag{7.9}$$

which is a sigmoidal function that fulfils the logistic equation

$$\begin{aligned} & -f' + f^2 - f = 0\,, \\ \text{or} \qquad & -f' = f(1-f)\,. \end{aligned} \tag{7.10}$$

This equation is like a rate equation with gain and loss terms and has been applied in such diverse fields as ecology, medicine, chemistry, linguistics, economics, for modelling population growth, modelling tumour growth, to describe autocatalytic reaction models, neural networks, and so on (see, e.g., [6], Sect. 10).

The same structure of equations as (7.10) is also found in different other areas of physics (and not only there). Another example discussed by Haken in his book on Synergetics ([6], Sect. 5.4) is a simplified model of a laser. In this case the photon production rate is determined by an equation for the temporal change of the photon number $n(t)$ having the form

$$\dot{n} = \text{gain} - \text{loss} \tag{7.11}$$

that can be specified to a basic laser equation

$$\dot{n} = -kn - k_1 n^2 \tag{7.12}$$

with appropriately-defined constants k and k_1 (for details, see [6]).

Of importance is the fact that, due to the nonlinearity of Eq. (7.12), this equation allows for bifurcations. This means small changes of a parameter can lead to drastic changes of the system like, e.g., phase transitions. This is demonstrated subsequently using a system that initially looks different from the ones discussed in this sub-section (due to a cubic nonlinearity) but can be brought into the same form and displays the property of a so-called Hopf bifurcation.

7.3 Nonlinear Dynamics with Hopf Bifurcation

A Hopf bifurcation generates a limit cycle starting from a fixed point ([7], Sect. 5.4). The initial equations may actually look different from the ones discussed in the previous sub-section. Starting from the set of coupled NL differential equations in rectangular Cartesian coordinates x and y,

$$\dot{x} = -\{\Gamma + \left(x^2 + y^2\right)\}x - \omega y \,, \tag{7.13}$$

$$\dot{y} = -\{\Gamma + \left(x^2 + y^2\right)\}y + \omega x \,, \tag{7.14}$$

by introducing polar coordinates r and θ with $\frac{d}{dt}\theta = \omega$, this can be reduced to a single equation for $r(t)$,

$$\dot{r} = -\Gamma r - r^3 \,, \tag{7.15}$$

i.e., a first-order differential equation with cubic nonlinearity where the parameter Γ can be expressed by the difference of another parameter a and its critical value a_c, $\Gamma = a - a_c$. The Riccati/Bernoulli form of (7.15) can be achieved simply by multiplying this equation by $2r$ and defining as a new function $R(t) = 2r^2(t)$ that now fulfils

$$\dot{R} + 2\Gamma R + R^2 = 0 \tag{7.16}$$

where 2Γ corresponds to $2\tilde{C}$ in our quantum mechanical case.

The solution of Eq. (7.16) can then be written down (in terms of $r(t)$) as

$$r^2(t) = \frac{\Gamma\, r_0^2\, e^{-2\Gamma t}}{r_0^2\left(1 - e^{-2\Gamma t}\right) + \Gamma} \tag{7.17}$$

with $r_0 = r(t = 0)$.

For $\Gamma \geq 0$ the trajectory approaches the origin (fixed point); for $\Gamma < 0$ it spirals towards a limit cycle with radius $r_\infty = |(a - a_c|^{\frac{1}{2}} = |\Gamma|^{\frac{1}{2}}$.

Equation (7.15) has the same structure as the Newtonian equation of motion for the motion in a double-well potential,

$$V(x) = \frac{m}{2}\omega_0^2 x^2 + \frac{1}{4}k_1 x^4 , \tag{7.18}$$

under the influence of a linear velocity dependent friction force $-m\gamma x$ (with $\gamma = 1$) if the acceleration term $\ddot{x}$ can be neglected, i.e.,

$$\dot{x} = -\omega_0^2 x - \frac{k_1}{m} x^3 . \tag{7.19}$$

This was used by Haken ([4], Sect. 5.1) as starting point for the discussion of dynamical processes including bifurcations and limit cycles that serves as basis for the simple laser model mentioned in the previous subsection.

Double-well potentials like (7.18) can also be related to phenomena like phase transitions, for instance those discussed by Landau. In the case of second-order phase transitions, this leads to equations of the form

$$\dot{q} = -\alpha q - \beta q^3 , \tag{7.20}$$

adopting the nomenclature of Haken ([4], Sect. 6.7).

7.4 Solitons and Riccati Equations

Solitary waves or solitons are waves travelling without changing their shape even in collision with other solitary waves, particularly not spreading in time, and can be described by NL differential equations. Historically, the observation of solitary waves was first reported by Scott-Russell in 1844 [8] and a theoretical description was given by Korteweg and de Vries in 1895 [9]. The interest in solitons began to grow in the 1960s with the development of modern computers and has shown to have applications in almost every area of physics and beyond, from the transition of light pulses in optical fibres, electrical pulses in nerve cells to tsunamis. The aim of this sub-section is to show the link between NL differential equations with soliton solutions and Riccati equations and (further via linearization of these equations) to corresponding SEs.

Connections between soliton theory and SEs are well-known in the literature [10, 11], e.g., in the Miura transformation [12] (changing the Korteweg–de Vries (KdV) equation into the modified KdV equation) which, via a logarithmic derivative, also leads to a TISE (as in the cases discussed above) and provides the basis for the inverse scatter theory.[2]

Here a *different* route is taken. It is shown how the Burgers equation and the KdV Equation (at least for cases with real one-soliton solutions) can be transformed into the same (real) Riccati equation that can be linearized to a TISE with well-known potential. Using a TI equivalent of Eq. (2.102), $\mathcal{K}(t) = -\tilde{\mathcal{C}}^2\mathcal{C}^{-1}$, a corresponding Riccati equation and TISE can also be found where the potential of this SE is simply the SUSY partner of the one in the above TISE.

As both soliton equations lead to the same Riccati equation, relations between them can also be established.

7.4.1 Burgers Equation

Starting point is the Burgers equation for the function $u(x,t)$ in the form

$$\frac{\partial}{\partial t}u + u\frac{\partial}{\partial x}u - D\frac{\partial^2}{\partial x^2}u = 0 \tag{7.21}$$

with diffusion constant D. As we are looking for spatially-localized solutions that propagate with constant speed c_{B} while maintaining their shape, we switch to a coordinate system travelling with the wave according to $X = x - c_{\text{B}}t$. Assuming no further explicit time-dependence, $\frac{\partial}{\partial t}$ can be replaced by $-c_{\text{B}}\frac{d}{dX}$ and $u(x,t)$ can be rewritten as $\hat{g}(X)$, now fulfilling the equation

$$-c_{\text{B}}\frac{d}{dX}\hat{g}_- + \hat{g}_-\frac{d}{dX}\hat{g}_- - D\frac{d^2}{dX^2}\hat{g}_- = 0 \tag{7.22}$$

(where $\frac{d}{dX} = \frac{d}{dx}$). The meaning of the subscript "minus" will become clear later on. The one-soliton solution [13] of this equation can be written as

$$\hat{g}_-(X) = c_{\text{B}}\left[1 - \tanh\left(\frac{c_{\text{B}}}{2D}X\right)\right]. \tag{7.23}$$

To link Eq. (7.22) to a Riccati equation, it is integrated with respect to X and, after being multiplied by $\frac{1}{2D}$, the function $g_-(X) = \frac{1}{2D}\hat{g}_-(X)$ fulfils the Bernoulli equation

$$-\frac{d}{dX}g_- + g_-^2 - \frac{c_{\text{B}}}{D}g_- = 0\,. \tag{7.24}$$

[2] A logarithmic derivate also occurs in the Hopf–Cole transformation that transforms the Burgers equation into a heat or diffusion equation; this will not be discussed further in this work.

Comparing it with the TD situation in Chap. 2, it is known that half of the coefficient of the term linear in g_-, here $\tilde{C}_- = -\frac{c_B}{2D}$, is simply the particular solution of the corresponding Riccati equation and, therefore,

$$K_- = g_-(X) + \tilde{C}_- = g_-(X) - \frac{c_B}{2D} = -\frac{c_B}{2D}\tanh\left(\frac{c_B}{2D}X\right) \tag{7.25}$$

then fulfils the Riccati equation

$$-\frac{d}{dX}K_- + K_-^2 - \left(\frac{c_B}{2D}\right)^2 = 0\,. \tag{7.26}$$

However, this equation is also compatible with the particular solution $\tilde{C}_+ = +\frac{c_B}{2D}$, i.e., a wave travelling in the opposite direction with the same velocity c_B. The Burgers equation (7.22) then changes into

$$c_B\frac{d}{dX}\hat{g}_+ + \hat{g}_+\frac{d}{dX}\hat{g}_+ - D\frac{d^2}{dX^2}\hat{g}_+ = 0 \tag{7.27}$$

and solution (7.23) into

$$\hat{g}_+(X) = -c_B\left[1 + \tanh\left(\frac{c_B}{2D}X\right)\right]. \tag{7.28}$$

After integrating Eq. (7.27), the resulting Bernoulli equation is now

$$-\frac{d}{dX}g_+ + g_+^2 + \frac{c_B}{D}g_+ = 0\,, \tag{7.29}$$

compatible with $\tilde{C}_+ = +\frac{c_B}{2D}$. Therefore, the variable K_+ fulfilling the corresponding Riccati equation is

$$K_+ = g_+(X) + \tilde{C}_+ = -\frac{c_B}{2D}\tanh\left(\frac{c_B}{2D}X\right) = K_- = K\,, \tag{7.30}$$

i.e., the same variable is valid for $+c_B$ and $-c_B$. So Eq. (7.26) (without any subscript of K) is valid for solitary waves travelling in opposite directions. This Riccati equation can now be linearized using $K = \frac{\frac{d}{dX}\Psi^{(1)}(X)}{\Psi^{(1)}(X)}$ to yield (after multiplication with $\frac{\hbar^2}{2m}$)

$$-\frac{\hbar^2}{2m}\frac{d^2}{dX^2}\Psi^{(1)} + \frac{\hbar^2}{2m}\left(\frac{c_B}{2D}\right)^2\left[1 - \frac{2}{\cosh^2\left(\frac{c_B}{2D}X\right)}\right]\Psi^{(1)} = 0\,. \tag{7.31}$$

This is the TISE for the well-known Rosen–Morse potential

$$V_1 = \frac{\hbar^2}{2m}\left(\frac{c_{\mathrm{B}}}{2D}\right)^2\left[1 - \frac{2}{\cosh^2\left(\frac{c_{\mathrm{B}}}{2D}X\right)}\right] \tag{7.32}$$

which has [14] one bound state Ψ_0 with energy $E_0 = 0$ (hence 0 on the rhs of Eq. (7.31)).

According to the definition of the potential energies in SUSY quantum mechanics in terms of logarithmic derivatives of the ground state wave function [14, 15], the (normalized) ground state wave function of Eq. (7.31) can be determined from Eq. (7.32) as

$$\Psi_0^{(1)} = \frac{1}{\sqrt{2}}\left(\frac{c_{\mathrm{B}}}{2D}\right)^{\frac{1}{2}} \frac{1}{\cosh\left(\frac{c_{\mathrm{B}}}{2D}X\right)} \,. \tag{7.33}$$

As shown in the TD case, if K fulfils a Riccati equation of type (7.26), the function $C = -\tilde{C}^2 K^{-1}$ also fulfils a Riccati equation, namely

$$\frac{d}{dX}C + C^2 - \tilde{C}^2 = 0 \tag{7.34}$$

(here the sign of the derivative term changes as $\tilde{C}^2$ is positive for real $\tilde{C}$). This definition of C leads to

$$C(X) = \frac{c_{\mathrm{B}}}{2D}\coth\left(\frac{c_{\mathrm{B}}}{2D}X\right) \tag{7.35}$$

which can again be written as the logarithmic derivative of a function $\Psi^{(2)}(X)$

$$C = \frac{\frac{d}{dx}\Psi^{(2)}}{\Psi^{(2)}} = \frac{\left(\frac{c_{\mathrm{B}}}{2D}\right)\cosh\left(\frac{c_{\mathrm{B}}}{2D}X\right)}{\sinh\left(\frac{c_{\mathrm{B}}}{2D}X\right)} = \frac{\frac{d}{dx}\sinh\left(\frac{c_{\mathrm{B}}}{2D}X\right)}{\sinh\left(\frac{c_{\mathrm{B}}}{2D}X\right)} \tag{7.36}$$

with[3] $\Psi^{(2)}(X) \propto \sinh\left(\frac{c_{\mathrm{B}}}{2D}X\right)$, linearizing Riccati equation (7.34) to

$$-\frac{\hbar^2}{2m}\frac{d^2}{dX^2}\Psi^{(2)} + \frac{\hbar^2}{2m}\left(\frac{c_{\mathrm{B}}}{2D}\right)^2\Psi^{(2)} = 0\,, \tag{7.37}$$

i.e., the TISE with potential

$$V_2 = \frac{\hbar^2}{2m}\left(\frac{c_{\mathrm{B}}}{2D}\right)^2 = \mathrm{const} \tag{7.38}$$

which is the SUSY partner potential of V_1.

For completeness it should be mentioned that, in addition to $\Psi^{(2)}(X)$, also $\Phi^{(2)}(X) \propto \cosh\left(\frac{c_{\mathrm{B}}}{2D}X\right)$ is an independent solution of Eq. (7.37) with the corresponding Riccati variable $C_\Phi = +\frac{c_{\mathrm{B}}}{2D}\tanh\left(\frac{c_{\mathrm{B}}}{2D}X\right) = -K$. Similarly, for the Rosen–Morse

[3] As $\Psi^{(1)}(X)$ is normalizable, $\Psi^{(2)}(X)$ is not.

potential V_1 the second solution would be $\Phi^{(1)}(X) \propto \frac{1}{\sinh(\frac{c_B}{2D}X)}$ and the corresponding Riccati variable $K_\Phi = -\frac{c_B}{2D}\coth\left(\frac{c_B}{2D}X\right) = -C$.

7.4.2 *Korteweg–de Vries Equation*

Applying the same type of coordinate transformation as in Sect. 7.4.1 to the KdV equation for the function $w(x,t)$,

$$\frac{\partial}{\partial t}w - 6w\frac{\partial}{\partial x}w + \frac{\partial^3}{\partial x^3}w = 0\,, \tag{7.39}$$

i.e., shifting to a system moving with constant velocity c_K according to $X = x - c_K t$, changes this equation into

$$-c_K\frac{d}{dX}f - 6f\frac{d}{dX}f + \frac{d^3}{dX^3}f = 0 \tag{7.40}$$

with $w(x,t)\hat{=}f(X)$.

Integration with respect to X now provides

$$\frac{d^2}{dX^2}f = 3f^2 + c_K f + A \tag{7.41}$$

with integration constant A.

Using $\frac{d^2}{dX^2} = \frac{1}{2}\frac{d}{df}\left(\frac{d}{dX}f\right)^2$ and integrating with respect to f then yields

$$\frac{1}{2}\left(\frac{d}{dX}f\right)^2 = f^3 + \frac{1}{2}c_K f^2 + Af + B\,. \tag{7.42}$$

Looking for solitary waves with boundary conditions f, $\frac{d}{dX}f$ and $\frac{d^2}{dX^2}f \to 0$, as $X \to \pm\infty$ (see [13]), the integration constants A and B can be chosen to be zero allowing one to write Eq. (7.42) as

$$-2f + \left(\frac{\frac{d}{dX}f}{f}\right)^2 - c_K = 0\,. \tag{7.43}$$

This would be the desired Riccati equation if $-2f \propto \frac{d}{dX}\left(\frac{\frac{d}{dX}f}{f}\right)$ would be valid.

As known [13], Eq. (7.40) possesses a soliton solution of the form[4]

---

[4]Actually, in $\cosh^2$ it should be more general $X - X_0$; but for the following this is not essential and $X_0 = 0$ will be assumed.

$$f(X) = -\frac{c_\mathrm{K}}{2} \frac{1}{\cosh^2\left(\frac{\sqrt{c_\mathrm{K}}}{2} X\right)} \tag{7.44}$$

which fulfils the condition

$$\frac{d}{dX}\left(\frac{\frac{d}{dX} f}{f}\right) = -\frac{c_\mathrm{K}}{2} \frac{1}{\cosh^2\left(\frac{\sqrt{c_\mathrm{K}}}{2} X\right)} = f\,. \tag{7.45}$$

Therefore, with

$$\tilde{f}(X) = -\sqrt{\frac{c_\mathrm{K}}{2}} \frac{1}{\cosh\left(\frac{\sqrt{c_\mathrm{K}}}{2} X\right)}\,, \tag{7.46}$$

Equation (7.43) can be rewritten as a Riccati equation

$$-\frac{d}{dX}\left(\frac{\frac{d}{dX} \tilde{f}}{\tilde{f}}\right) + \left(\frac{\frac{d}{dX} \tilde{f}}{\tilde{f}}\right)^2 - \frac{c_\mathrm{K}}{4} = 0 \tag{7.47}$$

that has the same form as Riccati equation (7.26) for $K(X)$, only replacing $\frac{c_\mathrm{B}}{D}$ with $\sqrt{c_\mathrm{K}}$ and $K = \frac{\frac{d}{dX}\Psi^{(1)}(X)}{\Psi^{(1)}(X)}$ with $\frac{\frac{d}{dX}\tilde{f}}{\tilde{f}}$. Therefore, the linearized TISE corresponding to Eq. (7.47) is simply Eq. (7.31) for the Rosen–Morse potential, only replacing $\Psi^{(1)}(X)$ with $\tilde{f}(X)$ and $\frac{c_\mathrm{B}}{D}$ with $\sqrt{c_\mathrm{K}}$. The relations to the corresponding TISE with SUSY partner potential are analogous to the case discussed for the Burgers equation.

7.4.3 Connections Between the Soliton Equations

As the Burgers equation (7.22) and the KdV (7.40) lead to the same kind of Riccati equation, (7.26) and (7.47), respectively, using K as given in (7.30), a relation can be established between the functions $\hat{g}_\pm(X)$ solving Eqs. (7.22), (7.27) and $f = \tilde{f}^2$, solving Eq. (7.40), in the form

$$2K = \frac{\hat{g}_\pm}{D} \pm \frac{c_\mathrm{B}}{D} \mathrel{\hat{=}} \frac{\frac{d}{dX} f}{f} \tag{7.48}$$

where only $\frac{c_\mathrm{B}}{D}$ in $\hat{g}$ must be replaced by $\sqrt{c_\mathrm{K}}$ in f or vice versa.

Only a short remark connecting the results of this work with another case shall follow where a Riccati equation is used in soliton theory, namely the Miura transformation [9, 10]. This transformation shows that if $w(x,t)$ fulfils the KdV equation (7.39), then a function $\hat{v}(x,t)$ fulfilling the Riccati equation

$$w = \frac{\partial}{\partial x}\hat{v} + \hat{v}^2 \tag{7.49}$$

obeys the modified KdV equation

$$\frac{\partial}{\partial t}\hat{v} - 6\hat{v}^2 \frac{\partial}{\partial x}\hat{v} + \frac{\partial^3}{\partial x^3}\hat{v} = 0\,, \tag{7.50}$$

or, using the moving coordinate system, $w(x, t)$ is replaced by $f(X)$, $\hat{v}(x, t)$ by $v(X)$ and $\frac{\partial}{\partial t}$ must be replaced by $-c_K \frac{d}{dX}$. Expressing $v(X)$ as a logarithmic derivative according to $v(X) = \frac{\frac{d}{dX}\Psi}{\Psi}$, this turns Riccati equation (7.49) into the TISE

$$-\frac{\hbar^2}{2m}\frac{d^2}{dX^2}\Psi + \frac{\hbar^2}{2m} f\Psi = 0\,. \tag{7.51}$$

The case where $f(X)$ is given in the form (7.44) shall be mentioned in connection with the results obtained previously. In this case, the TISE (7.51) acquires the form

$$-\frac{\hbar^2}{2m}\frac{d^2}{dX^2}\Psi + \frac{\hbar^2}{2m}\left(\frac{\sqrt{c_K}}{2}\right)^2 \frac{2}{\cosh^2\left(\frac{\sqrt{c_K}}{2}X\right)}\Psi = -\frac{\hbar^2}{2m}\frac{d^2}{dX^2}\Psi + V_{PT}\Psi = 0 \tag{7.52}$$

where V_{PT} is the well-known Pöschl–Teller potential (see, e.g., [16]).

As the KdV equation is Galilean invariant [13], the transformation

$$w(x, t) \to \Lambda + w(x + 6\Lambda t, t), \qquad -\infty < \Lambda < +\infty \tag{7.53}$$

leaves Eq. (7.40) unchanged for arbitrary (real) Λ. The x-dependence is not affected by this transformation. The only consequence for the Riccati equation (7.49) is that $w(x, t)$ must be replaced by $\tilde{w} = w + \Lambda$. This function $\tilde{w}$ now obeys the same KdV equation (7.49) as w but the linearization of Riccati equation (7.49) now yields a TISE where f is replaced by $f - \Lambda$,

$$-\frac{\hbar^2}{2m}\frac{d^2}{X^2}\Psi + \frac{\hbar^2}{2m}(f - \Lambda)\Psi = 0\,, \tag{7.54}$$

which is the basis for the inverse scattering method [13] and, for $f(X)$ again given in the form (7.44) and $\Lambda = \frac{c_K}{4}$, is identical to the TISE (7.31) for the Rosen–Morse potential.

As both soliton equations, the Burgers equation and the KdV equation, can be transformed into the same equation, the simple relation (7.48) can be established between the quantities fulfilling both soliton equations. Sharing the same Riccati equation also allows for further comparisons between the two equations (for further details, see also [17]).

After one integration, the Burgers equation in the moving coordinate frame turns into the homogeneous Bernoulli equation (7.24). A Bernoulli equation with a linear term in addition to the quadratic one also occurs in the quantum mechanical context; there, e.g., for the dissipative free motion in the TD as well as in the TI case.[5] In both cases the coefficient of this term depends on the friction constant γ, representing a property of the environment. The coefficient of the linear term in Eq. (7.24) is via the diffusion coefficient also proportional to some environmental parameter. (However, one must be careful with direct comparisons as, e.g., in the TI quantum system the coefficient is purely imaginary – which might be due to the fact that the Bernoulli variable, depending on the wave function, is complex – whereas the soliton solution is real.) This would be in agreement with the statement by Su and Gardner [18] that the Burgers equation is a limiting form for NL dissipative systems whereas the KdV equation is a limiting form for NL dispersive systems.

A different point of view concerning the Bernoulli equation (7.24) can be taken by assuming that the linear term originates from a transition between an inhomogeneous Riccati equation and the homogeneous Bernoulli equation via a particular solution of the former equation, as shown in the transition from Eq. (2.4) to (2.5). In this case, no dissipation is involved and the coefficient of the linear term in the Bernoulli equation simply represents (twice) this particular solution. However, this (non-dissipative) Riccati equation is basically the one obtained from the KdV equation after two integrations, also being in agreement with the statement by Su and Gardner [18].

The transition from the KdV to the modified KdV equation via the Miura transformation [12] – also involving a Riccati equation – provides the basis for the inverse scattering theory via Eq. (7.54). The same equation was also obtained by Malfliet [19] using a product ansatz for the solution of the KdV equation and its Galilean invariance. In this approach it turns out that the solution of the KdV equation is simply the square of the wave function of Eq. (7.54). In this case, the constant term $\frac{\hbar^2}{2m}\Lambda$ in Eq. (7.54) would correspond to a (kinetic-type) energy term proportional to c_{B}^2 (with velocity c_{B}) and the term proportional to $\frac{1}{\cosh^2}$ in Eqs. (7.31) and (7.32) would essentially be Ψ^2, turning Eq. (7.31) into a cubic NLSE (like equation (17) in [19]). This might lead to a way to find connections with further soliton equations, like the cubic NLSE (there, however, with a complex wave function Ψ).

7.5 Complex Riccati Equation in Classical Optics

In his second communication [20] Schrödinger tried to show some analogies between his wave equation "derived" from Hamiltonian mechanics and classical optics. In the following, analogies between the wave equation of classical optics and our complex Riccati formulation of quantum mechanics are shown.

---

[5] For potentials different from zero, e.g., quadratic in position variable, also linear terms appear in the Bernoulli equation already for the non-dissipative case with coefficients proportional to the particular solution of the Riccati equation.

For this purpose, the scalar wave equation of optics can be written (e.g., following Goldstein [21]) as

$$\Delta\phi - \frac{n^2}{c^2}\frac{\partial^2}{\partial t^2}\phi = 0 \tag{7.55}$$

with $\phi(\mathbf{r}, t)$ being a scalar quantity like a scalar electromagnetic potential, c the velocity of light in vacuum and $n = \frac{c}{v_{\text{med}}}$ the diffraction index (with v_{med} = velocity of light in the medium). In general, $n = n(\mathbf{r})$ depends on the medium and the position in space.

For the special case n = constant, plane wave solutions exist for Eq. (7.55),

$$\phi(\mathbf{r}, t) = \phi_0 \mathrm{e}^{\mathrm{i}(\mathbf{kr}-\omega t)} , \tag{7.56}$$

where the wave number $k = |\mathbf{k}|$ and the frequency ω are related via $k = \frac{2\pi}{\lambda} = \frac{n\omega}{c} = \frac{\omega}{v_{\text{med}}}$.

For $\mathbf{k}$ being parallel to the z-direction, this can be reduced to a one-dimensional problem with

$$\phi(z, t) = \phi_0 \mathrm{e}^{\mathrm{i}k_0(nz-ct)} \tag{7.57}$$

where k_0 = wave number in vacuum with $k = nk_0$.

In the case of geometric optics, n is *not* constant but varies (usually slightly) in space, i.e., $n = n(\mathbf{r})$. Then, plane waves are no longer solutions of the wave Eq. (7.55) but one can use a similar ansatz,

$$\phi(\mathbf{r}, t) = \mathrm{e}^{A(\mathbf{r})+\mathrm{i}k_0(L(\mathbf{r})-ct)} , \tag{7.58}$$

i.e., nz is replaced by $L(\mathbf{r})$, the optical wave length, also called phase or *Eikonal*. $A(\mathbf{r})$ and $L(\mathbf{r})$ are functions of the spatial variable $\mathbf{r}$ only. Inserting the complex function (7.58) into (7.55) leads to two equations for the real and imaginary parts, respectively,

$$\mathcal{R}e: \quad \Delta a + (\nabla A)^2 - (k_0\nabla L)^2 + (k_0 n)^2 = 0 \tag{7.59}$$

$$\mathcal{I}m: \quad \Delta(k_0 L) + 2(\nabla A)(k_0 \nabla L) = 0 . \tag{7.60}$$

Comparison with the TDSE discussed in Chap. 2 shows

$$\nabla A = \frac{\nabla \mathrm{e}^{A(\mathbf{r})}}{\mathrm{e}^{A(\mathbf{r})}} \mathrel{\hat{=}} \mathcal{C}_{\text{R}} = \frac{2\hbar}{m} y_{\text{R}} = \frac{\dot{\alpha}}{\alpha} , \tag{7.61}$$

$$k_0(\nabla L) \mathrel{\hat{=}} \mathcal{C}_{\text{I}} = \frac{2\hbar}{m} y_{\text{I}} = \frac{1}{\alpha^2} = \dot{\varphi} . \tag{7.62}$$

Combining real and imaginary parts to the complex variable

$$\mathcal{C}(\mathbf{r}) = \nabla A + \mathrm{i}k_0(\nabla L) , \tag{7.63}$$

corresponding to $\mathcal{C}(t) = \frac{2\hbar}{m} y(t) = \frac{\dot{\alpha}}{\alpha} + \mathrm{i}\dot{\varphi}$, allows the wave Eq. (7.55) to be rewritten as the *complex Riccati equation*

$$\nabla \mathcal{C}(\mathbf{r}) + \mathcal{C}^2(\mathbf{r}) + (k_0 n)^2 = 0 \,. \tag{7.64}$$

The usual approximation assumes that wavelengths are small in comparison with the magnitude of change in the medium, i.e., terms with $k_0^2 = \left(\frac{2\pi}{\lambda}\right)^2$ are dominant in (7.59); therefore, it is reduced to the *Eikonal equation* of geometric optics,

$$n^2 = (\nabla L)^2 \,, \tag{7.65}$$

thus, however, loosing all information contained in $A(\mathbf{r})$. The complete information can be obtained by solving the complex Riccati equation (7.64), as discussed in Chap. 2.

Further examples in nonlinear optics where Ermakov equations occur, corresponding to complex Riccati equations as shown in Sect. 2.3, can be found in [22, 23].

7.6 Ermakov Equation for Bose–Einstein Condensates

A Bose–Einstein condensate (BEC) is the ground state of a collection of interacting bosons trapped by an external potential. In the limit where the number of atoms is sufficiently large and the atomic interactions are sufficiently weak, the mean-field approximation can be applied where the effect felt by a particular atom due to the ensemble is approximated by the mean action of the entire fluid on the particle.

In this case, the macroscopic wave function for the BEC, $\Psi(\mathbf{r}, t)$, is determined by the Gross–Pitaevskii equation

$$\mathrm{i}\hbar \frac{\partial}{\partial t} \Psi(\mathbf{r}, t) = \left\{ -\frac{\hbar^2}{2m} \, \Delta + V(\mathbf{r}, t) + g|\Psi|^2 \right\} \Psi(\mathbf{r}, t) \,. \tag{7.66}$$

This equation is formally equivalent to a three-dimensional NLSE (with cubic nonlinearity, but a different interpretation of Ψ!) and, in general, it is non-integrable. However, in two dimensions it is possible to determine the dynamics by employing the "moment method" [24–30]. In this approach, integral relations are constructed directly from the wave function without solving the SE explicitly and the evolution of these physical quantities then parametrizes the dynamics of the WP solution of (7.66).[6]

Therefore, in the following, a two-dimensional cylindrically-symmetric BEC in a parabolic trapping potential

[6]Remark: the same model was also applied in NL optics to study the propagation of paraxial light beams in fibres (with applications in information technology) [30, 31].

$$V(\mathbf{r},t) = \frac{m}{2}\omega^2(t)r^2 \tag{7.67}$$

with TD frequency $\omega(t)$ is considered.

A derivation of the moment method (from a Lagrangian density) can be found, e.g., in [32]. Here, only a brief summary of this method is given. For the radially-symmetric problem with $\Psi(\mathbf{r}_\perp,\theta,t) = u(\mathbf{r}_\perp,t)\,\mathrm{e}^{\mathrm{i}k\theta}$, Eq. (7.66) (with $g = 1$) turns into

$$\mathrm{i}\hbar\frac{\partial}{\partial t}u = -\frac{\hbar^2}{2m}\left(\frac{1}{r}\frac{\partial}{\partial r}r\frac{\partial}{\partial r}u\right) + \left(\frac{\hbar^2}{2m}\frac{k^2}{r^2} + |u|^2 + \frac{m}{2}\omega^2(t)r^2\right)u\,. \tag{7.68}$$

Also this equation is non-integrable and has no analytic solutions, even for $\omega =$ constant. But one can define the following integral quantities, the so-called moments[7] M_i:

$$M_1 = \int d^2x\,|u|^2 \quad \text{, norm (intensity or number of particles)} \tag{7.69}$$

$$M_2 = \int d^2x\,r^2|u|^2 \quad \sim \text{width} \tag{7.70}$$

$$M_3 = \mathrm{i}\int d^2x\left(\frac{\frac{\partial u^*}{\partial r}}{u^*} - \frac{\frac{\partial u}{\partial r}}{u}\right)|u|^2 \tag{7.71}$$

$$= \mathrm{i}\int d^2x\left(u\frac{\partial u^*}{\partial r} - u^*\frac{\partial u}{\partial r}\right) \quad \sim \text{radial momentum}$$

$$M_4 = \frac{1}{2}\int d^2x\left(|u|^2 + \frac{k^2}{r^2}|u|^2 + |u|^4\right) \quad \sim \text{energy of WP} \tag{7.72}$$

where $d^2x = rdrd\theta$ and integration over θ yields 2π.

These M_i satisfy a simple and, most important, closed set of evolution equation (for details see [32]). From the conservation of norm or particle number follows $\dot{M}_1 = 0$. The other moments obey[8]

$$\dot{M}_2 = M_3\,, \tag{7.73}$$

$$\dot{M}_3 = 4M_4 - 2\omega^2 M_2\,, \tag{7.74}$$

$$\dot{M}_4 = -\frac{1}{2}\omega^2 M_3\,. \tag{7.75}$$

This system of equation has several invariants under time-evolution. The most important one is

$$Q = 2M_4M_2 - \frac{1}{4}M_3^2 = \text{const}\,. \tag{7.76}$$

---

[7] In optics these moments are used to calculate the beam-parameter evolution.

[8] Compare the similarity with the set of evolutions Eqs. (2.187)–(2.189) in Sect. 2.10.

With the help of (7.76), the system of Eqs. (7.73)–(7.75) can be reduced to one equation for $M_2(t)$,

$$\frac{1}{2}\frac{\ddot{M}_2}{M_2} - \frac{1}{4}\left(\frac{\dot{M}_2}{M_2}\right)^2 + \omega^2(t) = \frac{Q}{M_2^2} \,. \tag{7.77}$$

With $M_2 = \alpha^2$, this can be rewritten as Ermakov equation

$$\ddot{\alpha} + \omega^2(t)\,\alpha = \frac{Q}{\alpha^3} \,, \tag{7.78}$$

which is equivalent to a Riccati equation for the complex variable $C = \frac{\dot{\alpha}}{\alpha} + i\frac{1}{\alpha^2}$.

Comparison with the TDSE for the one-dimensional HO,

$$\mathrm{i}\hbar\frac{\partial}{\partial t}\Psi(x,t) = \left\{-\frac{\hbar^2}{2m}\frac{\partial^2}{\partial x^2} + \frac{m}{2}\omega^2(t)x^2\right\}\Psi(x,t) \tag{7.79}$$

leads to

$$M_1 = \int dx\ \Psi^*\Psi = \text{const.} = 1\ ,\ \text{norm},\ \dot{M}_1 = 0 \tag{7.80}$$

$$M_2 = \frac{2m}{\hbar}\int dx\ \Psi^*\tilde{x}^2\Psi = \frac{2m}{\hbar}\langle\tilde{x}^2\rangle\ \sim \text{position uncertainty, WP width} \tag{7.81}$$

$$M_3 = \frac{2}{\hbar}\int dx\ \Psi^*(\tilde{x}\tilde{p} + \tilde{p}\tilde{x})\Psi = \frac{2}{\hbar}\langle[\tilde{x},\tilde{p}]_+\rangle\ \sim \text{position-momentum uncertainty correlation} \tag{7.82}$$

$$M_4 = \frac{1}{m\hbar}\int dx\ \Psi^*\tilde{p}^2\Psi = \frac{1}{m\hbar}\langle\tilde{p}^2\rangle\ \sim \text{momentum uncertainty, quantum kinetic energy} \tag{7.83}$$

or, expressed in terms of α and $\dot{\alpha}$,

$$M_2 = \frac{2m}{\hbar}\langle\tilde{x}^2\rangle = \alpha^2 \,, \tag{7.84}$$

$$M_3 = \frac{2}{\hbar}\langle[\tilde{x},\tilde{p}]_+\rangle = 2\dot{\alpha}\alpha \,, \tag{7.85}$$

$$M_4 = \frac{1}{m\hbar}\langle\tilde{p}^2\rangle = \frac{1}{2}\left(\dot{\alpha}^2 + \frac{1}{\alpha^2}\right) \,. \tag{7.86}$$

These "moments" fulfil the same closed set of evolution equations as Eqs. (7.73)–(7.75) for the moments (7.69)–(7.72) and the invariant Q turns into $Q = 1$, thus turning the Ermakov equation (7.78) into the one obtained in Sect. 2.3, i.e., Eq. (2.16).

7.7 Ermakov Equation in Cosmology

In a paper by J.E. Lidsey [33], a correspondence was established between the BEC (as treated in the last subsection) and cosmology. The equations of motion for the BEC (as seen above) as well as a positively curved, perfect fluid cosmology (see below) can both be mapped onto a one-dimensional Ermakov equation.[9]

This comparison starts from the Einstein field equations and assumes a spatially-homogeneous and isotropic universe in agreement with the cosmological principle. In this case, the Robertson–Walker metric is applicable,

$$ds^2 = -c^2dt^2 + a^2(t)\left[\frac{dr^2}{1-kr^2} + r^2d\Theta^2 + r^2\sin^2\Theta\, d\phi^2\right], \tag{7.87}$$

where the scale factor $a(t)$ represents something like the "radius of the universe" and k, defining the curvature, can take the values $0, +1, -1$ for a flat, closed or open universe.

This leads to the Friedmann–Lemaître equations

$$\left(\frac{\dot{a}}{a}\right)^2 = H^2 = \frac{2}{3}\varrho - \frac{k}{a^2}, \tag{7.88}$$

$$\dot{\varrho} = -3H(\varrho + p), \tag{7.89}$$

with $H = \frac{\dot{a}}{a}$: Hubble parameter, overdot $^{\cdot} = \frac{d}{dt}$ with t: cosmological proper time, $4\pi G = c = 1$, $\Lambda = 0$ (cosmological constant).

Assuming the matter source as a self-interacting scalar field $\Phi(t)$, the energy density ϱ can be written as

$$\varrho = \frac{1}{2}\left(\frac{d\Phi}{dt}\right)^2 + U(\Phi) \tag{7.90}$$

with a potential $U(\Phi)$ and the pressure p as

$$p = \frac{1}{2}\left(\frac{d\Phi}{dt}\right)^2 - U(\Phi). \tag{7.91}$$

Taking the time-derivative of Eq. (7.88) and replacing $\dot{\varrho}$ using Eqs. (7.90) and (7.91) leads to

$$\frac{d}{dt}\left(\frac{\dot{a}}{a}\right) + \left(\frac{d\Phi}{dt}\right)^2 = \frac{k}{a^2}. \tag{7.92}$$

[9] Ermakov equations and corresponding invariants in a cosmological context are also discussed in [34–38], equivalent to a complex Riccati equation.

Introducing a new time-variable, the so-called conformal time τ, via $\frac{d}{dt} = a\frac{d}{d\tau}$, this equation turns into the Ermakov equation

$$\frac{d^2}{d\tau^2}a + \left(\frac{d\Phi}{d\tau}\right)^2 a = \frac{k}{a^3} \tag{7.93}$$

where $\frac{d\Phi}{d\tau}$ corresponds to $\omega(t)$ in the TDSE and for a closed universe, $k = 1$.

It is important that this (real) Ermakov equation is equivalent to a complex Riccati equation that can be linearized to a (complex) Newtonian equation. Considering only the Newtonian equation (where the imaginary part actually represents the physical Newtonian equation in one dimension), this has the same form as Eq. (7.93), only with $k = 0$, i.e., vanishing rhs, which, in the cosmological case, corresponds to a flat universe. Changing the real Newtonian equation to the complex one, in the relevant equation for the radius, i.e., Eq. (7.93), k changes from $k = 0$ to $k = 1$. This could mean that a description of the universe in terms of only real quantities could give the impression of a flat universe whereas a description in terms of complex quantities, as in quantum mechanics, could turn this into a closed-universe scenario. This is, however, only an idea originating from formal similarities, but might inspire further thoughts.

Knowing that a real NL Ermakov equation is equivalent to a complex NL Riccati equation allows to write the quantity fulfilling the Riccati equation, in this case, as

$$C = \frac{\frac{d}{d\tau}a}{a} + \mathrm{i}\frac{1}{a^2}\,. \tag{7.94}$$

Using the results from Sect. 2.11, this complex quantity enables to define corresponding creation and annihilation operators that could be used to create something like "coherent states" of the universe.

7.8 Complex Riccati Equation and Pythagorean Triples

In the Introduction, Plato's idea of "quantising" our material world in terms of right-angled triangles was presented and the question of finding such triangles was posed where the lengths of all three sides are integers (fulfilling the Pythagorean theorem), in other words, how are the so-called Pythagorean triples constructed?

For the answer let us return to the time-evolution of the complex quantity $C = \frac{2\hbar}{m}y$ that determines the spreading of the WP width, as discussed in Sect. 2.1. In the following, only the case $V = 0$ (i.e., $\omega = 0$) will be considered explicitly. As shown above, with a particular solution $\tilde{C}$ of the Riccati equation (2.4), its inhomogeneity can always be removed. The resulting additional linear term (at least for a constant coefficient $2\tilde{C}$) can also be removed. So we are dealing with a complex equation of the form

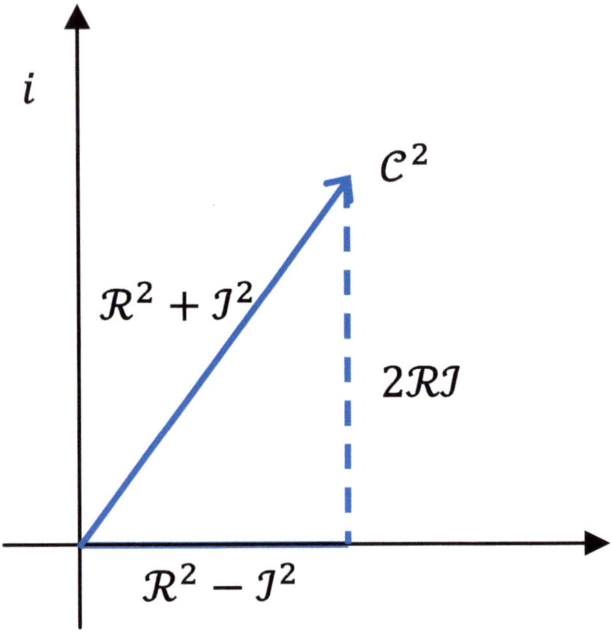

Fig. 7.1 Complex quantity $\mathcal{C}^2$ as a right-angled triangle defining the Pythagorean triples

$$\frac{d}{dt}\mathcal{C} + \mathcal{C}^2 = 0\,. \tag{7.95}$$

Then $-\frac{d}{dt}\mathcal{C}$ is also a complex quantity, $\mathcal{C}^2$, where its real and imaginary parts as well as its absolute value again define a right-angled triangle (in the complex plane) and each side contains contributions from the real part $\mathcal{R}$ and the imaginary part $\mathcal{I}$ of $\mathcal{C} = \mathcal{R} + \mathrm{i}\mathcal{I}$, i.e., $\mathcal{R}e\{\mathcal{C}^2\} = \mathcal{R}^2 - \mathcal{I}^2$, $\mathcal{I}m\{\mathcal{C}^2\} = 2\mathcal{R}\mathcal{I}$ and $|\mathcal{C}^2| = \mathcal{R}^2 + \mathcal{I}^2$ (see also Fig. 7.1).

If we now assume that $\mathcal{R}$ and $\mathcal{I}$ are no longer related to the WP width, but are simply numbers, in particular *integers* (with $\mathcal{R} > \mathcal{I}$), all three sides of the right-angled triangle created by real part, imaginary part and absolute value of $\mathcal{C}^2$ in the complex plane are also *integers*. As examples, we choose:

$$\begin{aligned} a)\ \mathcal{R} = 2,\ \mathcal{I} = 1: \qquad \mathcal{R}^2 - \mathcal{I}^2 &= 3 \\ 2\mathcal{R}\mathcal{I} &= 4 \\ \mathcal{R}^2 + \mathcal{I}^2 &= 5 \end{aligned} \tag{7.96}$$

with $9 + 16 = 25$,

$$b)\ \mathcal{R} = 3,\ \mathcal{I} = 2: \qquad \begin{aligned} \mathcal{R}^2 - \mathcal{I}^2 &= 5 \\ 2\mathcal{R}\mathcal{I} &= 12 \\ \mathcal{R}^2 + \mathcal{I}^2 &= 13 \end{aligned} \tag{7.97}$$

with $25 + 144 = 169$.

All possible (infinitely-many) Pythagorean triples can be obtained in this way by simply applying all integers $\mathcal{R}$ and $\mathcal{I}$ with $\mathcal{R} > \mathcal{I}$. In a physical context this means that whenever a physical quantity obeys a complex Riccati equation and this quantity can be "quantised" in the sense that its real and imaginary parts can be expressed as multiples of some units, its "evolution" (in time, space or any other variable, depending on the respective derivative) can also be expressed in terms of integer multiples of the same units.

This appears particularly interesting because it has been shown in the previous subsections that complex Riccati equations or corresponding Ermakov systems occur in many different areas of physics; not only in TD quantum mechanics, but also in TI quantum mechanics, classical optics, Bose–Einsten condensates, NL dynamics, cosmological models and many other areas not explicitly discussed here, like accelerator physics [39, 40], shallow water theory [41], magnetogasdynamics systems [42] and many more. So in all these systems, intrinsically, some kind of quantisation is possible. (Relations to the complex quadratic family as it occurs in the theory of fractals appear also obvious, but will not be discussed here.)

References

1. S. Hawking, *A Brief History of Time: From Big Bang to Black Holes* (Bantam Dell, New York, 1988)
2. R.P. Feynman, *Statistical Mechanics–A Set of Lectures* (Benjamin, Reading, 1972). Chap. 2.5
3. H.C. Rosu, F. Aveces de la Cruz, One-parameter Darboux-transformed quantum actions in Thermodynamics. Phys. Scripta **65**, 377–382 (2002)
4. H.C. Rosu, arXiv:math-ph/020813v1 (2002)
5. P.-F. Verhulst, Notice sur la loi que la population suit dans son accroissement. Correspondance Mathematique et Physique **10**, 113–121 (1838)
6. H. Haken, *Synergetics - An Introduction* (Springer, Berlin, 1978)
7. H.G. Schuster, *Deterministic Chaos: An Introduction* (Physik Verlag, Weinheim, 1984), p. 112
8. J.S. Russel, in Report on Waves, Report of the fourteenth meeting of the British Association for the Advancement of Science, York, Sept. 1844, pp. 311–390, Plates XLVII-LVII (London, 1845)
9. D.J. Korteweg, G. de Vries, On the change of form of long waves advancing in a rectangular canal, and on a new type of long stationary waves. Phil. Mag. **39**, 422–443 (1895)
10. R.K. Dodd, J.C. Eilbeck, J.D. Gibbon, H.C. Morris, *Solitons and Nonlinear Wave Equations* (Academic, London, 1982)
11. R. Meinel, G. Neugebauer, H. Steudel, *Solitonen: Nichtlineare Strukturen* (Akademie Verlag, Berlin, 1991)

12. R.M. Miura, Korteweg–de Vries equation and generalizations. I. A remarkable explicit nonlinear transformation. J. Math. Phys. **9**, 1202 (1968)
13. P.G. Drazin, R.S. Johnson, *Solitons: An Introduction* (Cambridge University Press, Cambridge, 1989)
14. H. Kalka, G. Soff, *Supersymmetrie* (Teubner, Stuttgart, 1997)
15. F. Cooper, A. Khare, U. Sukhatme, *Supersymmetry in Quantum Mechanics* (World Scientific, Singapore, 2001)
16. S. Flügge, *Practical Quantum Mechanics* (Springer, Berlin, 1974), p. 94
17. D. Schuch, From nonlinear Burgers and Korteweg–de Vries soliton equations via Riccati to linear Rosen-Morse and free particle Schrödinger equations. Phys. Scr. **90**, 074050 (2015)
18. C.H. Su, C.S. Gardner, Korteweg–de Vries equation and generalizations. III. Derivation of the Korteweg–de Vries equation and Burgers equation. J. Math. Phys. **10**, 536 (1969)
19. W. Malfliet, Simple solution method for the Korteweg–de Vries equation. Am. J. Phys. **49**, 666 (1981)
20. E. Schrödinger, Quantisierung als Eigenwertproblem (Zweite Mitteilung). Ann. d. Phys. **79**, 489–527 (1926)
21. H. Goldstein, *Classical Mechanics* (Addison-Wesley, Reading, 1951). Chap. 9.8
22. A.M. Goncharenko, Yu.A. Logvin, A.M. Samson, P.S. Shapovalov, S.I. Turovets, Ermakov Hamiltonian systems in nonlinear optics of elliptic Gaussian beams. Phys. Lett. A **160**, 138–142 (1991)
23. C. Rogers, B. Malomed, K. Chow, H. An, Ermakov–Ray–Reid systems in nonlinear optics. J. Phys. A: Math. Theor. **43**, 455214 (2010). (15 pp)
24. S.N. Vlasov, V.A. Petrischev, V.I. Talanov, Averaged description of wave beams in linear and nonlinear media (the method of moments). Radiophys. Quantum Electron. **14**, 1062–1070 (1971)
25. P.A. Belanger, Beam propagation and the ABCD ray matrices. Opt. Lett. **16**, 196–198 (1991)
26. M.A. Porras, J. Alda, E. Bernabeu, Nonlinear propagation and transformation of arbitrary laser beams by means of the generalized ABCD formalism. Appl. Opt. **32**, 5885–5892 (1993)
27. V.M. Perez-Garcia, M.A. Porras, L. Vazquez, The nonlinear Schrödinger equation with dissipation and the moment method. Phys. Lett. A **202**, 176–182 (1995)
28. V.M. Perez-Garcia, H. Michinel, J.I. Cirac, M. Lewenstein, P. Zoller, Low energy excitations of a Bose–Einstein condensate: a time-dependent variational analysis. Phys. Rev. Lett. **77**, 5320 (1996)
29. V.M. Perez-Garcia, H. Michinel, J.I. Cirac, M. Lewenstein, P. Zoller, Dynamics of Bose–Einstein condensates: variational solutions of the Gross–Pitaevskii equations. Phys. Rev. A **56**, 1424 (1997)
30. J.J. Garcia-Ripoll, V.M. Perez-Garcia, P. Torres, Extended parametric resonances in nonlinear Schrödinger systems. Phys. Rev. Lett. **83**, 1715 (1999)
31. V.M. Perez-Garcia, P. Torres, J.J. Garcia-Ripoll, H. Michinel, Moment analysis of paraxial propagation in a nonlinear graded index fibre. J. Opt. B: Quantum Semiclass. Opt. **2**, 353–358 (2000)
32. D. Anderson, Variational approach to nonlinear pulse propagation in optical fibers. Phys. Rev. A **27**, 3135 (1983)
33. J.E. Lidsey, Cosmic dynamics of Bose–Einstein condensates. Class. Quantum Grav. **21**, 777 (2004)
34. J.R. Ray, Cosmological particle creation. Phys. Rev. D **20**, 2632 (1979)
35. C. Bertoni, F. Finelli, G. Venturi, Adiabatic invariants and scalar fields in a de Sitter space-time. Phys. Lett. A **237**, 331–336 (1998)
36. F. Finelli, G.P. Vacca, G. Venturi, Chaotic inflation from a scalar field in nonclassical states. Phys. Rev. D **58**, 103514 (1998)
37. F. Finelli, A. Gruppuso, G. Venturi, Quantum fields in an expanding universe. Class. Quantum Grav. **16**, 3923–3935 (1999)
38. R.M. Hawkins, J.E. Lidsey, Ermakov–Pinney equation in scalar field cosmologies. Phys. Rev. D **66**, 023523 (2002)

39. E.D. Courant, H.S. Snyder, Theory of the alternating-gradient synchrotron. Ann. Phys. (N.Y.) **3**, 1–48 (1958)
40. J. Struckmeier, C. Riedel, Exact invariants for a class of three-dimensional time-dependent classical Hamiltonians. Phys. Rev. Lett. **85**, 3830–3833 (2000)
41. C. Rogers, H. An, Ermakov–Ray–Reid systems in (2+1)-dimensional rotating shallow water. Theory Stud. Appl. Math. **125**, 275–299 (2010)
42. H. An, C. Rogers, A 2+1-dimensional non-isothermal magnetogasdynamic system. Hamiltonian–Ermakov integrable reduction. SIGMA **8**, 057 (2012). (15 pp)

Chapter 8
Summary, Conclusions and Perspectives

The opening Chaps. (1–3) embark upon a nonlinear (NL) reformulation of quantum mechanics in terms of complex Riccati or the equivalent real Ermakov equations. To provide a solid foundation for the development of such a NL formulation of quantum theory, time-dependent (TD) quantum mechanics of systems with analytic solutions, i.e., Hamiltonians at most quadratic in position and momentum are used as starting point. The harmonic oscillator, the parametric oscillator and (in the limit $\omega \rightarrow 0$) the free motion are specifically considered. The corresponding Gaussian wave packet (WP) solutions of the conventional Schrödinger equation (SE) are completely determined by two parameters: their maximum and width; in these cases, both may be TD. The time-dependence of the WP-maximum, that is actually identical to the mean value of position $\langle x \rangle$, calculated with this WP is trivial. It is simply determined by the equation of motion for the trajectory $\eta(t)$ of a corresponding classical particle. The time-dependence of the width contains information pertaining to the quantum mechanical aspects, like tunnelling currents, ground-state energies and so on, making it the key point of interest. The equations of motion determining the time-evolution of the WP-width are complex, (quadratically) NL equations of Riccati-type. Different methods for treating the Riccati equation display distinct aspects and formulations of the quantum dynamics of the system [1–4].

In the direct treatment of the inhomogeneous Riccati equation, this can be transformed into a homogeneous NL Bernoulli equation, providing a particular solution of the Riccati equation is known. The examples possess particular solutions that are constants, thus leading to Bernoulli equations with solutions that can be given in analytic form. (For TD particular solutions, it is still possible to achieve solutions of the Bernoulli equation in closed form.) Characteristic of the solutions of the Bernoulli equation is their sensitivity to the initial conditions connected to the initial position uncertainty α_0 and initial temporal change of this quantity, $\dot{\alpha}_0$. This is not surprising for a NL differential equation but would not be expected in the linear formulation of quantum mechanics. An almost identical situation is known from supersymmetric quantum mechanics where families of isospectral potentials are obtained that have

D. Schuch, *Quantum Theory from a Nonlinear Perspective*,
Fundamental Theories of Physics 191, https://doi.org/10.1007/978-3-319-65594-9_8

rather different shapes of the potentials but same energy eigenvalues. Only there, they result from a real Riccati equation depending on a spatial variable instead of time.

In a different treatment, the complex NL Riccati equation is transformed into a real NL Ermakov equation. For this purpose, the imaginary part of the variable of the Riccati equation $\frac{2\hbar}{m}y = \mathcal{C}$ is set equal to the inverse square of a new variable $\alpha(t)$, i.e., $\mathcal{C}_I(t) = \frac{1}{\alpha^2(t)}$. The variable $\alpha(t)$ fulfils the Ermakov equation and is directly proportional to the WP width. From the coupled system of equations for $\eta(t)$ and $\alpha(t)$ a dynamical invariant, the Ermakov invariant, is derived. Not only is it still a constant of motion when the Hamiltonian no longer possesses this property, as in the case of the parametric oscillator with $\omega = \omega(t)$, but it also essentially has the dimension of action, the physical quantity that is quantized according to Planck in units of h or $\hbar = \frac{h}{2\pi}$.

Furthermore, the solution $\alpha(t)$ of the Ermakov equation (and hence the solution $\mathcal{C}(t) = \frac{\dot{\alpha}}{\alpha} + \mathrm{i}\frac{1}{\alpha^2}$ of the Riccati equation) and the solution of the Newtonian equation for $\eta(t)$ are not independent of each other. The quantity $\alpha(t)$ that describes the quantum uncertainty can be expressed in terms of the solutions of the corresponding Newtonian equation once they are known. There are different ways of accomplishing this.

Firstly, as demonstrated in Sect. 2.3.1 and Appendix A, if two linear-independent solutions of the Newtonian equation are known, $\alpha(t)$ can be determined with the help of the Ermakov invariant.

A second method involves a property of the NL Riccati equation that makes it particularly fitting for a NL description of quantum theory: its linearizability. Via a logarithmic derivative it sustains a kind of superposition principle that is essential for certain aspects of quantum theory. In this case, replacing $\mathcal{C}(t)$ with the logarithmic derivative of a new complex variable $\lambda(t)$, via $\mathcal{C}(t) = \frac{\dot{\lambda}}{\lambda}$, leads to a new complex Newtonian equation for $\lambda(t)$.

In polar coordinates $\lambda(t)$ can be written as $\lambda = \alpha\, \mathrm{e}^{\mathrm{i}\varphi}$ where its amplitude is identical to the Ermakov variable mentioned above.

Further, due to the nonlinearity of the Riccati equation, amplitude and phase of λ are not independent of each other but coupled via the conservation law $\dot{\varphi}\,\alpha^2 = const. = 1$.

In Cartesian coordinates, real and imaginary parts of $\lambda = u + \mathrm{i}z$ provide the time-dependent parameters $z(t)$ and $u(t)$ that completely determine the time-dependence of the TD Green function or Feynman kernel of the system. So, knowing the TD Green function enables one to determine the time-evolution of any initial state, not only of Gaussians (see, e.g., [5]).

Comparing the TD WP obtained via the TD Green function to the same Gaussian solution obtained via the direct method by inserting the Gaussian WP (2.1) into the TDSE (2.2) shows that the imaginary part of λ is directly proportional to the classical trajectory, $z(t) \propto \eta(t)$. From the conservation law, written in Cartesian coordinates as $\dot{z}u - \dot{u}z = 1$, the real part of λ, $u(t)$, can be calculated (as seen in Eq. (2.60)) once the imaginary part $z(t)$, the classical trajectory, is known. Knowing

z and u, the quantity α can now be obtained from $\alpha(t) = \sqrt{u^2 + z^2}$. Hence, quantum mechanical properties like tunnelling currents can be obtained, based essentially on the knowledge of two linear-independent solutions of the Newtonian equation. (This however involves the above-mentioned conservation law and depends on the initial uncertainties α_0 and $\dot{\alpha}_0$.) In this case, the two solutions of the Newtonian equations, z and u, are not only linearly independent of, but even orthogonal to, each other.

Knowing the relations between the solutions of the Newtonian equation of motion and the Ermakov variable $\alpha(t)$, as well as the relation between said variable and quantum properties like uncertainties, tunnelling currents, and so on, this formalism will be applied to cases where no analytic solutions of the Newtonian equation exist, such as for parametric oscillators with various time-dependent frequencies $\omega(t)$. As the relations derived above are also valid in this case, e.g., from numerical solutions of the Newtonian equation, also the corresponding time-dependence of the quantum properties can be calculated.

Seeing that the Newtonian equation is not only valid for the free motion and oscillators but actually for any potential, one could assume that also the relations between the solutions of the Newtonian equation and the corresponding quantum mechanical properties are valid, again via the Ermakov variable $\alpha(t)$. Therefore, solutions of the Newtonian equation shall be explored for arbitrary potentials (preferably in one dimension, to begin with) and the corresponding quantum properties calculated. Comparison with experimental data (where possible) could show to what extent this method could be applied.

From the TD Green function, insight can also be gained into the connection between Riccati equations and trigonometric/hyperbolic functions. Furthermore, the Ermakov invariant can also be written in the form of a Hamiltonian for a harmonic oscillator with unit frequency; now, however, the variable time t is replaced by the angle $\varphi(t)$. The corresponding equation of motion, written in a form where the variable φ is replaced by t, resembles a damped oscillator with "*damping coefficient*" $2\frac{\dot{\alpha}}{\alpha}$ (as seen in Eq. (2.74)). This seems somewhat surprising since, so far, only non-dissipative systems with time-reversible dynamics were dealt with. However, in the conventional damped motion, the damping coefficient is a constant thus breaking time-reversal symmetry for $t \to -t$ whereas here, due to the time-derivative in $\frac{\dot{\alpha}}{\alpha}$, also the "*damping coefficient*" changes sign for $t \to -t$. Therefore, the corresponding Newtonian equation is still invariant under time-reversal but it shows that dissipation might somehow be related to the temporal change of the amplitude $\alpha(t)$ of the complex quantity $\lambda(t)$.

Moreover, the Ermakov and Riccati variables, as well as $\lambda(t)$, can be used to rewrite the ground-state energy $\tilde{E}$ of the WPs in a form so that the usual Lagrange/Hamilton formalism of classical mechanics can also be applied to these quantum mechanical properties. From the conservation law mentioned earlier, it follows that the canonical momentum p_φ is not only a constant but also has the value

$\frac{\hbar}{2}$ which hints at a connection between the *angular momentum* for the motion in a *complex* plane and the quantum mechanical *spin*[1]!

Looking at the same TD quantum problems in momentum space shows that the information about the quantum mechanical properties and their dynamics is again expressed in terms of a complex Riccati equation. The differences, in comparison with the situation in position space, are simply that the variable of this Riccati equation is the inverse of the one in position space and the sign of the derivative term changes from plus to minus. Everything else said about Ermakov equations and invariants, TD Green functions, etc. applies as in position space.

Combining position and momentum space to form phase space and looking for an appropriate description of the quantum system in this space leads to the Wigner function. In the cases investigated in this work, the exponent of the Wigner function is essentially the Ermakov invariant (for shifted position and momentum variables). Different interpretations of this invariant are also given in this context [7].

Further light is shed on the nature of this invariant when considering the representation of canonical transformations in quantum mechanics in terms of the two-dimensional (real) symplectic group $Sp(2, R)$. This shows that the four elements of the 2×2 matrices representing the group $Sp(2, R)$ are simply the real and imaginary parts $u(t)$ and $z(t)$ of $\lambda(t)$ and their time-derivatives (with appropriate signs). In the case of TD quantum systems it turns out that these matrices not only transform initial position and momentum into their values at a later time t but, simultaneously, also the initial uncertainties into the ones at the later time.

The determinant of the 2×2 matrices reproduces the conservation law of angular momentum in the complex plane. However, expressing u, z and their derivatives in terms of η, $\dot{\eta}$, α and $\dot{\alpha}$ shows this determinant is actually also equivalent to the Ermakov invariant (divided by its constant value, thus reproducing again a determinant equal to one) [8].

And there is more to the discussion of the Ermakov invariant. It can also be derived via a dynamical Lie algebra [9]. This leads to a set of three coupled differential equations for the time-evolution of position- and momentum-uncertainties and their correlation. This, in turn, is identical to the one obtained by inserting the TD Green function into the TDSE (see Sect. 2.5).

Still further, factorization of the operator corresponding to the Ermakov invariant leads to a TD generalization of creation and annihilation operators analogous to the well-known treatment of the HO [10]. In these operators, the constant $\mathrm{i}\omega_0$ (corresponding to the imaginary particular solution C_I of the HO with constant WP width) is simply replaced by the full complex TD solution C of the complex Riccati equation. These operators can be applied to obtain coherent states with TD width and can also be used to create generalized excited states, starting from the corresponding ground state. In the course of this treatment, it is possible to create *harmonic states* for the

[1]In general, the transition between real and complex variables involves a factor $\frac{1}{2}$ when derivatives are considered, e.g., for $w = x + \mathrm{i}y$ follows $\frac{\partial}{\partial w^*} = \frac{1}{2}\left(\frac{\partial}{\partial x} + \mathrm{i}\frac{\partial}{\partial y}\right)$ or $\frac{\partial}{\partial w} = \frac{1}{2}\left(\frac{\partial}{\partial x} - \mathrm{i}\frac{\partial}{\partial y}\right)$; see [6].

free motion, i.e., solutions of the free motion SE in terms of a Gaussian function with complex exponent multiplied by Hermitian polynomials depending on a TD variable [11].

In the end, it has been shown that the operator corresponding to the Ermakov invariant and its eigenfunctions and eigenvalues can be used to transform TDSEs into formally TISEs where the time-dependence of the WP width is eliminated by a unitary transformation removing the term proportional to $\mathrm{i}\mathcal{C}_\mathrm{R}\tilde{x}^2$ in the exponent of the Gaussian function. This approach again confirms the TD Green function derived in Sect. 2.5 in a different way and offers possibilities of extending the Ermakov invariant to systems with terms additional to the quadratic ones in the potential.

However, considering TD quantum systems with exact, particularly analytic, solutions, the Riccati/Ermakov formalism is essentially still restricted to problems with at most quadratic potentials. Nevertheless, in quantum mechanics, there are more potentials allowing for exact or even analytic solutions, like the Coulomb problem, but in the time-independent version. A powerful tool for treating these systems is supersymmetry. In this case, the position-dependent part of the creation and annihilation operators of the HO, $\omega_0 x$, is replaced by a general function of position[2] $W(x)$, the so-called "*superpotential*". This is similar to the generalization in the TD case, by replacing $\mathrm{i}\omega_0$ with $\mathcal{C}$, but here the generalization concerns the position variable. The link to the Riccati treatment is established because the superpotential $W(x)$ also fulfils a (actually two) Riccati equation(s) (this time real) that provides the potential of the system. Similar to the TD case, $\omega_0 x$ (like $\mathrm{i}\omega_0$) represents only a particular solution of the (TI) HO-problem and the general solution can be obtained via a Bernoulli equation in exactly the same way as in the TD case. This leads to an entire family of distorted potentials depending on a parameter (corresponding to the initial uncertainties in the TD case) that all have different shapes but the same spectrum. The superpotential can be obtained essentially as a logarithmic derivative of the ground state wave function of the problem which is usually a real function.

An extension of this position-dependent Riccati formalism that also allows for complex wave functions, i.e., including excited states, was given by Reinisch [12]. His NL reformulation of TI quantum mechanics is based on Madelung's hydrodynamic formulation [13], leading to a NL Ermakov equation for the amplitude of the wave function which is equivalent to a (complex) NL Riccati equation for the logarithmic derivative of the complete complex wave function [14].

In this case, phase and amplitude of the wave function are coupled by a conservation law formally identical to the one that couples phase and amplitude of the complex TD function $\lambda(t)$ that fulfils a Newtonian equation of motion.

The formal similarity between this approach and supersymmetric quantum mechanics allows one to start with a problem with real potential (or even $V = 0$) and, via a Darboux transformation, introduce a complex partner potential with (essentially) the same real spectrum [15]. This opens the way for constructing a new class of non-Hermitian potentials with real spectrum. Further work in this direction is in

[2]This $W(x)$ should not be confused with the Wigner function $W(x, p, t)$.

progress. This striking similarity is even more obvious when comparing the different linear and NL formulations of the TD and TI theories, as shown in Fig. 3.1.

A problem that is still not satisfactorily solved is the treatment of open systems with irreversible dynamics and dissipative energetics, in classical as well as in quantum mechanics.

Aesthetic aspects aside, analytical mechanics in terms of Lagrangians and Hamiltonians has distinct advantages over "*vectorial mechanics*", like in Newton's laws of motion (see also [16]). The invariance of the first formulations under coordinate (or even canonical) transformations is already notable. More remarkable, however, is the fact that the Lagrangian and Hamiltonian forms can be derived from first (variational) principles and be used as starting points for either canonical quantization (by means of the Hamiltonian function) or path integral quantization [17] (by means of the Lagrangian function). In the Lagrangian or Hamiltonian formalism it is not possible to take into account the effect of a (possibly dissipative) environment acting on an open system by simply changing the equation(s) of motion of this system of interest without significantly modifying the formalism. The same naturally applies to the quantized version.

However, in the Newtonian form of classical mechanics, there are modified versions for the equation of motion of open systems in the trajectory picture, like in the form of the Langevin equation (containing a velocity-dependent friction and a fluctuating stochastic force). Equivalently, in a description in terms of (one-particle) distribution functions, this can be achieved in the form of Fokker–Planck equations containing irreversible diffusion terms. These equations are usually called "*phenomenological*" equations, implying that they are less fundamental because they are not derived from "*first principles*"– but isn't all of physics essentially a phenomenological theory?

In Chap. 4 it is shown that these phenomenological equations are quite compatible with other approaches based on the Lagrangian and Hamiltonian formalism, even (or particularly) after quantization. For this purpose, the well-accepted system-plus-reservoir approach is discussed that couples the system of interest to a (generally very large) number of environmental degrees of freedom, considering both, system plus environment (reservoir), as a closed Hamiltonian system. After assuming specific couplings between system and reservoir and eliminating the environmental degrees of freedom, an equation of motion is finally obtained that agrees with the above-mentioned phenomenological approaches (e.g., [18–20]).

On the other hand, why start with a large number of environmental degrees of freedom if they are eliminated in the end and simply makes the calculations more cumbersome, time-consuming and costly? The other extreme of the system-plus-reservoir approach, going from a large number (in the limit infinitely-many) of environmental degrees of freedom to only a single one, was also discussed in the form of the Bateman Hamiltonian [21].

As both approaches provide the same (averaged) equation of motion for the system of interest, including a linear velocity-dependent friction force, the challenge was how to connect these two formalisms.

The problem could be solved by considering another type of approach for open systems where the environmental degrees of freedom do not appear explicitly; only their effect on the system shows up in terms of the aforementioned friction force. These approaches also retain the canonical Lagrangian and/or Hamiltonian formalism but the connection between the physical position and momentum variables and the canonical ones is no longer given by a canonical, but by a *non-canonical* transformation. That means the canonical formalism of analytical mechanics can still be used for open dissipative systems if the class of transformations allowed for position and momentum variables is extended consistently from canonical to non-canonical transformations.

Two approaches of this kind were discussed in detail. One by Caldirola and Kanai [22, 23] using an explicitly TD Hamiltonian $\hat{H}_{\text{CK}}$, the other using exponentially-expanding coordinates leading to a Hamiltonian $\hat{H}_{exp}$ that is a constant of motion [24–27]. It was shown by Sun and Yu [28, 29] that the Hamiltonian by Caldirola and Kanai can be obtained starting from Caldeira and Leggett's conventional system-plus-reservoir approach using a large number of environmental degrees of freedom. On the other hand, the Bateman Hamiltonian with only one environmental degree of freedom can be transformed into the one depending on the expanding coordinates by imposing certain constraints, thus eliminating this additional environmental degree of freedom [30]. Furthermore, it could be shown that the two effective Hamiltonian approaches are connected via an explicitly TD *canonical* transformation leading from the explicitly TD Hamiltonian $\hat{H}_{\text{CK}}$ to the constant Hamiltonian $\hat{H}_{exp}$ (or vice versa) [31].

So far, the situation appears quite promising. Then the canonical quantization of the Hamiltonian approaches means replacing (in position space) the momentum with an appropriate operator that (apart from a factor $\frac{\hbar}{i}$) is essentially the derivative with respect to the canonical conjugate position variable. This should lead to the corresponding Hamiltonian operator and thus to the modified SE for the dissipative system. However, it turns out that this is not so simple because the procedure starting from $\hat{H}_{\text{CK}}$ seemingly leads to violation of the commutation relation and Heisenberg's uncertainty principle for *physical* momentum and position. For a solution of this problem, effective approaches were considered that work entirely on the physical level. In other words, applying operators of position and momentum that are identical to the definitions in conventional quantum mechanics and then adding some terms to the Hamiltonian *operator* that reflect the effect of the environment. As there are different environmental effects, like dissipation, irreversibility or both, there are also different approaches for this purpose.

There are models based on the Langevin equation and Ehrenfest's theorem assuming that, on an average, a Newtonian equation of motion including a linear velocity-dependent friction force (i.e., an averaged Langevin equation) should determine the time-evolution of the mean-value of position. However, this requirement is so vague that several (all NL) modifications of the Hamiltonian operator are possible, most of them with some unphysical results [32–34].

An approach by Gisin based on the system-plus-reservoir approach, but now for the complex wave function [35, 36] (instead of the density operator), breaks the time-reversal symmetry by introducing a non-Hermitian (and also NL) term that still allows normalization of the wave function. However, the explicit form of this term and its physical interpretation are ambiguous.

At last, an approach based on breaking the time-reversal symmetry on the level of the evolution equation for the probability density leads to the desired result. This is achieved by introducing an additional diffusion term, thus replacing the continuity equation of quantum mechanics with a Fokker–Planck-type equation (to be precise, a Smoluchowski equation). For this equation to be separable into two complex conjugate modified SEs, a corresponding separability condition is necessary, leading to a complex logarithmic NL term in the modified SE. The real part of the resulting complex logarithmic NL term provides the friction force in the Ehrenfest equation. Moreover, the imaginary part of the nonlinearity, directly corresponding to the diffusion term, makes allowance for the introduction of irreversibility like in the Gisin-approach but now with a physical motivation (that was also confirmed on group-theoretical grounds [37–40]). Furthermore, all unphysical results in the approaches based on Ehrenfest's theorem are eliminated [41–43].

There is only one missing link: the connection of this approach on the physical level with the ones discussed before on the canonical level. For this purpose Schrödinger's original definition of the wave function Ψ was used in terms of the action function S [44] according to $S = \frac{\hbar}{\mathrm{i}} \ln \Psi$. This shows that the additional complex nonlinearity on the physical level is essentially proportional to the action as defined by Schrödinger and allows for the formulation of a canonical action and Hamiltonian that are identical to the ones in the effective canonical approach by Caldirola and Kanai.

The critical point is that the physical action, $S = \frac{\hbar}{\mathrm{i}} \ln \Psi$, and the canonical action, $\hat{S} = \frac{\hbar}{\mathrm{i}} \ln \hat{\Psi}$ with $\hat{S} = e^{\gamma t} S$, are different and so are the wave functions $\Psi(x,t)$ and $\hat{\Psi}(x,t)$. Hence $\Psi(x,t)$ on the physical level and $\hat{\Psi}(x,t)$ on the canonical level are connected via a *non-unitary* transformation (corresponding to the non-canonical transformation in the classical case). This then solves the problem with the violation of the uncertainty principle mentioned above in connection with the Caldirola–Kanai approach. As long as one stays on the canonical level there is no problem with the Caldirola–Kanai approach involving $\hat{\Psi}_{\mathrm{CK}}(x,t)$ and the uncertainty principle involving the *canonical* momentum. Changing to the uncertainty product of position and *physical* momentum, it is not sufficient to transform the corresponding momentum operator. It is imperative that also the canonical wave function of Caldirola and Kanai, $\hat{\Psi}_{\mathrm{CK}}(x,t)$, is changed into the physical one, $\Psi_{\mathrm{NL}}(x,t)$, via the non-unitary transformation $\ln \hat{\Psi}_{\mathrm{CK}}(x,t) = \mathrm{e}^{\gamma t} \ln \Psi_{\mathrm{NL}}(x,t)$. The confusion leading to the apparent physical discrepancies has its origins in the specific form of the wave functions. The function $\hat{\Psi}_{\mathrm{CK}}(x,t)$ on the canonical level and $\Psi_{\mathrm{NL}}(x,t)$ on the physical level are both functions of the same variables x and t thus giving the impression that they have the same physical meaning (and analytical form) which is actually not the case [31, 45, 46].

With this in mind, the calculations concerning open (classical and quantum) systems can be performed on the canonical as well as on the physical level, whichever is more convenient. Important is that the results are converted consistently to the physical level in the end.

In Chap. 5 the discussion concentrates on the physical level where the position and momentum operators have the same form and meaning as in conventional quantum mechanics. Therefore, the modified NLSE with complex logarithmic nonlinearity (5.1) is considered. As in the non-dissipative case, this has analytic Gaussian wave packet solutions where the time-evolution of the maximum and width completely determine the dynamics of the (open) quantum systems.

The maximum of the Gaussian function in the dissipative case follows a Newtonian equation of motion including the above-mentioned linear velocity-dependent friction force as it occurs in the averaged Langevin equation (without stochastic force). The time-evolution of the width is again determined by a complex NL Riccati equation, similar to the one in the non-dissipative case but with an additional term linear in the complex variable $C(t)$ with friction parameter γ as coefficient. This does not affect the treatment of the Riccati equation much. As shown in the non-dissipative case, the direct solution of the Riccati equation via a particular solution and transformation into a (homogeneous) Bernoulli equation introduces such a linear term anyway, so the friction parameter only modifies the coefficient of this term. Similarly, also a formulation of the complex Riccati equation as a real (NL) Ermakov equation is possible and, together with the equation for the WP-maximum, a dynamical Ermakov invariant can also be obtained for the open dissipative system(s).

Furthermore, the Riccati equation can again be linearized to a complex Newtonian equation, now with a linear velocity-dependent friction term. The uncertainties of position, momentum and their correlation can also be expressed in terms of the Ermakov, Riccati or complex Newtonian variables. Even for these dissipative systems, a Lagrangian and Hamiltonian formalism can be obtained that provides the correct equations of motion for the uncertainties [47, 48].

Further details concerning the time-dependence of the uncertainties in the dissipative case are also given in Appendix B. However, there are not only similarities with the non-dissipative case but also interesting and important differences.

The interaction of the system with the environment causes quantitative effects like the change in frequency ω of an oscillator to the reduced frequency $\Omega = \sqrt{\omega^2 - \frac{\gamma}{4}}$. There are also qualitative effects. For the (undercritically) damped HO the environment does not reduce the ground-state energy from $\frac{\hbar}{2}\omega_0$ to $\frac{\hbar}{2}\Omega$ but supplies an additional contribution that can be expressed entirely in terms of the environmental parameters and lifts this energy above the ground-state energy of the undamped HO. This additional contribution to the energy can be interpreted in a similar manner as the energy contribution in the Langevin picture due to the stochastic force. As the additional energy contribution in this quantum case can be traced back to the imaginary part of the logarithmic nonlinearity, and thus to the diffusion term in the Smoluchowski equation, this term and the corresponding imaginary nonlinearity appear to represent something like the stochastic aspect of the open system.

The additional linear term in the complex Riccati equation does not only modify the corresponding one in the Bernoulli equation quantitatively, it can also lead to qualitative new effects. Despite the quadratic nonlinearity of the Bernoulli equation for the free motion problem, due to solutions ± 0, only one solution and thus only one set of uncertainties is possible in the non-dissipative case for the free motion. In the dissipative case, two different solutions exist with different time-evolution of the uncertainties and different values for the ground-state energies, resembling a bifurcation as known from NL dynamics [49]. This also looks like removing a degeneracy due to symmetry-breaking; in this case time-reversal symmetry. Furthermore, it is interesting that the difference in energy for the two ground states of the damped free motion also depends solely on the parameters of the environment, but not on quantum mechanical properties like $\hbar$. Assuming the diffusion coefficient D occurring in the Smoluchowski equation (that is the basis for separation into the NLSEs) can be equated with the Einstein relation $D = \frac{\mathsf{k_B T}}{m\gamma}$ (at least for an equilibrium situation at the initial time $t = 0$), then the energy gap between the two ground states is simply proportional to $\mathsf{k_B T}$.

The logarithmic NLSE for the damped free motion also has further interesting aspects. In this case, there are not only Gaussian-shaped solutions but also some resembling modified plane waves. By superposition of these plane waves the Gaussian WP can be constructed similarly as in the conservative case. The differences are only TD superposition coefficients and TD k-values, explaining a faster spreading of the damped free motion WP due to dephasing when compared to the conservative case. Nevertheless, the time-evolution of the uncertainty product for the damped free motion exhibits a surprisingly different behaviour. As the momentum uncertainty is no longer constant, but shrinking exponentially, the uncertainty product no longer diverges quadratically in time as in the conservative case. Instead, it approaches a finite value depending on the friction parameter and, thus, on the frequency of interaction with the environment. For infinitely-frequent interaction with (or observation of) the system, this shows that the uncertainty product approaches its minimum value for $\gamma \to \infty$ thus being similar to the quantum Zeno effect [42].

The damped HO provides new qualitative effects too [50]. So, for the undamped as well as for the damped HO, there are Gaussian WP solutions with constant width that, according to $\frac{\dot{\alpha}}{\alpha} = 0$, do not contribute to a tunnelling current. However, in the dissipative case the (reduced) frequency $\Omega = \sqrt{\omega^2 - \frac{\gamma}{4}}$ can be eliminated if $\omega_0 = \frac{\gamma}{2}$ is fulfilled. In this case, the WP solution of the dissipative HO with constant width turns into one with the width spreading quadratically in time, as known from the free motion without dissipation.

This is now similar to the situation in the Quantum Hall Effect (QHE). In this case, the environmental parameter γ is somehow fixed, e.g., via the temperature. However, the frequency ω can be changed by modifying the magnetic field. For certain values of this field, a current emerges that was not present before in this quantum device but disappears once the strength of the magnetic field is changed from this "*resonance*" condition. This effect will appear again in a different context.

Another interesting effect of this resonance condition is the behaviour of the ground-state energy. Whereas the classical energy of the system decays exponentially the quantum mechanical contribution, $\tilde{E} = \tilde{T} + \tilde{V} = \frac{1}{2m}\langle\tilde{p}^2\rangle_{\mathrm{NL}} + \frac{m}{2}\,\omega^2\langle\tilde{x}^2\rangle_{\mathrm{NL}}$, increases quadratically in time in the case $\omega_0 = \frac{\gamma}{2}$. This looks like a back-transfer of energy from the environment to the system. In the cases discussed, that seems to be a violation of the second law of thermodynamics. However, the energy from the environment is not transferred into the classical degrees of freedom; i.e., the WP-maximum does not start oscillating with increasing amplitude. Instead, the energy is increasing the quantum mechanical part $\tilde{E}$ of the system. Assuming an infinite heat bath as environment (as is usually done in a Markovian approximation), the increase of $\tilde{E}$ proportional to t^2 should not be a problem. For practical purposes it would be interesting to find out how to transform the additional quantum mechanical energy contribution into a classical one that could be used to perform mechanical, electrical or other "*macroscopic*" work.

Based on the linearized form of the complex Riccati equation, a TD Green function is also obtained for the dissipative systems [51].

Trying to apply this entire logarithmic NLSE formalism to momentum space seemed problematic. This is because the logarithmic term is particularly well-suited for position space as its derivative is proportional to velocity or momentum and thus able to provide the required friction force. In momentum space, an analogous procedure would lead to an *accelerating* force proportional to position – not what is required.

This problem could be overcome by reformulating the logarithmic NLSE in terms of a combined Schrödinger and diffusion equation with the help of the damped complex Newtonian equation for $\tilde{\lambda}(t)$. The same kind of WP solutions found in the non-dissipative case is also found in this circumstance. The WP width obeys also a complex Riccati equation where the variable is again the inverse of the one in position space [52].

Proceeding to phase space, the corresponding Wigner function for the dissipative system could be obtained using the dissipative Ermakov invariant where the difference between physical and canonical description must be taken into account [48]. Similarly, generalized creation and annihilation operators and the corresponding coherent states are obtained [10].

Comparing the complex linear Newtonian formulation of the dynamics for the non-dissipative and the dissipative case, the main difference is that the amplitude of the complex variable is damped (exponentially) in the latter case. How can this be transferred to the TI case? Then, there, the complex quantity is the wave function that must be normalizable, according to its interpretation; so a modification of its amplitude does not seem possible. What are the formal consequences of the environment in this dissipative case?

In the non-dissipative case, the formal similarity between the complex Newtonian equation for $\lambda(t)$ and the TISE for $\Psi(x)$ was shown. Therefore, the addition of the linear term $\gamma\dot{\tilde{\lambda}}$ in the TD case should correspond to the addition of a linear term proportional to $\nabla_x\Psi$ in the TI case. As Ψ is assumed to be complex, this would also lead to an additional contribution to the continuity equation for $\varrho = \Psi^*\Psi$. Assuming that this term should be compatible with the diffusion term in the TD case, it turns out that the desired term in the modified SE could be obtained, however now with a completely imaginary coefficient. This alone is not yet able to reproduce the desired friction force proportional to velocity as in the TD case; only half of it. The missing half is gained by taking into account the coupling of amplitude and phase of Ψ as established in the non-dissipative case and adjusting it according to the modified Hamilton–Jacobi equation in Madelung's hydrodynamic formulation of quantum mechanics. This final form is equivalent to the approach by Hasse [53], mentioned in Sect. 4.4.1, but without the problem of the non-vanishing mean-value of the friction term. The consequence is an additional term in the phase of the wave function corresponding to a unitary transformation. The resulting Ermakov equation is formally identical to the one in the TD case [31].

This essentially shows that, in the *linearized complex* formulation, a change of the amplitude of the complex quantity (in the TD case from $\lambda \propto \alpha$ to $\tilde{\lambda} \propto \alpha e^{-\frac{\gamma}{2}t}$) can equally well be represented by an additional phase factor while the amplitude (in the TI case the probability amplitude $|\Psi|$) remains unchanged[3] As a change of the amplitude of Ψ would correspond to a non-unitary transformation, the correlation between amplitude and phase will be the subject of further investigations.

Following the deliberations of quantum mechanical problems, selected examples were discussed to show the versatility of the Riccati equation. It occurs not only in the context of TD and TI quantum mechanics but also in many other fields of physics. Consequently, the various interrelations between different aspects of systems that can be described by Riccati equations and that were developed in the first chapters of this book might be transferrable (with possible modifications) to other physical problems.

As classical physics is usually expressed in terms of real quantities, examples with real Riccati equations were considered first. A certain qualitative change could be achieved by going from purely imaginary variables solving the NL Bernoulli equation (like for the HO in the non-dissipative case) to real variables (corresponding, e.g., to a change from time t to an imaginary "time" $\frac{\hbar}{k_B T}$), i.e., changing from trigonometric to hyperbolic functions. This leads to expressions well-known from statistical thermodynamics as well as to familiar problems dealt with in NL dynamics such as those that can be described by the logistic or Verhulst equation. Connections are also made to synergetics, established by Haken [55], and to systems that show a bifurcation phenomenon (Hopf bifurcation), a first step into the regime of chaotic systems [56].

A branch of NL systems that gained increasing interest towards the end of the last century was those with soliton solution, i.e., waves travelling without changing their shape. Solitons obeying the Burgers equation and the Korteweg–de Vries equation were chosen as they are particularly well-known examples. Although these two equations look quite different at first sight, it is possible to transform both of them into the same (real) Riccati equation that can be linearized to a TISE with well-known potential and exact solutions [57]. This reduction to the same Riccati

[3] A similar situation was already mentioned when comparing the complex log NLSE [54] with the approach by Doebner and Goldin [37–40], see Sect. 4.4.3.

equation means that relations previously not known can be established between the Burgers and Korteweg–de Vries equations.

As complex quantities are already used in classical optics (mainly for practical reasons) a first example for a *complex* Riccati equation outside the quantum mechanical context is discussed in connection with the scalar wave equation of optics.[4] This showed that it is possible to reformulate this problem in terms of a *complex* Riccati equation and that the usual Eikonal approximation loses important information (particularly associated with the "*centrifugal*" term of the corresponding Ermakov equation).

Moving from classical optics, quantum systems like Bose–Einstein condensates (BECs) are then considered. The dynamics of such a condensate can be described by the Gross–Pitaevskii equation, a cubic NLSE that cannot be solved exactly in this case. However it is possible to show that it can equally well be described by a set of so-called moments that cover all the dynamical properties of the condensate and fulfil a closed set of differential equations determining their time-evolution. It turned out that this set of equations can be condensed to one equation for the condensates width where this is an Ermakov equation and is thus equivalent to a complex Riccati equation [59].

By formal comparison, Lidsey [60] was able to transfer this microscopic problem to a really macroscopic, namely, a cosmological one. In this procedure it was possible to show that, based on a Robertson–Walker metric,[5] Friedmann–Lemaître equations can be transformed equation into an Ermakov equation. Again, as mentioned earlier via equivalence with a complex Riccati equation and further aspects as corresponding creation/annihilation operators, this provides various possibilities like the construction of "*coherent states of the universe*".

In the end, the squaring of a complex number, as it occurs in the complex Riccati or Bernoulli equation, provides the key to obtaining the Pythagorean triples. A complex number is chosen where real and imaginary parts are integers, the real part being larger than the imaginary. Squaring this number produces another complex number where the real part, imaginary part and absolute value again form a right-angled triangle (as in Fig. 7.1), i.e., all three sides can be expressed in terms of integers. So, for all possible integers for real and imaginary parts of the complex number, all possible Pythagorean triples can be obtained. This implies that any theory in which changes of a complex quantity are proportional to the square of this quantity can, in principle, be "*quantized*", no matter what kind of quantity this is [62, 63].

Up to this point, mainly one-dimensional problems were considered. Certainly the problems discussed in the quantum mechanical context can easily be extended to two or three dimensions. At first sight though, this would not provide any really new qualitative aspects. A different problem, also supplying analytical solutions, is the two-dimensional motion of a system in a magnetic field. The corresponding Hamil-

[4]Recent work by Gress and Cruz [58] has shown applications of the Ermakov formalism (where time t is replaced by spatial variables) to paraxial wave optics and Hermite–Gaussian modes.

[5]Even in the newly-developed approach of pseudo-complex general relativity [61], (real) Riccati equations are ubiquitous in Chap. 4 on pseudo-complex Robertson–Walker metric.

tonian is again quadratic in its variables and the motion is in a plane perpendicular to the direction of the magnetic field. This motion and its quantum mechanical description in terms of WPs without [64] and with dissipation should also be considered but is not presented here in detail. The dissipative quantum system with Gaussian WP solutions having constant width is discussed in [65, 66]; the same problem including TD widths of the WP solutions, again obeying complex Riccati equations, is considered in [67, 68]. Looking at coherent states in this magnetic field situation showed in the dissipative case that also the Hamiltonian density for the electric and magnetic fields, $H_{el} = \frac{1}{8\pi}(\mathbf{E}^2 + \mathbf{B}^2)$, must undergo a non-canonical transformation as in the transition between the Caldirola–Kanai system and the physical one. This is necessary in order to avoid unphysical gauge-dependencies in the equations of motion [69, 70]. This aspect of classical as well as quantum electrodynamics will be the subject of further studies.

As in the case of the damped HO, also certain cases have been studied [71] for the damped motion in a magnetic field with, e.g., resonance-like conditions but further investigations are planned.

The combination of motion in a magnetic field and dissipation naturally leads to problems that are well-established experimentally but, theoretically, (at least in certain aspects) not yet completely understood, namely, the QHE and, particularly, the Fractional Quantum Hall Effect (FQHE).

In a macroscopic electric circuit there is dissipation of the electromagnetic energy of the electric current into heat due to Ohmic damping. Going to a microscopic scale and very low temperatures, it turned out that the damping of the current is not continuous but occurs in steps that can be described by integers or fractions of integers. In a more speculative way, a connection shall be made with the Pythagorean quantization and the concept of quantized action, linking this to Sommerfeld's fine structure constant.

From the continuity equation for the probability density, or its equivalent Smoluchowski equation in the dissipative case, it follows that the probability current, or the velocity field, essentially depends on the relative change in time of the WP width $\alpha(t)$, i.e., $\frac{\dot{\alpha}}{\alpha}$. Here the total velocity field in both cases obeys $v_T = \dot{\eta} + \frac{\dot{\alpha}}{\alpha}\tilde{x}$ (again in one dimension) with the classical contribution $\dot{\eta} = \langle v \rangle$.

For $\dot{\alpha} = 0$, i.e., constant WP width, there is no quantum mechanical contribution to the current. However, for $\dot{\alpha} \neq 0$, typical quantum effects like tunnelling or QHEs may occur.

In the QHE, the current $j = \varrho\, v \propto \sigma E$ is proportional to the electric field strength E and the conductivity σ where $\sigma = n\, c \left(\frac{e^2}{c\hbar}\right) = n\, c\, \alpha_{\text{Som}}$ with $n =$ integer, $c =$ velocity of light, $e =$ electric elementary charge and $\alpha_{\text{Som}} = \left(\frac{e^2}{c\hbar}\right) \approx \frac{1}{137}$ is Sommerfeld's fine structure constant (not to be confused with $\alpha(t)$, the WP width as defined in Sect. 2!).

For fixed position of the systems centre of mass, $\dot{\eta} = 0$, the conductivity or current displays the following proportionality: $j \propto n \left(\frac{e^2}{c\hbar}\right) \propto \frac{\dot{\alpha}}{\alpha}$, therefore, changes of j by integers n would correspond to integer changes of $\frac{\dot{\alpha}}{\alpha}$, a possibility discussed in connection with the Pythagorean triples [62].

Asides from the QHE, the FQHE is also observed where $j \propto \frac{n}{k}\left(\frac{e^2}{c\hbar}\right)$ with n, k both being integers. There are different interpretations of this effect in the literature and not all possible combinations of n and k (being odd, even or both) have been observed. To my knowledge, no existing theory is able to explain all observed combinations of n and k or even to predict hitherto unobserved ones. So there might still be room for some unknown in the theory of the FQHE.

One interpretation [72] given some time ago (by the author and colleagues) assumes that the FQHE is a manifestation of a phenomenon where two effects are involved, each being quantized separately. Given that it is already stated in the introduction that *action* is the quantity usually quantized in nature, it is assumed that in the QHE one of the effects is quantized in terms of $\hbar$, the other in terms of $\frac{e^2}{c}$, so the ratio of both is just Sommerfeld's fine structure constant $\alpha_{\text{Som}} = \frac{e^2}{c\hbar}$.

The action $\hbar$ is certainly well known in quantum mechanics. The action $\frac{e^2}{c}$ was already mentioned by Einstein [73], Schrödinger [74] and Eddington [75], assuming its importance but not knowing its physical relevance. Here the interpretation of the FQHE is that this is the first physical effect to be observed that shows the relevance of a second quantum of action, $\frac{e^2}{c}$.

In [33], $\frac{e^2}{c}$ was called the "*quantum of electrostatic action*". Why this name? For two electrons interacting via Coulomb's law, the electrostatic action involved is *action* = *energy* × *time* = $\frac{e^2}{r} \times t$. As the distance between the electrons, r, and their interaction time, t, can be related via the interaction velocity $v = \frac{r}{t}$, the electrostatic action can be written as *action* = $\frac{e^2}{v}$. This action becomes minimum when the velocity attains its maximum value $v = c$ = velocity of light. So, the least Coulombic action is $\frac{e^2}{c}$.

Nature, however, is usually very economical, as expressed by the various extremal principles, e.g., in mechanics, optics, etc. Why, then, should it allow the luxury of two elementary quanta of action? Of course, no definite answer shall be given at this point; but, based on the facts stated earlier, the following remarks shall be permitted. Planck's constant (divided by 2π), $\hbar$, is well known, particularly in the quantum mechanics of oscillating systems where, e.g., the energy is quantized in terms of $\hbar\omega = \hbar\dot{\varphi}$, i.e., the frequency is related to some kind of angular velocity and thus to some kind of *angular* aspect. On the other hand, it has just been shown that $\frac{e^2}{c}$ is somehow related to a physical situation where a *radial* distance, $\frac{\dot{\alpha}}{\alpha}$, plays the important role.

As the radius and circumference of a circle cannot be expressed in terms of the same units ("*quanta*") because they are related via the irrational number π, perhaps also the "*radial*" *action* and the "*angular*" *action* cannot be expressed in terms of the same fundamental units.

As seen from the above discussion of integer values for $\frac{\dot{\alpha}}{\alpha}$ and $\dot{\varphi}$ the radial aspect for integer values of $\frac{e^2}{c}$ is connected to the transcendental number e $= 2.718\ldots$

while the angular aspect, $\dot{\varphi} = \frac{1}{\alpha^2}$, is connected to $\pi = 3.141\ldots$. Could this mean that $\frac{e^2}{c}$ and $\hbar$ simply offer the prospect of quantizing in a (complex?) two-dimensional (phase?) space radial changes as well as angular ones?!

References

1. D. Schuch, Connections between Newton- and Schrödinger-type equations in the description of reversible and irreversible dynamics. Int. J. Quant. Chem., Quant. Chem. Symp. **36**(S23), 59–72 (1989)
2. D. Schuch, On the complex relations between equations describing the dynamics of wave and particle aspects. Int. J. Quant. Chem. **42**, 663–683 (1992)
3. D. Schuch, M. Moshinsky, Connection between quantum-mechanical and classical time evolution via a dynamical invariant. Phys. Rev. A **73**, 062111, 10 pp (2006)
4. H. Cruz, D. Schuch, O. Castaños, O. Rosas-Ortiz, Time-evolution of quantum systems via a complex nonlinear Riccati equation. I. Conservative systems with time-independent Hamiltonian. Ann Phys. **360**, 44–60 (2015)
5. M. Moshinsky, D. Schuch, Diffraction in time with dissipation. J. Phys. A: Math. Gen. **34**, 4217–4226 (2001)
6. M.R. Spiegel, *Schaum's Outline Series; Theory and Problems of Complex Variables* (McGraw-Hill, New York, 1964), p. 69
7. D. Schuch, On the relation between the Wigner function and an exact dynamical invariant. Phys. Lett. A **338**, 225–231 (2005)
8. D. Schuch, M. Moshinsky, Wigner distribution functions and the representation of canonical transformations in time-dependent quantum mechanics. SIGMA **4**, 054, 12 pp (2008)
9. D. Schuch, R.S. Kaushal, Some remarks on dissipative Ermakov systems and damping in Bose–Einstein condensates. J. Phys. Conf. Ser. **306**, 012032, 10 pp (2011)
10. O. Castaños, D. Schuch, O. Rosas-Ortiz, Generalized coherent states for time-dependent and nonlinear Hamiltonian operators via complex Riccati equations. J. Math. Phys. A: Math. Theor. **46**, 075304, 20 pp (2013)
11. D. Schuch, O. Castaños, O. Rosas-Ortiz, Generalized creation and annihilation operators via complex nonlinear Riccati equations. J. Phys. Conf. Ser. **442**, 012058, 10 pp (2013)
12. G. Reinisch, Nonlinear quantum mechanics. Physica A **206**, 229–252 (1994)
13. E. Madelung, Quantentheorie in hydrodynamischer Form. Z. Phys. **40**, 322 (1927)
14. D. Schuch, Riccati and Ermakov equations in time-dependent and time-independent quantum systems. SIGMA **4**, 043, 16 pp (2008). arXiv:0805.1687
15. O. Rosas-Ortiz, O. Castaños, D. Schuch, New supersymmetry-generated complex potentials with real spectra. J. Phys. A: Math. Theor. **48**, 445302, 23 pp (2015)
16. C. Lanczos, *The Variational Principles of Mechanics*, 4th edn. (Dover, New York, 1970)
17. R.P. Feynman, A.R. Hibbs, *Quantum Mechanics and Path Integrals* (McGraw-Hill, New York, 1965)
18. A.O. Caldeira, A.J. Leggett, Influence of dissipation on quantum tunneling in macroscopic systems. Phys. Rev. Lett. **46**, 211 (1981)
19. A.O. Caldeira, A.J. Leggett, Quantum tunnelling in a dissipative system. Ann. Phys. (N.Y.) **149**, 374–456 (1983); ibid. **153**, 445 (E) (1984)
20. A.O. Caldeira, A.J. Leggett, Path integral approach to quantum Brownian motion. Phys. A **121**, 587–616 (1983)
21. H. Bateman, On dissipative systems and related variational principles. Phys. Rev. **38**, 815 (1931)
22. P. Caldirola, Forze non conservative nella meccanica quantistica. Nuovo Cimento **18**, 393–400 (1941)

23. E. Kanai, On the quantization of the dissipative systems. Prog. Theor. Phys. **3**, 440–442 (1948)
24. N.A. Lemos, Canonical approach to the damped harmonic oscillator. Am. J. Phys. **47**, 857 (1979)
25. H. Gzyl, Quantization of the damped harmonic oscillator. Phys. Rev. A **27**, 2297 (1983)
26. I.A. Pedrosa, Canonical transformations and exact invariants for dissipative systems. J. Math. Phys. **28**, 2662 (1987)
27. D. Schuch, A new Lagrange–Hamilton formalism for dissipative systems. Int. J. Quant. Chem., Quant. Chem. Symp. **24**, 767–780 (1990)
28. L.H. Yu, C.P. Sun, Evolution of the wave function in a dissipative system. Phys. Rev. A **49**, 592 (1994)
29. C.P. Sun, L.H. Yu, Exact dynamics of a quantum dissipative system in a constant external field. Phys. Rev. A **51**, 1845 (1995)
30. D. Schuch, J. Guerrero, F.F. López-Ruiz, V. Aldaya, Interrelations between different canonical descriptions of dissipative systems. Phys. Scr. **90**, 045209 (2015)
31. D. Schuch, Complex Riccati equations as a link between different approaches for the description of dissipative and irreversible systems. J. Phys. Conf. Ser. **380**, 012009, 22 pp (2012)
32. M.D. Kostin, On the Schrödinger-Langevin Equation. J. Chem. Phys. **57**, 3589 (1972)
33. G. Süssmann, unpublished, quoted in Albrecht, Ref. [34] and Hasse, Ref. [53]
34. K. Albrecht, A new class of Schrödinger operators for quantized friction. Phys. Lett. B **56**, 127–129 (1975)
35. N. Gisin, A simple nonlinear dissipative quantum evolution equation. J. Phys. A **14**, 2259–2267 (1981)
36. N. Gisin, Microscopic derivation of a class of non-linear dissipative Schrödinger-like equations. Phys. A **111**, 364–370 (1982)
37. H.-D. Doebner, G.A. Goldin, On a general nonlinear Schrödinger equation admitting diffusion currents. Phys. Lett. A **162**, 397–401 (1992)
38. H.-D. Doebner, G.A. Goldin, Properties of nonlinear Schrödinger equations associated with diffeomorphism group representations. J. Phys. A **27**, 1771 (1994)
39. H.-D. Doebner, J. Tolar, in *Symmetries in Science XI*, ed. by B.J. Gruber, G. Marmo, N. Yoshinaga (Kluwer Academic Publishers, Dortrecht, 2004), pp. 209–225
40. H.-D. Doebner, G.A. Goldin, Introducing nonlinear gauge transformations in a family of nonlinear Schrödinger equations. Phys. Rev. A **54**, 3764 (1996)
41. D. Schuch, K.-M. Chung, H. Hartmann, Nonlinear Schrödinger-type field equation for the description of dissipative systems. I. Derivation of the nonlinear field equation and one-dimensional example. J. Math. Phys. **24**, 1652–1660 (1983)
42. D. Schuch, K.-M. Chung, H. Hartmann, Nonlinear Schrödinger-type field equation for the description of dissipative systems. III. Frictionally damped free motion as an example for an aperiodic motion. J. Math. Phys. **25**, 3086–3090 (1984)
43. D. Schuch, K.-M. Chung, From macroscopic irreversibility to microscopic reversibility via a nonlinear Schrödinger-type field equation. Int. J. Quant. Chem. **29**, 1561–1573 (1986)
44. E. Schrödinger, Quantisierung als Eigenwertproblem (Erste Mitteilung). Ann. Phys. **79**, 361–376 (1926)
45. D. Schuch, Nonunitary connection between explicitly time-dependent and nonlinear approaches for the description of dissipative quantum systems. Phys. Rev. A **55**, 935–940 (1997)
46. D. Schuch, Effective description of the dissipative interaction between simple model-systems and their environment. Int. J. Quant. Chem. **72**, 537–547 (1999)
47. D. Schuch, Connection between quantum mechanical and classical time evolution of certain dissipative systems via a dynamical invariant. J. Math. Phys. **48**, 122701, 19 pp, 10 pp (2007)
48. H. Cruz, D. Schuch, O. Castaños, O. Rosas-Ortiz, Time-evolution of quantum systems via a complex nonlinear Riccati equation. II. Dissipative systems. Ann. Phys. **373**, 609–630 (2016)
49. D. Schuch, Can the breaking of time-reversal symmetry remove degeneracies in quantum mechanics? Phys. Lett. A **294**, 31–36 (2002)

50. D. Schuch, New energetic and dynamic quantum effects originating from the breaking of time-reversal symmetry. J. Phys. A: Math. Gen. **35**, 8615–8629 (2002)
51. D. Schuch, Green's function for dissipative quantum systems and its relation to nonlinear evolution equations, in *Group 24—Physical Applications and Mathematical Aspects of Geometry, Groups and Algebra*, ed. by J.P. Gazeau, R. Kerner (IOP Conference Series, Bristol, 2003), pp. 741–745
52. D. Schuch, On a form of nonlinear dissipative wave mechanics valid in position- and momentum-space. Int. J. Quant. Chem., Quant. Chem. Symp. **52**(S28), 251–259 (1994)
53. R.W. Hasse, On the quantum mechanical treatment of dissipative systems. J. Math. Phys **16**, 2005 (1975)
54. D. Schuch, Linearization of nonlinear Schrödinger equations – non-unitary transformations compared to nonlinear gauge transformations, in *Quantum Theory and Symmetries*, ed. by H.-D. Doebner, J.-D. Henning, W. Lücke, V.K. Dobrev (World Scientific, Singapore, 2000), p. 152
55. H. Haken, *Synergetics – An Introduction* (Springer, Berlin, 1978)
56. H.G. Schuster, *Deterministic Chaos; An Introduction* (Physik Verlag, Weinheim, 1984), p. 112
57. D. Schuch, From nonlinear Burgers and Korteweg–de Vries soliton equations via Riccati to linear Rosen–Morse and free particle Schrödinger equations. Phys. Scr. **90**, 074050 (2015)
58. Z. Gress, S. Cruz, A note on the off-axis Gaussian beams propagation in parabolic media. J. Phys. Conf. Ser. **839**, 012024 (2017)
59. J.J. Garcia-Ripoll, V.M. Perez-Garcia, P. Torres, Extended parametric resonances in nonlinear Schrödinger systems. Phys. Rev. Lett. **83**, 1715 (1999)
60. J.E. Lidsey, Cosmic dynamics of Bose–Einstein condensates. Class. Quantum Grav. **21**, 777 (2004)
61. P.O. Hess, M. Schäfer, W. Greiner, *Pseudo-Complex General Relativity* (Springer, Heidelberg, 2016)
62. D. Schuch, Pythagorean quantization, action(s) and the arrow of time. J. Phys. Conf. Ser. **237**, 012020, 20 pp (2010)
63. D. Schuch, Is quantum theory intrinsically nonlinear? Phys. Scr. **87** 038117 (10 pp) (2013) and Highlights 2013
64. A.D. Janussis, Ein neuer Beitrag zur Theorie der Gitterelektronen im äusseren homogenen Magnetfeld. Phys. Status Solidi B **6**, 217–226 (1964)
65. D. Schuch, *Theoretische Hilfsmittel zur Beschreibung der Ionenbewegung in der Ionen-Cyclotron-Resonance-Spectrometrie* (Inaugural-Dissertation, J.W. Goethe- University Frankfurt am Main, 1982)
66. D. Schuch, K.-M. Chung, H. Hartmann, Nonlinear Schrödinger-type field equation for the description of dissipative systems. II. Frictionally damped motion in a magnetic field. Int. J. Quant. Chem. **25**, 391–410 (1984)
67. D. Schuch, in *New Challenges in Computational Quantum Chemistry*, ed. by R. Broer, P.J.C. Aerts, P.S. Bagus (University of Groningen, 1994), pp. 255–269
68. C. Wecker, *Wellenmechanische Aspekte der Dynamik und Energetik der Bewegung ineinem Magnetfeld unter spezieller Berücksichtigung dissipativer Reibungskräfte* (Diplom-Thesis, J.W. Goethe-University Frankfurt am Main, 1993)
69. D. Schuch, M. Moshinsky, Coherent states and dissipation for the motion of a charged particle in a constant magnetic field. J. Phys. A: Math. Gen. **36**, 6571–6586 (2003)
70. D. Schuch, Irreversible wave packet dynamics in a magnetic field, in *Symmetries in Science XI*, ed. by B. Gruber (Kluwer, Dordrecht, 2004), pp. 471–490
71. M. Nuñez, P.O. Hess, D. Schuch, Quantum mechanics in dissipative systems with a strong magnetic field. Phys. Rev. A **70**, 032103, 5 pp (2004)
72. R.L.P. Wadlinger, G. Hunter, L. Kostro, D. Schuch, The Quantum of Electrostatic Action and the Quantum Hall Effect. Phys. Essays **3**, 194–197 (1990)
73. A. Einstein, Zum gegenwärtigen Stand des Strahlungsproblems. Z. Phys. **6**, 185–193 (1909)
74. E. Schrödinger, Über eine bemerkenswerte Eigenschaft der Quantenbahnen eines einzelnen Elektrons. Z. Phys. **12**, 13–23 (1922)
75. A.S. Eddington, *New Pathways in Science* (Cambridge University Press, Cambridge, 1935), p. 234

Appendix A
Method of Linear and Quadratic Invariants

According to classical Hamiltonian mechanics, the integration of the equations of motion becomes trivial if one can find a canonical transformation to variables that are constants. The same applies to the corresponding quantum mechanical operators. The linear time-independent operators $\hat{X}_{op}$ and $\hat{P}_{op}$ for the parametric oscillator are defined by the transformation

$$\begin{pmatrix} \hat{X}_{op} \\ \hat{P}_{op} \end{pmatrix} = \begin{pmatrix} g_1(t) & f_1(t) \\ g_2(t) & f_2(t) \end{pmatrix} \begin{pmatrix} x_{op} \\ p_{op} \end{pmatrix} \tag{A.1}$$

with $x_{op} = x$, $p_{op} = \frac{\hbar}{i}\frac{\partial}{\partial x}$ which satisfy the commutation relation

$$\left[\hat{X}_{op}, \hat{P}_{op}\right] = i\hbar(f_2 g_1 - f_1 g_2) = i\hbar, \tag{A.2}$$

therefore,

$$f_2 g_1 - f_1 g_2 = 1 \tag{A.3}$$

is valid.

The initial conditions are chosen so that $\hat{X}_{op}(t_0 = 0) = x_{op}$ and $\hat{P}_{op}(t_0 = 0) = p_{op}$, i.e.,

$$\begin{aligned} g_1(0) &= 1,\ f_1(0) = 0, \\ g_2(0) &= 0,\ f_2(0) = 1. \end{aligned} \tag{A.4}$$

Any invariant operator $\hat{K}$ must satisfy

$$\frac{d}{dt}\hat{K} = \frac{1}{i\hbar}\left[\hat{K}, \hat{H}_{op}\right]_{-} + \frac{\partial}{\partial t}\hat{K} = 0. \tag{A.5}$$

D. Schuch, *Quantum Theory from a Nonlinear Perspective*,
Fundamental Theories of Physics 191, https://doi.org/10.1007/978-3-319-65594-9

Inserting, for $\hat{K}$, the operators $\hat{X}_{\rm op}$ and $\hat{P}_{\rm op}$, leads to

$$\dot{f}_i(t) = -g_i(t), \tag{A.6}$$
$$\dot{g}_i(t) = \omega^2 f_i(t) \tag{A.7}$$

with $i = 1, 2$ which is equivalent to the second-order differential equation

$$\ddot{f}_i + \omega^2 f_i = 0 \tag{A.8}$$

for the Hamiltonian

$$\hat{H}_{\rm op} = \frac{1}{2m} p_{\rm op}^2 + \frac{m}{2}\omega^2 x^2. \tag{A.9}$$

With these invariant operators, the most general quadratic-invariant-operator can be proposed as

$$\hat{I} = \frac{1}{2}\left[A\hat{X}_{\rm op}^2 + B\hat{P}_{\rm op}^2 + C\left(\hat{X}_{\rm op}\hat{P}_{\rm op} + \hat{P}_{\rm op}\hat{X}_{\rm op}\right)\right] \tag{A.10}$$

where A, B and C are constants still to be determined. As a quadratic invariant for the system is already known, the Ermakov invariant, the invariant (A.10) can be assumed to be proportional to the operator corresponding to the Ermakov invariant

$$\hat{I} = \frac{1}{2}\left[\left(\alpha\frac{p_{\rm op}}{m} - \dot{\alpha}x\right)^2 + \left(\frac{x}{\alpha}\right)^2\right]. \tag{A.11}$$

Taking Eq. (A.10) to be equal to (A.11) and expressing $\hat{X}_{\rm op}$ and $\hat{P}_{\rm op}$ according to (A.1) in terms of $x_{\rm op}$, $p_{\rm op}$ and $x_{\rm op}p_{\rm op} + p_{\rm op}x_{\rm op}$, one obtains from the terms proportional to $p_{\rm op}^2$

$$\alpha^2 = Af_1^2 + Bf_2^2 + 2Cf_1f_2, \tag{A.12}$$

from the ones proportional to $x_{\rm op}^2$

$$\dot{\alpha}^2 + \frac{1}{\alpha^2} = Ag_1^2 + Bg_2^2 + 2Cg_1g_2, \tag{A.13}$$

and from the ones proportional to $x_{\rm op}p_{\rm op} + p_{\rm op}x_{\rm op}$

$$-\dot{\alpha}\alpha = Af_1g_1 + Bf_2g_2 + C(f_1g_2 + f_2g_1). \tag{A.14}$$

Expressing $\dot{\alpha}^2$ using (A.14) and inserting it into (A.13) finally allows the writing of the constant C in terms of the other two constants A and B as

$$|C| = \sqrt{AB - 1}. \tag{A.15}$$

Using the initial conditions for $f_1(t)$ and $f_2(t)$ and requiring that $\alpha_0 = \alpha(t_0 = 0)$ and $\dot{\alpha}_0 = \dot{\alpha}(t_0 = 0)$, the three constants take the form

$$A = \frac{m}{\hbar}\left(\dot{\alpha}_0^2 + \frac{1}{\alpha_0^2}\right), \quad B = \frac{1}{m\hbar}\alpha_0^2, \quad |C| = \frac{1}{\hbar}\dot{\alpha}_0\alpha_0. \tag{A.16}$$

So the WP width, or amplitude of the complex quantity $\lambda = \alpha\, \mathrm{e}^{i\varphi}$ can be obtained by taking the square-root of (A.12) to yield

$$\alpha(t) = \pm\sqrt{\left(\dot{\alpha}_0^2 + \frac{1}{\alpha_0^2}\right)f_1^2(t) + \alpha_0^2 f_2^2(t) \mp \dot{\alpha}_0\alpha_0 f_1(t) f_2(t)}. \tag{A.17}$$

As $\alpha(t)$ is related to the WP width, only the positive sign of the square-root is physically reasonable. That means, knowing two linear-independent solutions of the classical Newtonian equation (A.8), the quantum mechanical uncertainty $\alpha(t)$ can be determined if the initial uncertainty α_0 and the initial time-derivative of it, $\dot{\alpha}_0$, are known.

Appendix B
Position and Momentum Uncertainties in the Dissipative Case

The uncertainties of position and momentum, $\langle\tilde{x}^2\rangle_{\rm NL}(t)$ and $\langle\tilde{p}^2\rangle_{\rm NL}(t)$, and their correlation, $\langle[\tilde{x},\tilde{p}]_+\rangle_{\rm NL}(t)$, can be expressed in terms of $\alpha_{\rm NL}(t)$ and $\dot{\alpha}_{\rm NL}(t)$ as shown in Eqs. (5.17–5.19) where, in the last case, actually $+\dot{\alpha}\alpha$ and $-\dot{\alpha}\alpha$ must be considered[1] if probability currents (see below) travelling in opposite directions are also to be included. So, the expressions to be determined are

$$\langle\tilde{x}^2\rangle_{\rm NL} = \frac{\hbar}{2m}\,\alpha_{\rm NL}^2 \tag{B.1}$$

$$\langle\tilde{p}^2\rangle_{\rm NL} = \frac{\hbar m}{2}\left[\left(\dot{\alpha}_{\rm NL}^2 + \frac{1}{\alpha_{\rm NL}^2}\right) \mp \gamma\dot{\alpha}_{\rm NL}\alpha_{\rm NL} + \frac{\gamma^2}{4}\alpha_{\rm NL}^2\right] \tag{B.2}$$

$$\langle[\tilde{x},\tilde{p}]_+\rangle_{\rm NL} = \hbar\left[\pm\dot{\alpha}_{\rm NL}\,\alpha_{\rm NL} - \frac{\gamma}{2}\alpha_{\rm NL}^2\right]. \tag{B.3}$$

This can be traced back to the determination of three quantities corresponding to Eqs. (A.12–A.14) in the non-dissipative case,[2]

$$\alpha_\pm^2(t) = \left[\left(\dot{\alpha}_0^2 + \frac{1}{\alpha_0^2}\right)\xi_1^2(t) + \alpha_0^2\,\xi_2^2(t) \mp 2\dot{\alpha}_0\alpha_0\,\xi_1(t)\xi_2(t)\right], \tag{B.4}$$

$$\left(\dot{\alpha}^2 + \frac{1}{\alpha^2}\right)_\pm(t) = \left[\left(\dot{\alpha}_0^2 + \frac{1}{\alpha_0^2}\right)g_1^2 + \alpha_0^2\,g_2^2 \mp 2\dot{\alpha}_0\alpha_0\,g_1g_2\right], \tag{B.5}$$

$$(\dot{\alpha}\alpha)_\pm(t) = \left[-\left(\dot{\alpha}_0^2 + \frac{1}{\alpha_0^2}\right)\xi_1g_1 - \alpha_0^2\,\xi_2g_2 \pm \dot{\alpha}_0\alpha_0\,(\xi_1g_2 + \xi_2g_1)\right]. \tag{B.6}$$

[1] The same also applies already in the non-dissipative case but has there no really interesting consequences, particularly for the initial condition $\dot{\alpha}_0 = \langle[\tilde{x},\tilde{p}]_+\rangle_0 = 0$.

[2] In the following, the subscript "NL" is dropped as $\alpha_{\rm NL}$ and $\dot{\alpha}_{\rm NL}$ cannot be confused with $\alpha_{\rm L}$ and $\dot{\alpha}_{\rm L}$ as the latter do not occur in Appendix B.

D. Schuch, *Quantum Theory from a Nonlinear Perspective*,
Fundamental Theories of Physics 191, https://doi.org/10.1007/978-3-319-65594-9

The subscript "±" indicates that, for $\dot{\alpha}_0 \neq 0$, two different signs must be considered for the third terms in each expression.

In comparison with the non-dissipative case, the major difference now is that $f_1(t)$ is replaced by $\xi_1(t)$ with

$$\xi_1(t) = -\frac{1}{v_0}\eta(t)\,e^{\frac{\gamma}{2}t} \tag{B.7}$$

(also like the second linear independent solution $\xi_2(t)$) fulfilling the Newtonian equation

$$\ddot{\xi}_i + \left(\omega_0^2 - \frac{\gamma^2}{4}\right)\xi_i = \ddot{\xi}_i + \Phi^2\xi_i = 0, \quad i = 1, 2 \tag{B.8}$$

with the initial conditions

$$\xi_1(t_0) = 0, \quad \xi_2(t_0) = 1, \quad g_1(t_0) = 1, \quad g_2(t_0) = 0, \tag{B.9}$$

where, like in the non-dissipative case, $g_i = -\dot{\xi}_i$ is valid. The choice $\xi_2(t) = g_1(t)$ is in agreement with the initial conditions and provides a solution of (B.8) that is linearly independent of $\xi_1(t)$.

For the four damped cases under consideration, the free motion ($V = 0$) and the HO with $\omega_0 = \frac{\gamma}{2}$ and $\omega_0 \gtrless \frac{\gamma}{2}$, the expressions for $\xi_1(t)$, $g_1(t)$, $\eta(t)$ and Φ^2 are given in Table B.1 (due to $\xi_1(0) = 0$, it follows that $\eta(0) = \eta_0 = 0$).

The WP width, or $\alpha(t)$, the solution of the Ermakov equation (5.15), can be obtained from Eq. (B.4) by simply taking its square-root. In the case of the WP width, only the positive sign of the two possible ones of the square-root is physically reasonable, still permitting two possible values due to the plus-minus-sign of the term proportional to $\dot{\alpha}_0\alpha_0$. Considering the time-derivative of $\alpha(t)$, which actually

Table B.1 In this table the values of $-\xi_1(t)$, $g_1(t)$ and $\eta(t)$ are given for the initial conditions (B.9) and the four cases under consideration. Further, the values that Φ^2, appearing in Eq. (B.8), attains in these cases is indicated where $\Theta^2 = \frac{\gamma^2}{4} - \omega_0^2 = -\Omega^2 > 0$

	$V = 0$	$\omega_0 = \frac{\gamma}{2}$	$\omega_0 > \frac{\gamma}{2}$	$\omega_0 < \frac{\gamma}{2}$
$-\xi_1$	$\frac{1}{\frac{\gamma}{2}}\sinh\frac{\gamma}{2}t$	t	$\frac{1}{\Omega}\sin\Omega t$	$\frac{1}{\theta}\sinh\theta t$
g_1	$\cosh\frac{\gamma}{2}t$	1	$\cos\Omega t$	$\cosh\theta t$
Φ^2	$-\frac{\gamma^2}{4}$	0	$\Omega^2 = \omega_0^2 - \frac{\gamma^2}{4}$	$-\theta^2 = \left(\frac{\gamma^2}{4} - \omega_0^2\right)$
$\eta(t)$	$\frac{v_0}{\gamma}(1 - e^{-\gamma t})$	$v_0 e^{-\frac{\gamma}{2}t}t$	$\frac{v_0}{\Omega}e^{-\frac{\gamma}{2}t}\sin\Omega t$	$\frac{v_0}{\Omega}e^{-\frac{\gamma}{2}t}\sinh\theta t$

enters the expression for the tunnelling currents (see below) in the Smoluchowski equation (5.84), both signs for $\dot{\alpha}(t)$ are possible, corresponding to currents flowing in opposite directions. Therefore, for $\dot{\alpha}(t)$, or $\dot{\alpha}\alpha$, that enter $\langle[\tilde{x},\tilde{p}]_+\rangle$ and $\langle\tilde{p}^2\rangle$, four possibilities exist for $\dot{\alpha}_0 \neq 0$ (but only two for $\dot{\alpha}_0 = 0$ as α_+ and α_- then coincide).

As is obvious from Eqs. (B.4–B.6), the actual expressions depend, apart from $\xi_i(t)$ and $g_i(t)$, also strongly on the initial uncertainties α_0 and $\dot{\alpha}_0$ (particularly on the latter).

Different choices for α_0 in our case simply correspond to different initial widths of the WPs. The possible choices for $\dot{\alpha}_0$ can be related to different physical situations, particularly concerning the probability currents and the uncertainty product.

So is the choice $\dot{\alpha}_0 = 0$ related to a vanishing initial contribution of the total probability current to tunnelling, i.e., for

$$j_{\mathrm{T}} = j_- + j_{\mathrm{D}} = \varrho\,(v_- + v_{\mathrm{D}}) = \varrho\left(\left[\dot{\eta} + \left(\frac{\dot{\alpha}}{\alpha} - \frac{\gamma}{2}\right)\tilde{x}\right] + \frac{\gamma}{2}\tilde{x}\right) = \varrho\left(\dot{\eta} + \frac{\dot{\alpha}}{\alpha}\tilde{x}\right) \tag{B.10}$$

the tunnelling contribution

$$\tilde{j}_{\mathrm{tun}} = j_{\mathrm{T}} - \varrho\dot{\eta} = \varrho\left(\frac{\dot{\alpha}}{\alpha}\tilde{x}\right) \tag{B.11}$$

vanishes for $t = 0$.

However, this consequently means that the initial WP is not a minimum uncertainty WP as this is only obtained for $\langle[\tilde{x},\tilde{p}]_+\rangle_0 = 0$ which can be reached, according to Eq. (B.3), for $\dot{\alpha}_0 = \frac{\gamma}{2}\alpha_0$. In this case, also the different signs of the third terms in Eqs. (B.4–B.6) matter.

Taking all of this into account, it would lead to 72 (in part bulky) expressions. However, due to structural similarities, these can be reduced to a smaller number of compact relations that essentially depend on $\xi_1(t)$ and its (negative) time-derivative g_1. By means of Table B.1, explicit expressions for all cases under consideration can be obtained. (The left subscripts at the brackets correspond to the positive or negative sing of $\dot{\alpha}\alpha$ and $\beta_0 = \frac{1}{\alpha_0^2}$ is used like in the non-dissipative case.)

Position uncertainties:

$\dot{\alpha}_0 = 0$:

$$\langle\tilde{x}^2\rangle = \frac{\hbar}{2m}\alpha_0^2\left[\beta_0^2\xi_1^2 + g_1^2\right] \tag{B.12}$$

$\dot{\alpha}_0 = \frac{\gamma}{2}$:

$$\langle\tilde{x}^2\rangle_\pm = \frac{\hbar}{2m}\alpha_0^2\left[\beta_0^2\xi_1^2 + \left(g_1 \mp \frac{\gamma}{2}\xi_1\right)^2\right] \tag{B.13}$$

Momentum uncertainties:
$\dot{\alpha}_0 = 0$:

$$(+\dot{\alpha}\alpha):\ {}_{+}\langle\tilde{p}^2\rangle = \frac{m\hbar}{2}\alpha_0^2\left[\beta_0^2\left(g_1+\frac{\gamma}{2}\xi_1\right)^2+\left(\Phi^2\xi_1-\frac{\gamma}{2}g_1\right)^2\right] \quad (B.14)$$

$$(-\dot{\alpha}\alpha):\ {}_{-}\langle\tilde{p}^2\rangle = \frac{m\hbar}{2}\alpha_0^2\left[\beta_0^2\left(g_1-\frac{\gamma}{2}\xi_1\right)^2+\left(\Phi^2\xi_1+\frac{\gamma}{2}g_1\right)^2\right] \quad (B.15)$$

$\dot{\alpha}_0 = \frac{\gamma}{2}\alpha_0$:

$$(+\dot{\alpha}\alpha):\ {}_{+}\langle\tilde{p}^2\rangle_{\pm} = \frac{m\hbar}{2}\alpha_0^2\left[\beta_0^2\left(g_1+\frac{\gamma}{2}\xi_1\right)^2+\left\{\left(\Phi^2\pm\frac{\gamma^2}{4}\right)\xi_1-\frac{\gamma}{2}(1\mp 1)g_1\right\}^2\right] \quad (B.16)$$

$$(-\dot{\alpha}\alpha):\ {}_{-}\langle\tilde{p}^2\rangle_{\pm} = \frac{m\hbar}{2}\alpha_0^2\left[\beta_0^2\left(g_1-\frac{\gamma}{2}\xi_1\right)^2+\left\{\left(\Phi^2\mp\frac{\gamma^2}{4}\right)\xi_1+\frac{\gamma}{2}(1\pm 1)g_1\right\}^2\right] \quad (B.17)$$

Position-momentum uncertainties correlation:
$\dot{\alpha}_0 = 0$:

$$(+\dot{\alpha}\alpha):\quad {}_{+}\langle[\tilde{x},\tilde{p}]_+\rangle = \hbar\alpha_0^2\left[-\beta_0^2\xi_1\left(g_1+\frac{\gamma}{2}\xi_1\right)+\left(\Phi^2\xi_1-\frac{\gamma}{2}g_1\right)g_1\right] \quad (B.18)$$

$$(-\dot{\alpha}\alpha):\quad {}_{-}\langle[\tilde{x},\tilde{p}]_+\rangle = \hbar\alpha_0^2\left[\beta_0^2\xi_1\left(g_1-\frac{\gamma}{2}\xi_1\right)-\left(\Phi^2\xi_1+\frac{\gamma}{2}g_1\right)g_1\right] \quad (B.19)$$

$\dot{\alpha}_0 = \frac{\gamma}{2}\alpha_0$:

$$\begin{aligned}(+\dot{\alpha}\alpha):\quad {}_{+}\langle[\tilde{x},\tilde{p}]_+\rangle_{\pm} = \hbar\alpha_0^2\Big[&-\beta_0^2\xi_1\left(g_1+\frac{\gamma}{2}\xi_1\right)\\ &+\left\{\left(\Phi^2\pm\frac{\gamma^2}{4}\right)\xi_1-\frac{\gamma}{2}(1\mp 1)g_1\right\}\left(g_1\mp\frac{\gamma}{2}\xi_1\right)\Big]\end{aligned} \quad (B.20)$$

$$\begin{aligned}(-\dot{\alpha}\alpha):\quad {}_{-}\langle[\tilde{x},\tilde{p}]_+\rangle_{\pm} = \hbar\alpha_0^2\Big[&\beta_0^2\xi_1\left(g_1-\frac{\gamma}{2}\xi_1\right)\\ &-\left\{\left(\Phi^2\mp\frac{\gamma^2}{4}\right)\xi_1+\frac{\gamma}{2}(1\pm 1)g_1\right\}\left(g_1\mp\frac{\gamma}{2}\xi_1\right)\Big]\end{aligned} \quad (B.21)$$

Checking that these uncertainties fulfil the Schrödinger–Robertson relation

$$\langle\tilde{x}^2\rangle\langle\tilde{p}^2\rangle-\left(\frac{1}{2}\langle[\tilde{x},\tilde{p}]_+\rangle\right)^2=\frac{\hbar^2}{4} \quad (B.22)$$

shows that this is given providing the condition

$$g_1^2 + \Phi^2 \xi_1^2 = \frac{1}{v_0^2}\left(\dot{\eta}^2 + \gamma\eta\dot{\eta} + \omega_0^2\eta^2\right) e^{\gamma t} = 1 \tag{B.23}$$

holds; which is the case if $\eta(t)$ obeys the Newtonian equation (4.41) including the friction force. For the chosen initial condition $\eta_0 = 0$, Eq. (B.23) is also in agreement with the constant of motion (4.48), discussed in connection with the expanding coordinate system.

From the explicit expressions for the uncertainties, i.e., after inserting $\xi_1(t)$ and $g_1(t)$ from Table B.1, one discovers a certain symmetry under time reversal.

For $\dot{\alpha}_0 = 0$, the change from $+\dot{\alpha}\alpha$ to $-\dot{\alpha}\alpha$ leads to the time-reversed expression, i.e., ${}_+\langle\ldots\rangle(t) = {}_-\langle\ldots\rangle(-t)$.

For $\dot{\alpha}_0 = \frac{\gamma}{2}\alpha_0$, two different expressions are possible for the terms given in (B.4–B.6), indicated by the right subscript of the corresponding uncertainties. Changing from + to −, i.e., $\langle\ldots\rangle_+ \to \langle\ldots\rangle_-$, leads to additional terms for the momentum uncertainties $\langle\tilde{p}^2\rangle_-$ and the position-momentum correlations $\langle[\tilde{x},\tilde{p}]_+\rangle_-$. Changing the sign of $\dot{\alpha}\alpha$ simultaneously then leads to the same, but time-reversed, expression, i.e., ${}_+\langle\ldots\rangle_+(t) = {}_-\langle\ldots\rangle_-(-t)$ and ${}_-\langle\ldots\rangle_+(t) = {}_+\langle\ldots\rangle_-(-t)$. So, in general, half of the expressions are just the time-reversed version of the other half.

With the help of the initial conditions (B.9), the initial uncertainty products $\langle\tilde{x}^2\rangle(0)\langle\tilde{p}^2\rangle(0) = \frac{\hbar^2}{4} + \left(\frac{1}{2}\langle[\tilde{x},\tilde{p}]_+\rangle(0)\right)^2 \geq \frac{\hbar^2}{4}$ can be determined straightforwardly where only the equal-sign corresponds to a minimum uncertainty WP.

For $\dot{\alpha}_0 = 0$, one obtains in all possible cases

$$\langle\tilde{x}^2\rangle(0)\langle\tilde{p}^2\rangle(0) = \frac{\hbar^2}{4}\left[1 + \left(\frac{\gamma}{2}\alpha_0\right)^2\right] > \frac{\hbar^2}{4}, \tag{B.24}$$

which is expected as $\dot{\alpha}_0 = 0$ is incompatible with $\langle[\tilde{x},\tilde{p}]_+\rangle(0) = 0$. But also for $\dot{\alpha}_0 = \frac{\gamma}{2}\alpha_0$, which is compatible with this requirement, a minimum uncertainty product is not always obtained. So the combinations where the two subscripts of the uncertainties are different

$${}_+\langle\tilde{x}^2\rangle_-(0)\,{}_+\langle\tilde{p}^2\rangle_-(0) = {}_-\langle\tilde{x}^2\rangle_+(0)\,{}_-\langle\tilde{p}^2\rangle_+(0) = \frac{\hbar^2}{4}\left[1 + (\gamma\alpha_0)^2\right] > \frac{\hbar^2}{4}, \tag{B.25}$$

leads to a larger value than $\frac{\hbar^2}{4}$. Only when the subscripts are the same,

$${}_+\langle\tilde{x}^2\rangle_+(0)\,{}_+\langle\tilde{p}^2\rangle_+(0) = {}_-\langle\tilde{x}^2\rangle_-(0)\,{}_-\langle\tilde{p}^2\rangle_-(0) = \frac{\hbar^2}{4}, \tag{B.26}$$

the minimum product is obtained.

Also the contributions to the tunnelling currents, originating from $\frac{\dot{\alpha}}{\alpha}(t)$, can be expressed completely with the help of ξ_1, g_1 and Φ.

In particular, for $\dot{\alpha}_0 = 0$ one obtains

$$\left(\frac{\dot{\alpha}}{\alpha}\right)(t) = \frac{\left(\Phi^2 - \beta_0^2\right)}{\beta_0 \frac{\xi_1}{g_1} + \frac{g_1}{\xi_1}} = \frac{\left(\Phi^2 - \beta_0^2\right)\xi_1 g_1}{\beta_0^2 \xi_1^2 + g_1^2} \tag{B.27}$$

and for $\dot{\alpha}_0 = \frac{\gamma}{2}\alpha_0$

$$\left(\frac{\dot{\alpha}}{\alpha}\right)_{\pm}(t) = \frac{\left(\Phi^2 \xi_1 \pm \frac{\gamma}{2} g_1\right)\left(g_1 \mp \frac{\gamma}{2}\xi_1\right) - \beta_0^2 \xi_1 g_1}{\beta_0^2 \xi_1^2 + \left(g_1 \mp \frac{\gamma}{2}\xi_1\right)^2}. \tag{B.28}$$

In the latter case, the terms proportional to β_0^2 remain unchanged, like for $\dot{\alpha}_0 = 0$, but g_1 and $\Phi^2\xi_1$ change according to $g_1 \to g_1 \mp \frac{\gamma}{2}\xi_1$ and $\Phi^2\xi_1 \to \Phi^2\xi_1 \pm \frac{\gamma}{2} g_1$.

The explicit TD expressions are given below.

For $\dot{\alpha}_0 = 0$ one obtains:

$$V = 0: \quad \frac{\dot{\alpha}}{\alpha} = \frac{\left(\beta_0^2 + \frac{\gamma^2}{4}\right)\frac{\sinh\frac{\gamma}{2}t}{\frac{\gamma}{2}}\cosh\frac{\gamma}{2}t}{\beta_0^2\left(\frac{\sinh\frac{\gamma}{2}t}{\frac{\gamma}{2}}\right)^2 + \cosh^2\frac{\gamma}{2}t} \tag{B.29}$$

$$\omega_0 = \frac{\gamma}{2}: \quad \frac{\dot{\alpha}}{\alpha} = \frac{\beta_0^2 t}{\beta_0^2 t^2 + 1} \tag{B.30}$$

$$\omega_0 > \frac{\gamma}{2}: \quad \frac{\dot{\alpha}}{\alpha} = \frac{\left(\beta_0^2 - \Omega^2\right)\frac{\sin\Omega t}{\Omega}\cos\Omega t}{\beta_0^2\left(\frac{\sin\Omega t}{\Omega}\right)^2 + \cos^2\Omega t} \tag{B.31}$$

$$\omega_0 < \frac{\gamma}{2}: \quad \frac{\dot{\alpha}}{\alpha} = \frac{\left(\beta_0^2 + \Theta^2\right)\frac{\sinh\Theta t}{\Theta}\cosh\Theta t}{\beta_0^2\left(\frac{\sinh\Theta t}{\Theta}\right)^2 + \cosh^2\Theta t}. \tag{B.32}$$

For $\dot{\alpha}_0 = \frac{\gamma}{2}\alpha_0$ one obtains:

$$V = 0: \quad \left(\frac{\dot{\alpha}}{\alpha}\right)_\pm = \frac{\beta_0^2 + \frac{\sinh\frac{\gamma}{2}t}{\frac{\gamma}{2}} \cosh\frac{\gamma}{2}t \pm \frac{\gamma}{2}\mathrm{e}^{\pm\gamma t}}{\beta_0^2 \left(\frac{\sinh\frac{\gamma}{2}t}{\frac{\gamma}{2}}\right)^2 + \mathrm{e}^{\pm\gamma t}}. \tag{B.33}$$

$$\omega_0 = \frac{\gamma}{2}: \left(\frac{\dot{\alpha}}{\alpha}\right)_\pm = \frac{\beta_0^2 t \pm \frac{\gamma}{2}\left(1 \pm \frac{\gamma}{2}t\right)}{\beta_0^2 t^2 + \left(1 \pm \frac{\gamma}{2}t\right)^2} \tag{B.34}$$

$$\omega_0 > \frac{\gamma}{2}: \left(\frac{\dot{\alpha}}{\alpha}\right)_\pm = \frac{\beta_0^2 \frac{\sin\Omega t}{\Omega}\cos\Omega t \pm \left(\frac{\gamma}{2}\cos\Omega t \mp \Omega\sin\Omega t\right)\left(\cos\Omega t \pm \frac{\gamma}{2}\frac{\sin\Omega t}{\Omega}\right)}{\beta_0^2\left(\frac{\sin\Omega t}{\Omega}\right)^2 + \left(\cos\Omega t \pm \frac{\gamma}{2}\frac{\sin\Omega t}{\Omega}\right)^2} \tag{B.35}$$

$$\omega_0 < \frac{\gamma}{2}: \left(\frac{\dot{\alpha}}{\alpha}\right)_\pm = \frac{\beta_0^2 \frac{\sinh\Theta t}{\Theta}\cosh\Theta t \pm \left(\frac{\gamma}{2}\cosh\Theta t \pm \Theta\sinh\Theta t\right)\left(\cosh\Theta t \pm \frac{\gamma}{2}\frac{\sinh\Theta t}{\Theta}\right)}{\beta_0^2\left(\frac{\sinh\Theta t}{\Theta}\right)^2 + \left(\cosh\Theta t \pm \frac{\gamma}{2}\frac{\sinh\Theta t}{\Theta}\right)^2}. \tag{B.36}$$

Appendix C
Classical Lagrange–Hamilton Formalism in Expanding Coordinates

In Sect. 4.3.2 it has been shown that a dissipative harmonic oscillator (i.e., $V = \frac{m}{2}\omega^2 x^2$) with a linear velocity-dependent friction force can be described either on the canonical level in the usual formalism or on the physical level, with a modification of this formalism. The time-dependence of any mechanical property, either in canonical or physical variables, can be expressed equivalently in either of these two formalisms. This is demonstrated in the following for the *physical* position and momentum and for the (non conserved) energy $E = T + V = E(t)$.

(1) Determination of the time-dependence of mechanical quantities with the help of the **canonical** relation

$$\begin{aligned} \frac{d}{dt}\hat{F}(\hat{Q},\hat{P},t) &= -\left(\frac{\partial \hat{H}_{exp}}{\partial \hat{Q}}\frac{\partial \hat{F}}{\partial \hat{P}} - \frac{\partial \hat{H}_{exp}}{\partial \hat{P}}\frac{\partial \hat{F}}{\partial \hat{Q}}\right) + \frac{\partial}{\partial t}\hat{F} \\ &= -\{\hat{H}_{exp},\hat{F}\}_{(Q,P)-} + \frac{\partial}{\partial t}\hat{F} \end{aligned} \tag{C.1}$$

with

$$\hat{H}_{exp}(\hat{Q},\hat{P}) = \frac{1}{2m}\hat{P}^2 + \frac{m}{2}\left(\omega^2 - \frac{\gamma^2}{4}\right)\hat{Q}^2. \tag{C.2}$$

(a) $\hat{F} = x = \hat{Q}\,\mathrm{e}^{-\frac{\gamma}{2}t} = \hat{F}(\hat{Q},t)$:

$$\frac{\partial}{\partial t}\hat{F} = -\frac{\gamma}{2}\hat{Q}\,\mathrm{e}^{-\frac{\gamma}{2}t} = -\frac{\gamma}{2}x, \tag{C.3}$$

$$\begin{aligned} \frac{d}{dt}x &= \frac{\partial \hat{H}_{exp}}{\partial \hat{P}}\frac{\partial\left(\hat{Q}e^{-\frac{\gamma}{2}t}\right)}{\partial \hat{Q}} + \frac{\partial}{\partial t}\left(\hat{Q}e^{-\frac{\gamma}{2}t}\right) \\ &= \frac{\partial \hat{H}_{exp}}{\partial \hat{P}}\,\mathrm{e}^{-\frac{\gamma}{2}t} - \frac{\gamma}{2}\hat{Q}\,\mathrm{e}^{-\frac{\gamma}{2}t} = \left(\frac{1}{m}\hat{P} - \frac{\gamma}{2}\hat{Q}\right)\mathrm{e}^{-\frac{\gamma}{2}t} = \dot{x}. \end{aligned} \tag{C.4}$$

D. Schuch, *Quantum Theory from a Nonlinear Perspective*,
Fundamental Theories of Physics 191, https://doi.org/10.1007/978-3-319-65594-9

(b) $\hat{F} = p = m\dot{x} = \left(\hat{P} - m\frac{\gamma}{2}\hat{Q}\right)\, \mathrm{e}^{-\frac{\gamma}{2}t} = \hat{F}(\hat{Q}, \hat{P}, t):$

$$\frac{\partial}{\partial t}\hat{F} = -m\frac{\gamma}{2}\left(\dot{\hat{Q}} - \frac{\gamma}{2}\hat{Q}\right)\, \mathrm{e}^{-\frac{\gamma}{2}t} = -m\frac{\gamma}{2}\dot{x} = -\frac{\gamma}{2}p, \tag{C.5}$$

$$\begin{aligned} \frac{d}{dt}p &= -\left[\frac{\partial \hat{H}_{exp}}{\partial \hat{Q}} + m\frac{\gamma}{2}\frac{\partial \hat{H}_{exp}}{\partial \hat{P}}\right]\, \mathrm{e}^{-\frac{\gamma}{2}t} - m\frac{\gamma}{2}\left(\dot{\hat{Q}} - \frac{\gamma}{2}\hat{Q}\right)\, \mathrm{e}^{-\frac{\gamma}{2}t} \\ &= -\left[\dot{\hat{P}} + m\frac{\gamma}{2}\dot{\hat{Q}} + \frac{\gamma}{2}\hat{P} - m\frac{\gamma^2}{4}\hat{Q}\right]\, \mathrm{e}^{-\frac{\gamma}{2}t} \\ &= -m\left(\omega^2 x + \gamma\dot{x}\right) = -m\omega^2 x - \gamma p. \end{aligned} \tag{C.6}$$

(c) $\hat{F} = T + V = E(t) = \left(\frac{1}{2m}\hat{P}^2 + \frac{\gamma}{2}\hat{Q}\hat{P} + \frac{m}{2}\left(\omega^2 + \frac{\gamma^2}{4}\right)\hat{Q}^2\right)\mathrm{e}^{-\gamma t} = \hat{F}(\hat{Q}, \hat{P}, t):$

$$\frac{\partial}{\partial t}\hat{F} = -\gamma(T + V) = -\gamma E(t), \tag{C.7}$$

$$\begin{aligned} \frac{d}{dt}(T + V) &= \left[\dot{\hat{P}}\left(\frac{1}{m}\hat{P} - \frac{\gamma}{2}\hat{Q}\right) + \dot{\hat{Q}}\left(\frac{\gamma}{2}\hat{P} - m\left(\omega^2 + \frac{\gamma^2}{4}\right)\hat{Q}^2\right]\mathrm{e}^{-\gamma t} - \gamma E(t) \\ &= -\gamma\left[\frac{1}{m}\hat{P}^2 - \gamma\hat{Q}\hat{P} + m\frac{\gamma^2}{4}\hat{Q}^2\right]\mathrm{e}^{-\gamma t} = -2\gamma T. \end{aligned} \tag{C.8}$$

(2) Determination of the time-dependence of mechanical quantities with the help of the modified formalism on the **physical** level, i.e.,

$$\begin{aligned} \frac{d}{dt}F(x, p, t) &= -\{H, F\}_{(x,p)-} + \left\{F, \frac{\gamma}{2}px\right\}_{(x,p)-} - \left\{F, \frac{\gamma}{2}px\right\}_{(x,p)+} + \frac{\partial}{\partial t}F \\ &= -\{(T + V), F\}_{(x,p)-} - \gamma p\frac{\partial}{\partial p}F + \frac{\partial}{\partial t}F \end{aligned} \tag{C.9}$$

with

$$H = T + V = \frac{1}{2m}p^2 + \frac{m}{2}\omega^2 x^2. \tag{C.10}$$

(a) $F = x:$

$$\frac{\partial}{\partial t}F = 0, \ \frac{\partial}{\partial p}F = 0, \tag{C.11}$$

$$\frac{d}{dt}x = -\{(T + V), x\}_{(x,p)-} = \frac{\partial T}{\partial p} = \frac{\partial}{\partial p}\left(\frac{p^2}{2m}\right) = \frac{p}{m}. \tag{C.12}$$

(b) $F = p$:

$$\frac{\partial}{\partial t}F = 0, \ \frac{\partial}{\partial p}F = 0, \tag{C.13}$$

$$\frac{d}{dt}p = -\{(T+V), p\}_{(x,p)-} - \gamma p\frac{\partial p}{\partial p} = -\frac{\partial V}{\partial x} - \gamma p. \tag{C.14}$$

(c) $F = T + V = E(t)$:

$$\frac{\partial}{\partial t}F = 0, \tag{C.15}$$

$$\frac{d}{dt}(T+V) = -\{H, H\}_{(x,p)-} - \gamma p\frac{\partial T}{\partial p} = -\gamma\frac{p^2}{m} = -2\gamma T. \tag{C.16}$$

Appendix D
On the Connection Between the Bateman Hamiltonian and the Hamiltonian in Expanding Coordinates

In Sect. 4.5 it has been shown that the connection between the Bateman approach and the effective canonical one using expanding coordinates is not unique. This is so because there are three parameters involved but only two equations to determine them. Three choices for the third parameter and their consequences are discussed in detail here.

D.1 The Case $c = 0$

For this choice of c it follows that $a = \frac{m}{2}$ and $b = m\,\frac{\gamma}{4}$. In the ensuing discussions, for the Bateman system $\hat{x} = x = \hat{Q}\,\mathrm{e}^{-\frac{\gamma}{2}t}$ and $\hat{p}_y = m\,(\dot{x} + \frac{\gamma}{2}x) = \hat{P}\,\mathrm{e}^{-\frac{\gamma}{2}t}$ are always valid, only $\hat{y}$ and $\hat{p}_x$ expressed in terms of of x and $\dot{x}$ are changing. Therefore, $\hat{y}$ and $\hat{p}_x$ are supplied with a (second) subscript indicating which parameter is set equal to zero.

One obtains in this case

$$\hat{y}_c = \frac{1}{2}x\,\mathrm{e}^{\gamma t} = \frac{1}{2}\,\hat{Q}\,\mathrm{e}^{\frac{\gamma}{2}t} \tag{D.1}$$

$$\hat{p}_{x,c} = \frac{m}{2}\left(\dot{x} + \frac{\gamma}{2}\,x\right)\,\mathrm{e}^{\gamma t} = \frac{1}{2}\,\hat{P}\,\mathrm{e}^{\frac{\gamma}{2}t}. \tag{D.2}$$

Inserting this into $\hat{H}_{\mathrm{B}}$ (Eq. (4.19)) turns it into

$$\hat{H}_{\mathrm{B},c} = \frac{1}{m}\,\hat{p}_{x,c}\,\hat{p}_y + m\left(\omega^2 - \frac{\gamma^2}{4}\right)\hat{x}\,\hat{y}_c = \hat{H}_{\Omega} \tag{D.3}$$

as

$$\hat{D} = \frac{\gamma}{2}\,(\hat{y}_c\,\hat{p}_y - \hat{x}\,\hat{p}_{x,c}) = 0. \tag{D.4}$$

D. Schuch, *Quantum Theory from a Nonlinear Perspective*,
Fundamental Theories of Physics 191, https://doi.org/10.1007/978-3-319-65594-9

$\hat{H}_{\mathrm{B},c}$ in (D.3) when expressed in terms of x and $\dot{x}$ is identical to $\hat{H}_{exp}$ as given in (4.51). However, $\hat{H}_{\mathrm{B},c}$ is no longer a Hamiltonian that provides the correct equations of motion. The reason is that the constraints contain an explicit time-dependence.

On the other hand, $\hat{y}_c$ as defined in (D.1) now fulfils the equation of motion for x, i.e.,

$$\ddot{\hat{y}}_c - \gamma\,\dot{\hat{y}}_c + \omega^2\,\hat{y}_c = \frac{1}{2}\left(\ddot{x} + \gamma\dot{x} + \omega^2 x\right)\mathrm{e}^{\gamma t} = 0. \tag{D.5}$$

D.2 The Case $a = 0$

For this choice of a it follows that $c = \frac{1}{\gamma}$ and $b = -\frac{m}{\gamma}\left(\omega^2 - \frac{\gamma^2}{4}\right)$. The canonical variables that are still missing attain in this case the values

$$\hat{y}_a = \frac{1}{\gamma}\left(\dot{x} + \frac{\gamma}{2}\,x\right)\mathrm{e}^{\gamma t} = \frac{1}{m\gamma}\,\hat{P}\,\mathrm{e}^{\frac{\gamma}{2}t} \tag{D.6}$$

$$\hat{p}_{x,a} = -\frac{m}{\gamma}\left(\omega^2 - \frac{\gamma^2}{4}\right)x\,\mathrm{e}^{\gamma t} = -\frac{m}{\gamma}\,\Omega^2\,\hat{Q}\,\mathrm{e}^{\frac{\gamma}{2}t}. \tag{D.7}$$

Inserted into $\hat{H}_{\mathrm{B}}$, this now yields

$$\hat{H}_{\mathrm{B},a} = \frac{\gamma}{2}\left(\hat{y}_a\,\hat{p}_y - \hat{x}\,\hat{p}_{x,a}\right) = \hat{D} \tag{D.8}$$

with

$$\hat{H}_\Omega = \frac{1}{m}\,\hat{p}_{x,a}\,\hat{p}_y + m\left(\omega^2 - \frac{\gamma^2}{4}\right)\hat{x}\,\hat{y}_a = 0, \tag{D.9}$$

i.e., just the opposite situation of the case $c = 0$.

Again, $\hat{H}_{\mathrm{B},a}$ is no longer a proper Hamiltonian function that provides the correct equations of motion (see the comments in the previous case).

In this case, the equation of motion for $\hat{y}_a$ leads to

$$\ddot{\hat{y}}_a - \gamma\dot{\hat{y}}_a + \omega^2\hat{y}_a = \left[\frac{1}{\gamma}\frac{d}{dt}\left(\ddot{x} + \gamma\dot{x} + \omega^2 x\right) + \frac{1}{2}\left(\ddot{x} + \gamma\dot{x} + \omega^2 x\right)\right]\mathrm{e}^{\gamma t} = 0. \tag{D.10}$$

D.3 The Case $b = 0$

Now one obtains $c = \frac{\frac{\gamma}{4}}{\omega^2}$ and $a = \frac{m}{2\omega^2}\left(\omega^2 - \frac{\gamma^2}{4}\right)$, leading to

$$\hat{y}_b = \left(\frac{\frac{\gamma}{4}}{\omega^2}\, \dot{x} + \frac{1}{2}\, x \right) \mathrm{e}^{\gamma t} = \frac{1}{2\omega^2} \left(\frac{\gamma}{2m} \hat{P} + \Omega^2 \hat{Q} \right) \mathrm{e}^{\frac{\gamma}{2} t} \tag{D.11}$$

$$\hat{p}_{x,b} = \frac{m}{2\omega^2} \left(\omega^2 - \frac{\gamma^2}{4} \right) \dot{x}\, \mathrm{e}^{\gamma t} = \frac{m\Omega^2}{2\omega^2} \left(\frac{1}{m} \hat{P} - \frac{\gamma}{2} \hat{Q} \right) \mathrm{e}^{\frac{\gamma}{2} t}. \tag{D.12}$$

Comparison with $\hat{H}_{\mathrm{B}}$ (Eq. (4.19) or (4.51)) shows that now

$$\hat{H}_\Omega = \frac{1}{m}\, \hat{p}_{x,b}\, \hat{p}_y + m \left(\omega^2 - \frac{\gamma^2}{4} \right) \hat{x}\, \hat{y}_b = \left(1 - \frac{\frac{\gamma^2}{4}}{\omega^2} \right) \hat{H}_{\mathrm{B}} = \frac{\Omega^2}{\omega^2}\, \hat{H}_{\mathrm{B}} \tag{D.13}$$

$$\hat{D} = \frac{\gamma}{2} \left(\hat{y}_b\, \hat{p}_y - \hat{x}\, \hat{p}_{x,b} \right) = \frac{\frac{\gamma^2}{4}}{\omega^2}\, \hat{H}_{\mathrm{B}} \tag{D.14}$$

is valid.

The equation of motion for $\hat{y}_b$ now takes the form

$$\ddot{\hat{y}}_b - \gamma \dot{\hat{y}}_b + \omega^2 \hat{y}_b = \left[\frac{\frac{\gamma}{2}}{\omega^2} \frac{d}{dt} \left(\ddot{x} + \gamma \dot{x} + \omega^2 x \right) + \frac{1}{2} \left(\ddot{x} + \gamma \dot{x} + \omega^2 x \right) \right] \mathrm{e}^{\gamma t} = 0. \tag{D.15}$$

Appendix E
Logarithmic Nonlinear Schrödinger Equation via Complex Hydrodynamic Equation of Motion

Using Schrödinger's original definition of the wave function Ψ, via the action function S and taking into account that Ψ is generally complex, leads to the *complex action*

$$S_c = \frac{\hbar}{\mathrm{i}} \ln \Psi \tag{E.1}$$

as mentioned in Sect. 4.5.

A corresponding complex momemtum p_c (or velocity v_c) can then be defined according to

$$p_c = m v_c = \frac{\partial}{\partial x} S_c = \frac{\hbar}{\mathrm{i}} \frac{\frac{\partial}{\partial x}\Psi}{\Psi}. \tag{E.2}$$

With these quantities, the TDSE can be rewritten as a modified complex Hamilton–Jacobi equation where only a kind of complex "quantum potential" appears on the rhs instead of zero, as in the classical case,

$$\frac{\partial}{\partial t} S_c + \frac{1}{2m} p_c^2 + V = \mathrm{i} \frac{\hbar}{2m} \frac{\partial}{\partial x} p_c. \tag{E.3}$$

Taking the spatial derivative of this equation, with $v_c = \frac{1}{m} p_c$, leads to a complex Newtonian equation in a moving coordinate frame, expressed by the Lagrangian time-derivative $\frac{D}{Dt} = \left(\frac{\partial}{\partial t} + v_c \frac{\partial}{\partial x}\right)$ as

$$m \left(\frac{\partial}{\partial t} v_c + v_c \frac{\partial}{\partial x} v_c \right) = m \frac{D}{Dt} v_c = -\frac{\partial}{\partial x} V - \mathrm{i} \frac{\hbar}{2} \frac{\partial^2}{\partial x^2} v_c. \tag{E.4}$$

For Gaussian WPs, as considered in our case, $\frac{\partial^2}{\partial x^2} v_c = 0$ is valid.

Equation (E.4) is a hydrodynamic form of a Newtonian equation of motion for a complex velocity v_c and a complex "quantum force" $F_{qu} = -\mathrm{i}\frac{\hbar}{2}\frac{\partial^2}{\partial x^2} v_c$ which vanishes for Gaussian WPs.

D. Schuch, *Quantum Theory from a Nonlinear Perspective*,
Fundamental Theories of Physics 191, https://doi.org/10.1007/978-3-319-65594-9

In the same way as one can consider a linear velocity-dependent friction force $-m\gamma v$ in the Newtonian equation of motion, one could think of adding a corresponding force depending on the complex velocity leading, in this case, to

$$m\frac{D}{Dt}v_c = -\frac{\partial}{\partial x}V - m\gamma v_c = -\frac{\partial}{\partial x}V - m\gamma\left(\frac{\hbar}{\mathrm{i}m}\frac{\partial}{\partial x}\ln\Psi\right), \tag{E.5}$$

and providing, on an average, the correct equation of motion including the friction force

$$m\frac{d}{dt}\langle v_c\rangle = -\left\langle\frac{\partial}{\partial x}V\right\rangle - m\gamma\langle v_c\rangle, \tag{E.6}$$

where $\langle v_c\rangle = \langle v_{op}\rangle = \dot{\eta}$ in our notation.

Following the path taken above in the opposite direction, i.e., integrating instead of differentiating, leads to $-V - \gamma\frac{\hbar}{\mathrm{i}}(\ln\Psi - f(t))$ on the rhs of Eq. (E.5) where, as mentioned before for normalization purposes, $f(t) = -<\ln\Psi>$ is necessary. The lhs remains unchanged as in the conservative case, finally leading to the logarithmic NLSE

$$\mathrm{i}\hbar\frac{\partial}{\partial t}\Psi_{\mathrm{NL}} = \left\{-\frac{\hbar^2}{2m}\Delta + V + \gamma\frac{\hbar}{\mathrm{i}}(\ln\Psi_{\mathrm{NL}} - \langle\ln\Psi_{\mathrm{NL}}\rangle)\right\}\Psi_{\mathrm{NL}}. \tag{E.7}$$

Index

D. Schuch, *Quantum Theory from a Nonlinear Perspective*,
Fundamental Theories of Physics 191, https://doi.org/10.1007/978-3-319-65594-9

MIX
Papier aus verantwortungsvollen Quellen
Paper from responsible sources
FSC® C105338

If you have any concerns about our products,
you can contact us on
ProductSafety@springernature.com

In case Publisher is established outside the EU,
the EU authorized representative is:
Springer Nature Customer Service Center GmbH
Europaplatz 3, 69115 Heidelberg, Germany

Printed by Libri Plureos GmbH
in Hamburg, Germany